SDGs and Textiles

AF895263

Editor-in-Chief

Hafeezullah Memon, International Institute of Silk, College of Textile Science and Engineering, Zhejiang Sci-Tech University, Hangzhou, China

The book series "SDGs and Textiles" addresses the strategies to achieve sustainable development goals (SDGs) in the present, past, and future. It presents books about the present and future policies of textile ministries of different countries, and books related to sustainability education around different parts of the world in the textile sector. Moreover, it would welcome the conference proceeding related to SDGs and Textiles. The series would cover books comparing the sustainability and SDGs of different institutions and countries. The individual book volumes in the series are thematic. The goal of each book is to give readers a comprehensive overview of a different area of sustainability in the textile sector. As a collection, the series provides valuable resources to a broad audience in academia, the research community, industry, and anyone looking to expand their knowledge of SDGs and Textiles.

Textiles and life are together – life cannot be separated from textiles as it is the most important need for human beings after food. In 2015, the United Nations General Assembly proposed 17 interlinked global goals to be achieved by 2030. Since then, academia and industry have paid much attention to achieving these goals. Textile found its close relation with almost all of these 17 goals.

SDG 1 - No Poverty: Poverty would never be overcome by a charity only; it is essential to develop people's skills to have a better and wealthy life. Thus, the textile can be considered an excellent discipline to achieve this goal by creating jobs and small and medium businesses.

SDG 2 - Zero Hunger: Through the effective utilization of advanced application of Agrotech Textiles, it is possible to have higher crop yields and save crops from rough weather, unexpected rains, floods, insects, etc.; thus, geotextiles play an essential in achieving this goal of sustainable development.

SDG 3 - Good Health & well-being: There has been much health consciousness after Covid19, and medical textiles assist in getting good health and well-being.

SDG 4 - Learning & Education: Textile or fashion has remained a significant discipline for societies for ages, and there has always remained much to explore in this field. Textile-related universities may play a vital role by offering free access to their education resources, training and spreading information among the locals.

SDG 5 - Gender Equality: The textile sector is one of the industrial sectors that accepted gender equality long ago; in particular, the garment sector has more females than males. Thus, the textile sector has been doing gender equality. Moreover, there has been a recent trend for Gender Neutral Clothing, which need worth studying and may further assist gender equality.

SDG 6 - Clean Water & Sanitation: Textiles could be achieved through filtration, and of course, textile is one of the critical materials for filtration.

SDG 7 - Affordable & Clean Energy: With the recent advancement in material science and engineering, the textile sector has come on the front for, not only by using this clean energy during textile production but also by assisting the production of this clean energy, either in the form of wind turbines blades made of textile composites or by energy harvesting from T-Shirts, etc.

SDG 8 - Decent Work: Recently, there has been much attention that the textile workers are not paid well, labor rights are not cared about, etc.

SDG 9 - Industry and innovation: Textile Industry always follows innovation; the textile companies that do not chase innovation cannot survive in the market.

SDG 10 - Reduced Inequalities: Getting better life and well-being would help reduce inequalities in the textile industry.

SDG 11 - Sustainable Cities: Sustainable Textile Cities through Buildtech and transport textiles.

SDG 12 - Consumption and Production: Textile and garment consumption and production all come under.

SDG 13 - Climate Action: Oekotech or Ecotech Textile, waste management of textiles are upfront to achieve this goal of sustainable development.

SDG 14 - Life Below Water: Mitigating microfiber waste in rivers and oceans may come under the context of it. There has been much attention on this subject after passing the bill at the parliament level of the UK.

SDG 15 - Life on Land: Geotech or Geotextiles studies life on land.

SDG 16 - Peace, Justice, and Strong Institutions: Protective textiles are doing their best to achieve peace, justice, and strong institutions.

SDG 17 - Partnerships for the Goals: The application of textiles to achieve sustainable development goals is only an example. In all textiles sectors, combined efforts of all the goals are essential to achieve true sustainability.

José Moleiro Martins
Editor

Industry and Innovation: Textile Industry

Springer

Editor
José Moleiro Martins
Department of Management
Lisbon Accounting and Business School
Polytechnic Institute of Lisbon
Lisbon, Portugal

ISSN 2948-1236 ISSN 2948-1244 (electronic)
SDGs and Textiles
ISBN 978-3-031-57806-9 ISBN 978-3-031-57804-5 (eBook)
https://doi.org/10.1007/978-3-031-57804-5

This Springer imprint is published by the registered company Springer Nature Switzerland AG
The registered company address is: Gewerbestrasse 11, 6330 Cham, Switzerland

Paper in this product is recyclable.

Preface

Globalization provides access to practically all textiles traded worldwide. While beneficial for consumers, it poses challenges for the textile manufacturing industry. The textile industry is one of the most polluting sectors. Therefore, in addition to the impact of global competition demanding continuous innovation, the adoption of a sustainability strategy is necessary to reduce its negative impact on the environment. Competitiveness in the textile industry hinges on sustainable innovation.

In 2010, the implementation of Industry 4.0 began, a term introduced at the Hanover Fair in 2011. It represents an intelligent industry with a new way of organizing production methods and utilizing resources more efficiently. Its flexibility and adaptability to production needs aligned with consumer demands are based on new technologies such as blockchain and the Internet of Things (IoT).

In this new scenario, traditional manufacturing sectors, such as the textile industry, are facing times of change. It is imperative for textile companies to transition from machine-dominated manufacturing to digital manufacturing. This sector relies on product customization and short manufacturing cycles. However, the textile industry is mainly composed of small and medium-sized enterprises (SMEs) with limited capacity to invest and adopt new production technologies.

Interoperability can occur in the textile industrial ecosystem through the integration of companies via value networks. Interoperability involves the ability to communicate between cyber-physical systems, IoT devices, production units, and collaborators. In this context, computational intelligence techniques can also be applied in cutting, sewing, finishing, and packaging clothing. The new paradigm includes productive modular structures composed of devices in the Internet of Things environment. Robotic tasks can be used in the autonomous or collaborative assembly of clothing sets, as well as in the unfolding of garment pieces.

SMEs in the textile sector can benefit from technological innovations in cloud computing for smart clothing, fabric customization, production line control, mass customization, store programming, efficiency improvement, and pollution reduction in the environment. This new production approach requires new business models, increased management capabilities, and the updating of physical infrastructure and information technologies in companies, enabling the management of distributed resources (employees and equipment) in remote interaction with each other.

In this panorama, this book includes various subjects that help understand and deepen knowledge in the field of textile industry and innovation, in a cross-cutting thematic approach, as follows.

Chapter "Risk and Innovation Management of Small Businesses in the Textile Industry". This study offers a strategic compass for businesses, researchers, and policymakers, underscoring the multidimensional and global endeavors to bolster the textile industry's innovative and risk management capacities.

Chapter "Impact of Effective Leadership Styles to Achieve the Sustainable Growth of Manufacturing Firms". This research examines the crucial role of leadership in driving sustainable growth within the manufacturing sector, with a specific focus on the textile industry. It contributes to the understanding of the dynamic interplay between leadership styles and sustainable manufacturing practices.

Chapter "The Role of the Supply Chain in Developing Innovation Processes in the Textile Industry". This chapter explores the vital interconnection between supply chain management and innovation in the textile manufacturing industry. The emergence of smart textiles and eco-friendly materials has further underscored the need for a supply chain capable of sourcing novel materials while maintaining quality and cost-efficiency.

Chapter "Communication and Market Orientation as Part of an Innovation Strategy in Smart Textile Companies". This study addresses the reasons why people are becoming increasingly entrepreneurial in the European textile and clothing industries, and how innovation and communication have become so important in the strategies for entrepreneurship in these industries.

Chapter "The Influence of the Marketing Orientation of Textile Companies in Increasing Their Competitiveness". The study's main aim is to examine the influence of the marketing orientation on the competitive advantage of textile firms in terms of market innovation and global strategy. The study found that market turbulence moderates the association between marketing orientation, strategy, and competitive advantage.

Chapter "Knowledge and Innovation Management as a Driver of Competitiveness in Textile Firms". This study aims to comprehensively investigate the role of knowledge and innovation management as crucial drivers of competitiveness in textile firms. The analysis highlights a pivotal interplay between knowledge and innovation management in shaping competitiveness, with a growing scholarly focus on the subject.

Chapter "How Knowledge Management Capabilities Enhance Textile Firms' Innovative Performance". According to this study, innovation is an indispensable factor in bolstering the operational effectiveness of textile companies. It enhances the organization's capacity to operate effectively. Therefore, innovation possesses the capacity to enhance the efficiency of businesses. It is of utmost importance to assist businesses with employee training and research and development expenditures, as doing so has a direct positive impact on output.

Chapter "Innovation, the Clothing Industry, and Textile Manufacturing in Portugal: An Exploratory Analysis of the Sector Between 2003 and 2022". The study points to the need to define strategies aimed at increasing the efficiency of economic activities and promoting a strategy based on competitiveness, adaptability, and greater efficiency in the textile and clothing industry in Portugal.

Chapter "Innovation and Green Finance Management in the Textile Industry: Emerging Trends and Challenges". This research aims to conduct a comprehensive bibliometric analysis to examine the evolving landscape of innovation and green finance in the textile industry. It informs researchers, industry stakeholders, and policymakers about emerging trends and opportunities to advance sustainable practices within the sector.

Chapter "The Innovative Influence of Industry 4.0 on the Textile Industry". This study aims to perform a bibliometric analysis of the burgeoning field of research at the intersection of Industry 4.0 and the textile industry. The article concludes with a call for future research to broaden the scope to encompass sustainability, economic, and policy perspectives, thereby enriching the current understanding of this industrial transformation.

Chapter "Textile Innovation, Sustainability, Survival, and the Future". This research describes the future of textiles in the world, especially in Portugal, is undoubtedly challenging. It seems that the market of the future will face significant concerns related to the environmental and social impact of fashion. For humans and the textile industry to survive, the mission must demonstrate economic viability and combine truly environmentally sustainable innovation with a better life.

Chapter "The Role of Financing in the Development of Innovation Processes in Textile Companies". This research delves into the intricate relationship between financing and innovation in the textile industry, exploring its historical evolution, current practices, and future trends. The shift toward diverse and sophisticated financing strategies reflects the industry's need to adapt to changing market demands and foster innovation, ensuring sustainability and technological progression in the highly competitive global textile market.

Chapter "Management Control Systems and Innovation in Textile Industry". The study reveals the essential role of MCS in shaping organizational culture and driving innovation, emphasizing the need for integrating sustainability and technological advancements. It identifies gaps and potential areas for future research, stressing the importance of international collaboration and exploring under-represented topics to enrich the textile industry's innovation and adaptability.

Chapter "Enhancing Textile Enterprises' Integrated Development: Innovative Management Optimization and Monitoring System". This study introduces an innovative monitoring system for the integrated development of textile enterprises, addressing a significant gap in the current literature. The research focuses on the integration of management practices with technological advancements, essential for the textile industry's dynamic and evolving market.

Chapter "Export Performance of the Textile Industry: Influence of Financial Resources and Productivity". This study aims to examine the effect of certain factors on the export performance of the Portuguese textile industry. The study provides several findings for export-oriented firms, revealing that financial resources, productivity, and size are significant determinants of export performance.

Chapter "Sustainability Innovation in the Textile Industry". The aim of this chapter is to identify key themes of sustainability innovation in the textile industry and to map research in this area. These include basic themes consisting of sustainability, innovation, circular economy, and nanotechnology; niche themes consisting of green innovation, smart textiles, and electrospinning; as well as emerging themes, namely competitiveness and technological innovation.

Finally, I would like to express gratitude to all the authors who contributed to the completion of the book.

Estoril, Portugal
February 2024

José Moleiro Martins

Contents

Risk and Innovation Management of Small Businesses in the Textile Industry

António Abreu

Abstract This study aims to conduct a bibliometric analysis of the literature at the nexus of risk management, innovation, and small businesses in the textile industry. The study conducted a bibliometric analysis of publications in the Web of Science and Scopus scientific databases for 1985–2022 using RStudio and Biblioshiny software. Drawing from the Web of Science and Scopus databases, the study maps the evolution of academic discourse, highlighting thematic trends, seminal works, and existing research gaps. The temporal analysis indicates heightened interest during global disruptions, notably the 2008 financial crisis and the COVID-19 pandemic. Despite an upward trajectory in publications, citation metrics stress the primacy of quality over sheer volume. A geographic examination reveals a diversified global input, with emerging economies like Brazil and India accentuating their academic footprint. Institutional trends spotlight the sustained commitment from universities such as Zhejiang University and Lodz University of Technology. Lotka's Law underpins a significant author productivity skewness, indicating a broad spectrum of perspectives, yet spotlighting a select group with higher output potential. The findings urge closer academia-industry collaboration, an emphasis on research quality, and a watchful eye on new academic entrants. This study offers a strategic compass for businesses, researchers, and policymakers, underscoring the multi-dimensional and global endeavors to bolster the textile industry's innovative and risk management capacities.

A. Abreu (✉)
Unit for Innovation and Research in Engineering, Polytechnic University of Lisbon, 1959-007 Lisbon, Portugal
e-mail: ajfa@dem.isel.ipl.pt

Center of Technology and Systems (UNINOVA-CTS) and Associated Lab of Intelligent Systems (LASI), 2829-516 Caparica, Portugal

Lisbon Accounting and Business School, Polytechnic Institute of Lisbon (ed.), *Industry and Innovation: Textile Industry*, SDGs and Textiles,
https://doi.org/10.1007/978-3-031-57804-5_1

1 Introduction

The textile industry, one of the oldest sectors of human civilization, remains a key pillar of global commerce and economic progress. With roots reaching back to prehistoric eras, textiles have played an indispensable role in socio-cultural evolution, technological advancement, and economic development (Araújo et al., 2019). Over the centuries, it has evolved from hand-woven materials produced for local consumption to a highly mechanized, globalized industry characterized by vast production scales and intricate supply chains. Today, it stands as a representation of a complex web of activities encompassing the design, production, distribution, and consumption of textile products (Shafi et al., 2021).

The dynamic interplay of rapidly evolving global markets, changing consumer preferences, sustainability concerns, and technological innovations presents both opportunities and threats to businesses within the textile industry (Ikram, 2022). These drivers of change are especially salient for small businesses, which often operate with limited resources, making them particularly susceptible to external shocks and the risk of market turbulence.

Risk management, in this context, has emerged as a critical imperative for ensuring the sustained competitiveness and longevity of these businesses. Without appropriate risk mitigation strategies, small textile enterprises may find it challenging to navigate the turbulent waters of the global market. From supply chain disruptions and volatile raw material prices to stringent environmental regulations and competitive pressures, the litany of potential risks is extensive (Luque and Herrero-García, 2019). However, it is not enough for these businesses to merely manage risks; to thrive, they must also embrace innovation.

Innovation, typically perceived as the lifeblood of business survival and growth, plays a quintessential role in shaping the textile industry's future trajectory. For small businesses, innovation can be the distinguishing factor that propels them ahead of competitors, allows them to tap into new market segments, or even reinvents their business models for unprecedented success. Yet, innovation is not without its risks. The paradox lies in the fact that while innovation is necessary for competitive advantage, it simultaneously exposes businesses to new uncertainties.

The dynamics between risk and innovation, especially within the context of small businesses in the textile industry, is an intriguing area of study. Despite its significance, there exists a noticeable gap in the literature pertaining to how these two domains intersect and influence one another. Bibliometric analysis offers a robust methodological tool to delve into this lacuna. By mapping the existing literature landscape, it enables researchers to identify prevailing trends, seminal works, key contributors, and potential avenues for future research. Such analyses not only bring clarity to the current state of knowledge but also catalyze further exploration by highlighting understudied areas.

This study aims to conduct a bibliometric analysis of the literature at the nexus of risk management, innovation, and small businesses in the textile industry. By offering a holistic view of the current research landscape, it seeks to (1) elucidate the major themes and trends that have emerged over the years, (2) pinpoint the seminal works that have shaped the discourse, (3) identify gaps and areas that warrant deeper investigation, and (4) provide a foundation upon which future researchers can build. The study conducted a bibliometric analysis of publications in the Web of Science and Scopus scientific databases for 1985–2022 using RStudio and Biblioshiny software.

As global economies become increasingly interconnected and industries more complex, understanding the multifaceted relationship between risk and innovation is crucial. For the textile industry, which holds cultural, economic, and historical significance, the stakes are even higher. Small businesses, often hailed as the backbone of economies worldwide, stand at the crossroads of these challenges. Ensuring their survival and prosperity entails a thorough comprehension of how they navigate the intertwined pathways of risk and innovation. This study is a step in that direction, offering insights that hold implications not just for academia but also for practitioners, policymakers, and stakeholders within the textile industry.

2 Systematic Literature Review

The span of nearly four decades from 1985 to 2022 provides a rich tapestry of academic contributions, marking distinct shifts in thought processes, methodologies, and paradigms in diverse research areas. An analysis of the publications indexed in two of the most esteemed scientific databases, Scopus and Web of Science, yields a compelling narrative of academic endeavors during this period.

According to the data presented in Table 1, the studied timespan encompasses 132 documents sourced from an array of 112 different journals, books, and other platforms. This corpus of academic material showcases steady growth, with an annual growth rate pegged at 3.72%. The average age of these documents is 10.3 years, suggesting a mix of both seminal, foundational papers and newer, emergent ideas.

Interestingly, the average number of citations per document stands at 11.87, highlighting the significance and impact of these works within the academic community. Such a number underscores the importance of these documents in furthering understanding, sparking debate, and laying the groundwork for subsequent research.

Diving deeper into authorship metrics, a total of 313 authors have contributed to this collection of documents. While collaborative efforts dominate, 42 authors have singularly penned documents, accounting for 56 single-authored pieces. This reflects a healthy balance between individual scholarship and collaborative explorations. Collaboration is further exemplified by the fact that each document averages about 2.58 co-authors. Significantly, international collaborations constitute 12.88% of the total, indicating the global nature of discussions and the cross-border exchanges enriching academic discourse.

Breaking down the 132 documents by type provides insights into the modes of scholarly communication favored over the period. Articles lead the way with 79 entries, followed by conference papers (24), book chapters (10), reviews (9), and conference reviews (6). Books (3) and notes (1) round off the list. The predominance

Table 1 Main information about data

Description	Results
Timespan	1985:2022
Sources (Journals, Books, etc.)	112
Documents	132
Annual growth rate %	3.72
Document average age	10.3
Average citations per doc	11.87
Authors	–
Authors	313
Authors of single-authored docs	42
Authors collaboration	–
Single-authored docs	56
Co-authors per doc	2.58
International co-authorships %	12.88
Document types	–
Articles	79
Books	3
Book chapters	10
Conference papers	24
Conference reviews	6
Notes	1
Reviews	9

of articles and conference papers underscores the importance of peer-reviewed journals and conferences as primary platforms for disseminating research findings and engaging with the academic community.

The bibliometric data spanning 1985–2022, as sourced from Scopus and Web of Science, offers a panoramic view of academic contributions during this period. From the steady growth of publications to the balance between individual and collaborative efforts, and from the global nature of discussions to the variety of document types, the narrative is one of dynamic scholarship and continuous evolution. In order to conduct the research and identify the key features of the conducted research, scientists performed a cluster analysis of publications by the keywords "risk", "innovation management", "small business", and "textile industry" using the VOSviewer program. The results of this analysis are visualized in Fig. 1.

The two most frequent terms, "textile industry" and "textiles", highlight the core theme and focus of the analyzed documents. Their predominance underlines the sector-specific nature of the content. "Risk assessment", "health risks", and "occupational health" are clearly prevalent. This underscores the strong emphasis on risk management in the textile industry, encompassing both generalized and specific

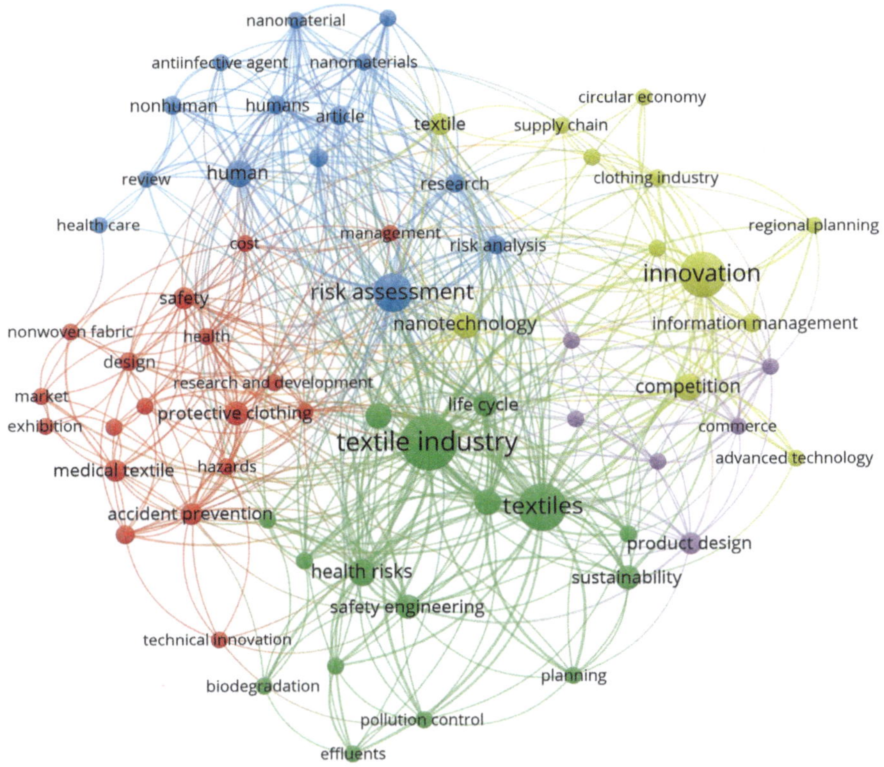

Fig. 1 Cluster analysis of publications in the context of risk and innovation management of small businesses in the textile industry in the scopus and web of science databases from 1985 to 2022

threats. The mention of "innovation" indicates a focus on forward-thinking strategies, improvements, and technological advancements that businesses in the textile industry are either adopting or considering. "Nanotechnology" appearing in the list suggests a growing interest in integrating advanced technologies into textiles. Nanotechnology offers potential benefits like improved fabric properties, but it may also introduce new risks, emphasizing the need for risk assessment. The mention of "competition" emphasizes the challenges faced by small businesses in the textile sector. This may pertain to both local market competition and global competitive pressures, underscoring the need for innovative solutions to maintain or enhance market position. Words like "human" and "occupational health" bring attention to the human factor in the textile industry. This suggests an awareness of the direct impacts on workers, stressing the importance of safe practices, well-being, and perhaps even ethical considerations. "Protective clothing" is a term that integrates both risk (protection from harm) and innovation (developing advanced protective textiles). This could refer to clothing designed for workers within the textile industry or for end consumers in specific environments.

Figure 2 provides a breakdown of the most frequent words that appear in the literature or articles discussing risk and innovation management in small businesses within the textile industry.

Over the past few decades, the textile industry has experienced unprecedented changes influenced by socio-economic dynamics, technological advancements, and global trends. One of the key areas of interest within this realm has been risk and innovation management (Fig. 3).

Recent literature underscores the importance of sustainability in the textile sector, addressing environmental implications and advocating for eco-friendly innovations (Behera et al., 2021; Gu et al., 2021; Levanen et al., 2021; Phan et al., 2020). Such contributions shed light on the transition from traditional processes to more sustainable models, emphasizing the need for businesses to integrate sustainable practices not just for environmental benefits but also to mitigate associated risks (Fu et al.,

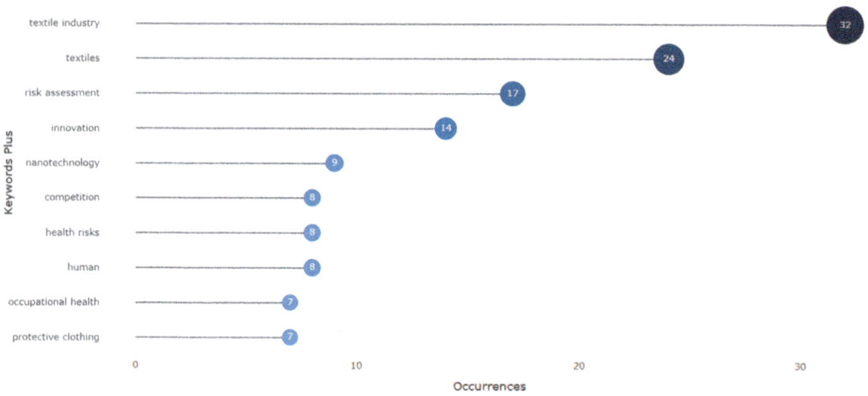

Fig. 2 Most frequently occurring words in the researched publications

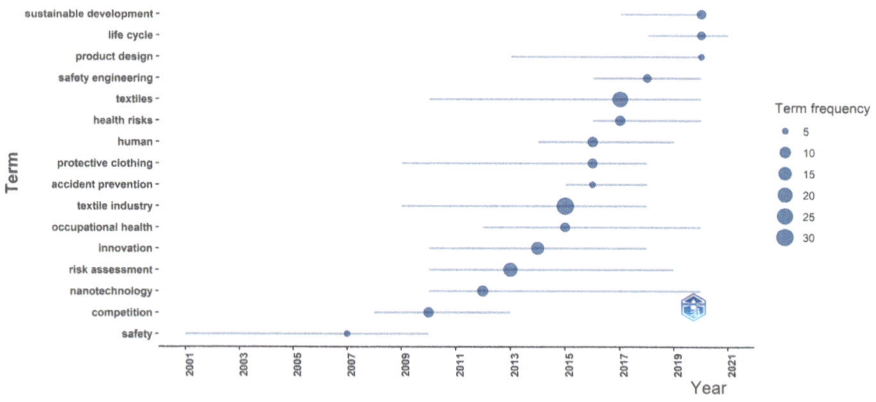

Fig. 3 Trend topics in the researched publications

2023; Seidu et al., 2023). Indeed, in examining the broader ecosystem, sustainability is viewed as both a challenge and an opportunity for small businesses in the textile sector (Denuwara et al., 2019).

The advancement of digital technology has also profoundly impacted the industry. A significant number of studies have been dedicated to the exploration of technological innovations and their implications on risk management. The integration of intelligent systems, for instance, offers new avenues for improving efficiency, customization, and supply chain management (Ding et al., 2020; Ledda & Palomba, 2020; Romao & Simoes, 2020). While these advancements open up unprecedented opportunities for businesses, they also introduce unique risks, emphasizing the need for an effective risk management framework tailored to the digital age (Ho & Watanabe, 2020).

Historical analyses provide a broader perspective on the textile industry's evolution, shedding light on past challenges and strategies, thereby offering valuable insights into managing contemporary risks and embracing new innovations (Donges & Selgert, 2019). Furthermore, as businesses within the textile industry are becoming increasingly global, international collaborations and interactions have gained prominence. This international dimension adds layers of complexity in managing risks and fostering innovation, as highlighted by the percentage of international co-authorships observed in the bibliometric data.

The importance of collaboration, both at the individual and institutional levels, is further echoed in the literature. Collaborative endeavors are often seen as a conduit for fostering innovation, pooling resources, sharing risks, and navigating complex challenges that single entities might find overwhelming (Kafuku, 2019; Pfoser et al., 2021).

Literature delving into the nuances of the textile industry's unique risks also explores the cultural, socio-economic, and market-specific dimensions. There are studies dedicated to understanding the cultural implications of textiles, suggesting that understanding cultural shifts and preferences can be pivotal for risk management and innovative product development (del Solar, 2019). Studies have also emphasized the importance of understanding consumer behavior, market dynamics, and the broader societal implications of textiles. An understanding of these dimensions is crucial for businesses to identify potential risks and innovate in ways that resonate with their target audiences (Sugg, 2022; Tsai & Su, 2022).

Some research highlights the intricate relationship between the environment, materials, and textile production. The interactions between various materials, especially in the context of microbial processes, have profound implications for product quality, sustainability, and associated business risks (Bockmühl et al., 2019; Lugani et al., 2021). In addition to the aforementioned themes, a set of studies has also provided valuable insights into niche areas of the textile industry. These range from specific types of materials such as silk and its properties (Shah et al., 2022; Wang et al., 2019) to more overarching discussions on the economic, historical, and social aspects of textiles and their significance in global trade and culture (Vanegas-López et al., 2021; Wescoat & Rawoot, 2021).

The literature spanning offers a comprehensive view of the multifaceted domain of risk and innovation management within the textile industry. The intersection of sustainability, technology, collaboration, and socio-cultural dimensions presents both challenges and opportunities for small businesses in this sector. As the industry continues to evolve, understanding these dynamics, as evidenced by the rich tapestry of academic contributions, will be pivotal for businesses to navigate future uncertainties and harness the power of innovation.

The main themes emerging from the bibliometric analysis related to risk and innovation management of small businesses in the textile industry are discerned by identifying clusters and examining the number of occurrences, centrality measures, and PageRank values for the most frequent words (Fig. 4).

Safety cluster. The words most frequently associated with safety are "safety", "medical textile", "testing", "waste disposal", and "conference", among others."Safety" has the highest betweenness centrality, highlighting its role as a bridge in the network, while also commanding a considerable PageRank centrality value. Other terms in this cluster like "fabric" and "silk" have very high betweenness centrality values, suggesting they too might play significant intermediary roles within the theme of safety.

Textile industry cluster. The dominant terms in this cluster include "textile industry", "textiles", "innovation", "nanotechnology", and "competition". These terms appear most frequently and thus form the core of the discussions. "Textile industry" and "textiles" have remarkably high betweenness centrality values, emphasizing their essential position within this thematic discussion. Their PageRank centrality

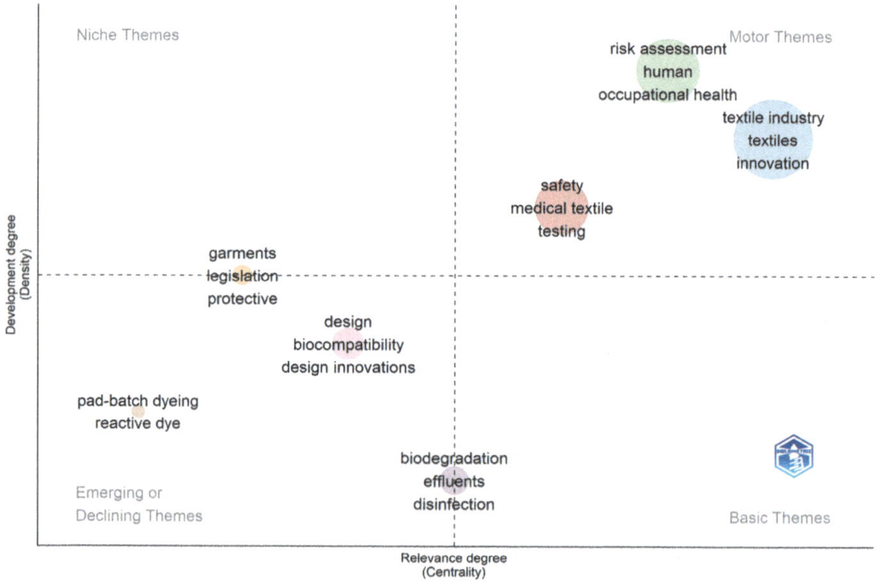

Fig. 4 Thematic map for the researched publications

Table 2 Cluster analysis according to the thematic map

Cluster	Callon centrality	Callon density	Rank centrality	Rank density	Cluster frequency
Safety	6.5417	84.5029	5	5	49
Textile industry	15.9478	115.1425	7	6	258
Risk assessment	9.1724	163.1192	6	7	93
Biodegradation	1.8611	61.1111	4	1	10
Garments	0.25	83.3333	2	4	6
Pad-batch dyeing	0	75	1	2	4

values are significantly higher than others, indicating their dominant influence in the network. Words like "innovation", "nanotechnology", and "health risks" underline the areas of focus within the textile industry, highlighting the areas of innovation and concerns related to the industry.

High betweenness centrality suggests that a term acts as a bridge or connection between different parts of the network (Table 2). In our data, words like "textile industry" and "safety" have high betweenness centrality, making them crucial in connecting different discussions. High closeness centrality means a term is closer to all other terms in the network. However, given our data, most terms have relatively similar closeness centrality values. PageRank centrality gives an indication of the most influential terms in the network. Here, "textile industry", "textiles", and "innovation" stand out, suggesting they are central toto the discourse.

The vast range of terms indicates the diverse areas of interest when discussing risk and innovation management in the textile industry. From focusing on technological aspects (e.g., "nanotechnology", "Internet of Things", "Industry 4.0") to environmental and health concerns (e.g., "sustainable development", "pollution control", "health risks"), the discourse appears comprehensive. There's also a considerable focus on business and management strategies, as evidenced by terms like "risk management", "information management", "human resource management", and "production management".

The textile industry, from this bibliometric analysis, is observed to be a complex field with multiple facets. Safety remains a prominent theme, with emphasis also on the industry's innovations, risks, and the interplay between different aspects of the business, technology, and environmental factors. For small businesses in the textile sector, understanding these key themes is essential for both risk mitigation and harnessing innovation opportunities.

2.1 *The Common Research Trends and Meta-Analysis*

Figure 5 provided sheds light on the annual scientific production concerning risk and innovation management in the textile industry over the span of approximately four

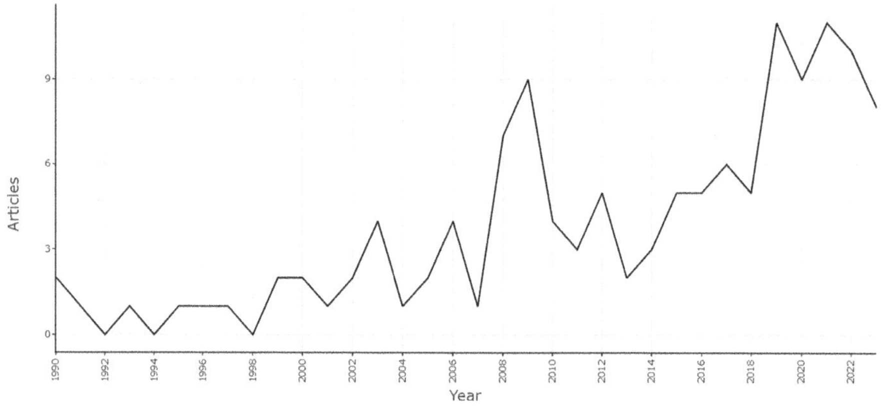

Fig. 5 Annual scientific production for the direction of risk and innovation management in the textile industry

decades, from 1985 to 2022. Over the 38-year period, there are noticeable fluctuations in the number of articles produced annually. The overall trend, however, shows an upward trajectory especially when comparing the early years to the more recent ones. The decade 1985–1994 began with a minimal interest in the topic, with only 2 articles in 1985. The number of articles remained low, with several years seeing no publications at all. 1995–2004—the production remained fairly consistent, with an average of about 1.4 articles per year. A significant spike was observed in 2003 with 4 articles. The decade 2005–2014 saw an overall increase in the annual scientific production, reaching up to 9 articles in 2009. The decade averaged around 4 articles per year.

The period 2015–2022 has the highest consistent production, peaking at 11 articles in both 2019 and 2021. The least number of articles in this period is still a notable 5, suggesting heightened interest and research in the field. The years 2008, 2009, 2019, 2021, and 2022 stand out with a significant number of articles, suggesting periods of heightened research interest or perhaps responses to global events or industry shifts that required more attention to risk and innovation management in textiles.

The global financial crisis around 2008–2009 might have prompted businesses to delve deeper into risk management strategies, explaining the surge in publications during this period. The latter peaks around 2019–2022 might be influenced by the disruptions caused by the COVID-19 pandemic, which affected global supply chains and demanded innovative solutions, especially for industries like textiles. From the bibliometric data presented, it is evident that the importance of understanding risk and fostering innovation in the textile industry has grown over time. While the initial years showed sporadic interest, the past two decades, especially the 2015–2022 period, highlight an increased awareness and emphasis on the subjects of risk and innovation in the textile business sector.

Table 3 provides insights into the average citations per year in the context of research articles related to risk and innovation management for small businesses in the textile industry.

The year 2013 stands out with an average of 87 citations per article, and a Mean Total Citation (MeanTCperYear) of 7.91. This suggests that the literature published in this year may have had a significant impact or presented groundbreaking insights. The number of publications (N) has generally seen an increasing trend in the 2010s,

Table 3 Average citations per year for the researched publications

Year	MeanTCperArt	N	MeanTCperYear	Citable years
1990	40.5	2.00	1.19	34
1991	13	1.00	0.39	33
1993	0	1.00	0.00	31
1995	0	1.00	0.00	29
1996	0	1.00	0.00	28
1997	0	1.00	0.00	27
1999	1	2.00	0.04	25
2000	58	2.00	2.42	24
2001	1	1.00	0.04	23
2002	0	2.00	0.00	22
2003	0	4.00	0.00	21
2004	0	1.00	0.00	20
2005	33.5	2.00	1.76	19
2006	2.25	4.00	0.12	18
2007	1	1.00	0.06	17
2008	0	7.00	0.00	16
2009	10.33	9.00	0.69	15
2010	15.75	4.00	1.12	14
2011	2	3.00	0.15	13
2012	1.6	5.00	0.13	12
2013	87	2.00	7.91	11
2014	19.67	3.00	1.97	10
2015	6.4	5.00	0.71	9
2016	22.8	5.00	2.85	8
2017	11	6.00	1.57	7
2018	2.2	5.00	0.37	6
2019	25.73	11.00	5.15	5
2020	10.89	9.00	2.72	4
2021	18.18	11.00	6.06	3
2022	6.6	10.00	3.30	2

with 2019 and 2021 both having 11 publications, the highest in the dataset. A peak in citations is observed in the early 1990s and early 2000s, followed by a few years of minimal citations. Another peak appears around 2013, followed by relatively consistent citations until 2022. The data suggests periods of increased research interest and impact, notably in the early 1990s, early 2000s, and around 2013. The reasons behind these spikes could be multifaceted, possibly involving industry changes, global events, or advances in technology and methodology.

The increasing number of publications in recent years may indicate growing interest or evolving challenges and opportunities in the field of risk and innovation management for small textile businesses. Citations do not always correlate with the number of publications. Some years with fewer publications had a higher average citation rate, suggesting the quality or significance of research can be more impactful than the quantity. Figure 6 tracks the publication count of various journals over the years, specifically in relation to risk and innovation management in small businesses in the textile industry.

Based on the total number of publications over the years, "Technical Textiles International" and "Technovation" are the dominant sources with the highest consistent contributions. While "Technical Textiles International" and others have been consistent, journals like "Advances in Intelligent Systems and Computing" show a rapid increase in recent years, indicating its emerging role in the field. The latter part of the figure, especially post-2019, shows more diversity in journals contributing to the topic, indicating a broader academic interest and more comprehensive coverage of the subject. The trends suggest a growing and diversifying interest in the topic, with both traditional and emerging journals contributing to the discourse.

Figure 7 provides an overview of authorship production over time within the context of the analyzed research.

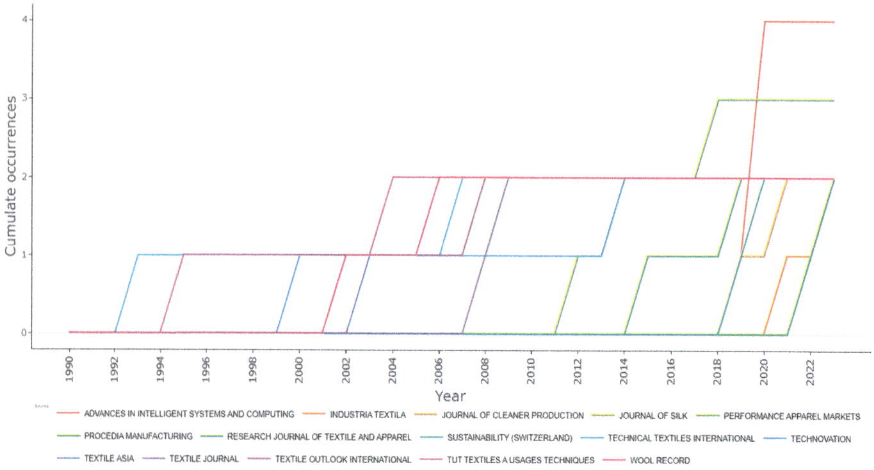

Fig. 6 Sources' production over time

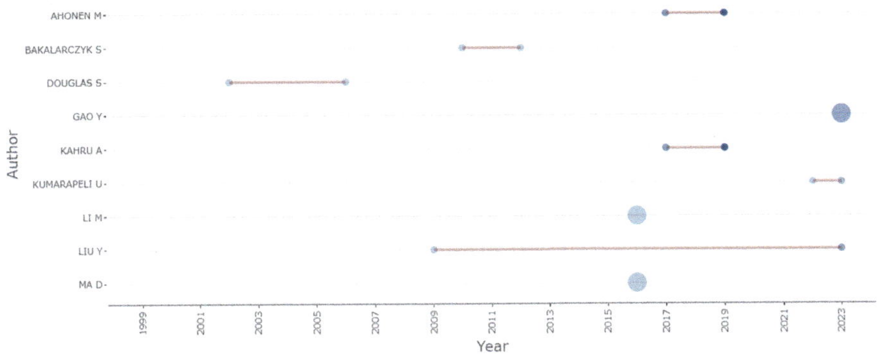

Fig. 7 Authors' production over time

The papers of Ahonen and Kahru (Dunne et al., 2020; Rosenberg et al., 2019) from 2019 have the highest citation count at 39. Average yearly citations provide insights into the relevance and importance of the work over time. Papers with high TCpY suggest they continue to be influential and are frequently referenced in ongoing research. All other authors have relatively low citation counts, ranging from 0 to 2 for their respective papers. Li and Ma (2016), despite publishing twice in 2016, have relatively low average yearly citations of 0.250. Authors like Liu Y. (from 2009) and Li M. (from 2016) have publications with the lowest TCpY, with 0.133 and 0.250 respectively. The bibliometric data provides a comprehensive overview of the authors' contributions and the impact of their research in the textile industry. It can serve as a foundation for further analysis to identify influential authors, emerging trends, and the trajectory of research in this sector.

In the evolving field of risk and innovation management within the textile industry, several significant works have emerged that have shaped the discourse. The examination of these pivotal contributions illuminates the direction and scope of current research trends (Fig. 8).

One of the most influential works was presented by Dangelico et al. (2013), which has received consistent attention, amassing 167 citations over the past decade, highlighting its lasting impact. This paper's normalized citation rate is 1.92, indicating its broad influence in the discipline relative to its publication year. Similarly, a study from Ayres (1990) has maintained relevance with a normalized citation rate. Both studies offer valuable insights into the nuanced complexities of innovation in the textile realm.

Recent literature has exhibited a heightened emphasis on sustainable and environmental aspects. de Oliveira Neto's et al. (2019) work has been highly cited, emphasizing the growing attention to cleaner production methods. This document, along with Behera et al. (2021) contribution, both possess an impressive, normalized citation rate of 5.67, which underscores the urgency and significance of environmental concerns in modern textile innovation.

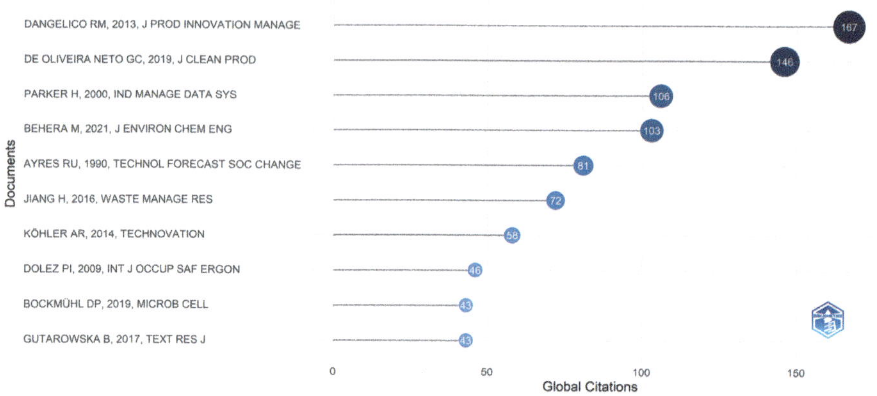

Fig. 8 Most global cited documents

The industry's adaptation to technological shifts can be traced back to Parker's (2000) publication, which has maintained relevance over the years, given its normalized citation rate of 1.83. On the other hand, Köhler and Som (2014) study focuses on the challenges and opportunities brought about by technological innovations in the textile domain, as indicated by its normalized citation rate of 2.95.

Waste management, another crucial aspect of the textile industry, was detailed in Jiang et al. (2016) research, which offers a perspective on the industry's approach to minimizing waste, supported by its 3.16 normalized citation rate. Meanwhile, safety remains a pivotal concern, as evidenced by Dolez and Vu-Khanh (2009) investigation, which garnered a high normalized citation rate of 4.45.

Recent studies by Bockmühl et al. (2019) and Gutarowska et al. (2017) have delved into the microbial and chemical aspects of textiles. Their respective normalized citation rates, 1.67 and 3.91, demonstrate the multifaceted nature of textile research, which encompasses both innovative and risk management domains. The cited works collectively paint a comprehensive picture of the textile industry's trajectory, with a keen emphasis on sustainability, technological adaptation, waste management, and safety.

The data underscores the principle of Lotka's law (Fig. 9).

The vast majority of authors (94.9%) have produced only a single document. This can be attributed to the challenges faced in producing multiple research works, lack of resources or interest, or simply the nature of academic publishing where many contribute once or twice but do not continue in the long term. On the other hand, a minuscule percentage of authors (those who've written 2 or more documents, adding up to 5.1%) have shown a higher productivity rate. The single author who has written 13 documents stands out, representing the pinnacle of productivity within this dataset.

Figure 10 provides the frequency (number of publications or occurrences) of different universities in the textile industry's risk and innovation management. The general trend across all universities shows a rising interest in risk and innovation

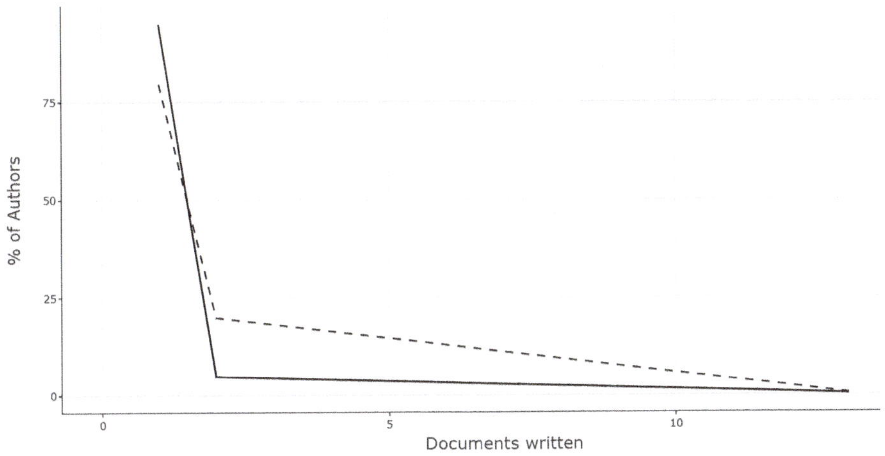

Fig. 9 Author productivity through Lotka's law

management in the textile industry over the years. This might reflect the growing importance of the topic in the textile sector or increased academic attention to the domain. Institutions like RMIT University and the University of Moratuwa have seemingly begun contributing later compared to others but have shown a rapid increase in their contributions once they started. Zhejiang University and Lodz University of Technology have consistently contributed over a longer period. Their steady numbers suggest a long-term commitment to research in this domain.

The affiliation of Zhejiang University started having relevant publications or mentions from the year 2008, with a steady rise to 3 by the year 2022. This indicates a growing interest or contribution from this institution in the specific domain over

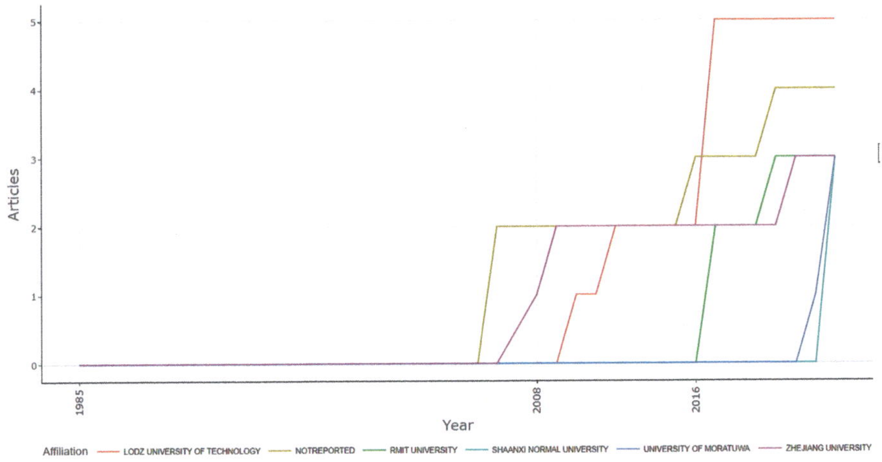

Fig. 10 Affiliations' production over time

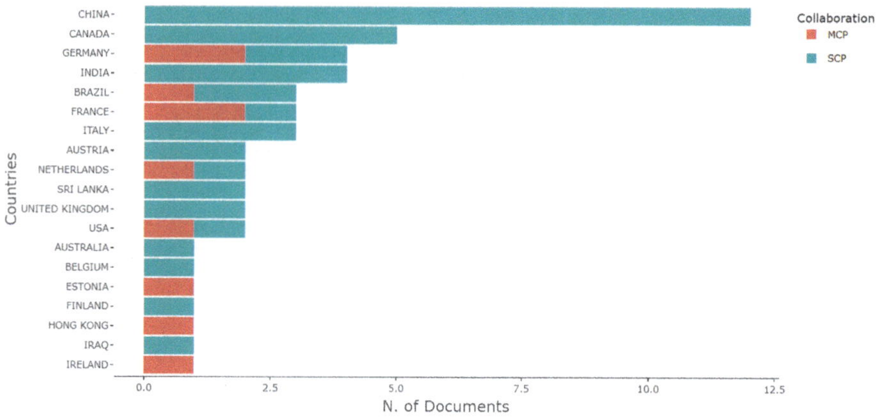

Fig. 11 Corresponding author's countries

the years. For Lodz University of Technology the activity started a bit later, with the first mention in 2010. However, from 2017 onwards, there's a sharp increase to 5 publications or mentions annually, suggesting a consistently high level of contribution or interest in this field. RMIT University shows no activity in the domain until 2017. From 2017 to 2020, there's a steady increase in the number of publications or mentions, reaching 3 by 2020 and maintaining this through 2022. University of Moratuwa showed no activity in the domain for a significant duration, only recording its first mention in 2022. However, there's a notable jump from 1 in 2022 to 3 in 2022, indicating a possible growing interest or burgeoning research area for the institution.

Figure 11 presents a bibliometric analysis regarding the distribution of articles in the domain of risk and innovation management of small businesses in the textile industry. This distribution is based on the country of the corresponding author.

China is the leading country with 12 articles, accounting for a frequency (Freq) of 9.1% among the countries listed. Canada and Germany follow with 5 and 4 articles respectively. India, Brazil, France, and Italy each have contributed 3 articles. Every article with a corresponding author from China, Canada, India, Italy, and Austria is a Single Country Publication (SCP). This implies these articles had authors only from the specified country and not from international collaborations. Germany, Brazil, France, and the Netherlands have a mix of SCP and Multi-Country Publications (MCP), indicating some level of international collaboration in their research articles. France has the highest MCP ratio at 0.667. Out of the 3 articles, 2 of them involved international collaboration. Germany and the Netherlands follow closely with an MCP ratio of 0.5. China, which has the highest number of articles, shows no signs of international collaboration as the MCP ratio stands at 0. This might indicate a more insular or self-sufficient research environment in this specific domain for China. Similarly, Canada, India, Italy, and Austria also have an MCP ratio of 0.

A higher MCP ratio suggests a broader collaboration across borders in the research domain. Countries with higher MCP ratios may benefit from diverse perspectives

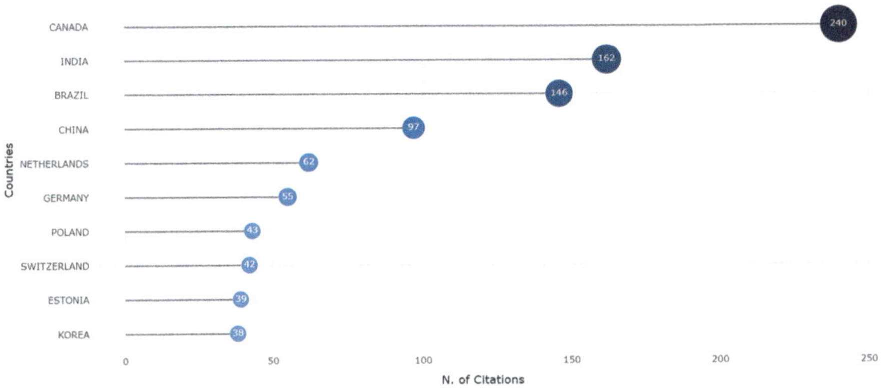

Fig. 12 Most cited countries

and expertise, which can enhance the depth and quality of research. On the other hand, countries with lower MCP ratios might be relying on local expertise or may have strong domestic institutions and research environments that facilitate more localized research. While China leads in terms of sheer numbers, countries like France, Germany, and the Netherlands demonstrate a more collaborative approach in the domain of risk and innovation management for small businesses in the textile industry. The figure provides valuable insights into the global research landscape and highlights the balance between domestic and collaborative research across countries.

Figure 12 presents the most cited countries in the context of risk and innovation management of small businesses in the textile industry based on two main metrics—Total Citations (TC) and Average Article Citations.

At first glance, Canada leads in the total number of citations (TC) with 240 citations. However, when we dive into the average article citations, Brazil outpaces Canada with an average of 48.70 citations per article, even though Brazil has fewer total citations. This indicates that while Canada produces a significant volume of research, the average quality or impact of individual articles from Brazil is slightly higher.

Countries such as India and Brazil, which are often seen as emerging markets, have made substantial contributions in terms of citations, rivaling established Western countries. This underscores the increasing importance and influence of research from non-traditional hubs in the global academic landscape.

China, a global powerhouse in many research domains, has a relatively low average of 8.10 citations per article in this particular field. This number is quite low when compared to other countries in the list, suggesting that while China may be producing a significant number of papers, they might not be as influential or widely recognized in the global textile industry's risk and innovation management domain.

The presence of countries like the Netherlands, Germany, Poland, Switzerland, and Estonia highlights the diverse research output from Europe. Each of these nations, despite their varying sizes and economic scales, has made noteworthy contributions to the field. Their average article citations, hovering around the 30–40 mark (except for Germany), indicate a good balance between quantity and quality.

Countries such as Poland, Switzerland, Estonia, and Korea have average article citations that closely match their total citation numbers, suggesting that these countries have produced a relatively consistent number of articles, each of which has garnered similar attention. While traditional research giants continue to make their mark, emerging markets are making significant inroads, suggesting a shift in the balance of academic power and influence. It also emphasizes the importance of looking beyond sheer volume to understand the true impact and quality of research output.

Figure 13 provides a quantitative breakdown of the research output across various countries. China and India are leading contributors, with China accounting for the highest number of publications at 29. This highlights Asia's significant role in the textile industry and its investment in research related to innovation and risk management. The prominence of China and India could be attributed to their established textile manufacturing hubs, vast raw material bases, and increasing emphasis on innovation and technology.

European countries, namely Germany, Switzerland, the UK, France, and the Netherlands, collectively contribute 53 publications. Their combined contribution surpasses even that of China, underscoring Europe's continued importance in the global textile industry and its focus on ensuring that small businesses are equipped to handle risks and foster innovation.

The USA and Canada collectively account for 20 publications. While the USA has traditionally been a significant player in various research areas, its modest contribution (relative to China and India) suggests a possible shift in the textile industry's epicenter. Canada's equal contribution with countries like Australia and France suggests an active engagement in research despite its smaller population. The USA, often a global leader in scientific research, trails behind Germany and India in

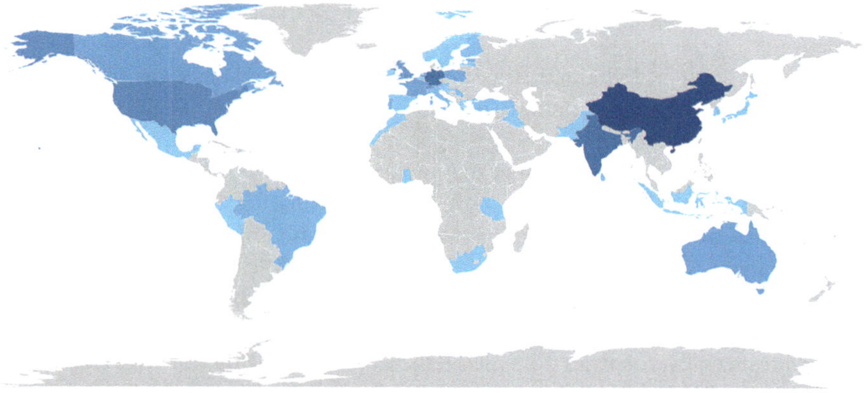

Fig. 13 Countries' scientific production

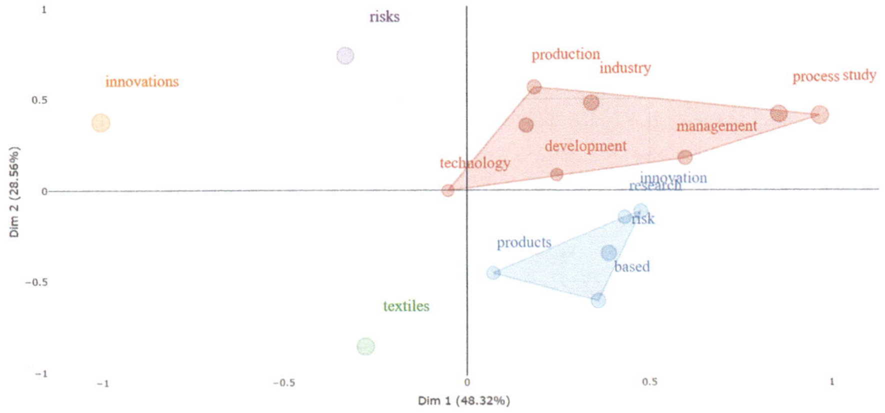

Fig. 14 Word map of the researched publications factorial analysis

this specific area. This may indicate a relative decline in the importance of the textile sector in the US or a shift in research priorities.

Countries like Switzerland, the UK, Australia, Canada, France, and the Netherlands, though having fewer publications compared to the leaders, demonstrate consistent dedication to research. Their contributions can be seen as pivotal, providing diverse perspectives and insights, especially from regions with different industry challenges and dynamics.

Figure 14 presents the results of a factorial analysis which maps out selected words based on their scores in two dimensions (Dim.1 and Dim.2) and then groups them into clusters. Most of the words are positive in Dimension 1 (Dim.1) with "innovations" being the only outlier having a score of –1. The words are more spread out in Dimension 2 (Dim.2), ranging from –0.86 to 0.74.

There are five distinct clusters. Cluster 1 is the most populated and contains words primarily associated with the core theme of the textile industry. Words such as"textile","industry","development","technology","production","process","study", and"management" are part of this cluster. The words in this cluster are mostly positive in both dimensions, suggesting these concepts may be closely related in the literature and are central to discussions on the textile industry.

Cluster 2 focuses on aspects related to research, risks, and the basis of innovations. Words include"risk","innovation","research","based", and"products". These words, especially"based","risk", and"products", have a negative score in Dim.2, hinting that they might be set in contrast or opposition to some of the concepts in cluster 1.

"Textiles" is the only word in cluster 3, and it is distinctly different from"textile" found in cluster 1. It has negative scores in both dimensions, which suggests it might represent a different context or perspective in the literature compared to cluster 1's"textile".

Cluster 4 contains only the word "risks", which stands out due to its high positive score in Dim.2. This suggests that while"risk" (cluster 2) is closely related to research

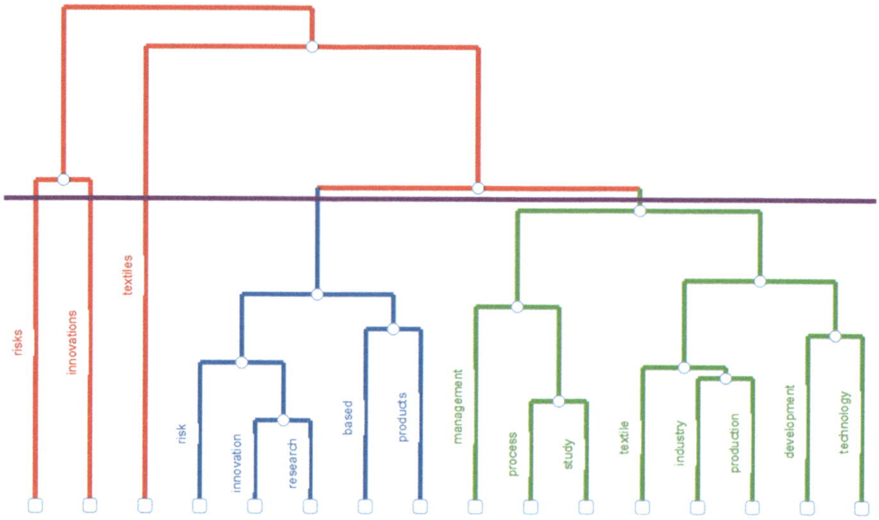

Fig. 15 Factorial analysis dendrogram of the researched publications

and innovation,"risks" might be discussed in a different context, possibly with respect to broader industry challenges or externalities.

Cluster 5 "Innovations" is the sole word in this cluster and has a notably negative score in Dim.1. This indicates that"innovations", as a plural, might have a different or more specialized discourse in the literature compared to "innovation" in cluster 2 (Fig. 15).

Discussions around the textile industry (cluster 1) emphasize not just the core industry but also its processes, technological development, and management. Research, risks, and innovation (cluster 2) are central themes, but their context differs when terms shift from singular to plural (like"risk" vs."risks" and"innovation" vs."innovations"). The differentiation between "textile" and"textiles" might suggest varying focuses on the industry as a whole versus specific products or aspects (Table 4).

The factorial analysis provides a structured visualization of the core themes in risk and innovation management of small businesses in the textile industry. The multi-dimensional approach offers a comprehensive overview, revealing the intricate balance between practical operations, innovations, risks, and broader industry aspects. Understanding these clusters can help industry players and researchers direct their focus effectively for enhanced innovation and risk management.

Table 4 Factorial analysis indicators of the researched publications

Word	Dim.1	Dim.2	Cluster
Textile	0.16	0.36	1
Technology	–0.05	0	1
Study	0.97	0.41	1
Risks	–0.33	0.74	4
risk	0.39	–0.35	2
Research	0.43	–0.15	2
Products	0.07	–0.45	2
Production	0.18	0.57	1
Process	0.85	0.41	1
Management	0.6	0.17	1
Innovations	–1	0.37	5
Innovation	0.48	–0.11	2
Industry	0.34	0.48	1
Development	0.25	0.08	1
Based	0.36	–0.61	2

3 Conclusion

The trajectory of research publications on risk and innovation management in the textile industry over the past four decades indicates a growing academic and practical interest in the subject, with significant spikes possibly connected to global events such as the 2008 financial crisis and the COVID-19 pandemic. Despite an increasing volume of research articles over time, citation metrics reveal that the impact or significance of a paper is not always related to the quantity of output, emphasizing the need to focus on quality. A notable diversity in research contributions from countries worldwide underscores the global importance of risk and innovation management in the textile sector. Emerging markets, like Brazil and India, are making notable strides in academic influence. The trend among institutions, such as Zhejiang University and Lodz University of Technology, suggests a long-term commitment and steady growth in contributions to the domain. Lotka's law's application reveals a skewness in author productivity, with the majority contributing once, while a small percentage are significantly more productive.

Given the increasing importance of the domain, academic institutions and industry stakeholders should collaborate closely, leveraging the knowledge of the past and present to craft future innovation strategies.

Emerging markets should be encouraged and supported to continue their research efforts, possibly opening doors to new perspectives and innovative solutions in risk management. Journals and researchers should emphasize quality over quantity, given that impactful papers can lead to greater advancements in the field. Academic institutions newly venturing into the domain, like RMIT University and the University of Moratuwa, should be observed for potential fresh insights and methodologies.

The vast number of authors contributing singular documents might indicate a wide range of perspectives, innovations, and ideas. For small businesses, this could mean a plethora of approaches to risk and innovation management. The few authors producing multiple papers can be considered as potential thought leaders or experts in their domain. Businesses can benefit by closely following their works, partnering, or consulting with them for specialized insights. Small businesses can identify and collaborate with the more productive authors, leveraging their expertise for business growth, innovative projects, or risk mitigation strategies.

The analysis provides a snapshot of the academic landscape of risk and innovation management in the textile industry. Different institutions have varying levels of engagement in the domain, with some being early adopters and consistent contributors, while others are just beginning to delve into the topic. The increasing numbers in more recent years might be indicative of the growing relevance and importance of the subject in both academic and industry circles.

For businesses, policymakers, and researchers, this bibliometric analysis can serve multiple purposes. It can guide collaboration opportunities, identify leading research hubs, and provide a roadmap for future investment in research and development. While Asia emerges as a significant player, the collective strength of European nations and consistent contributions from North America and other regions highlight the collaborative and widespread efforts to advance the industry.

References

Araújo, R., Santos, G., da Costa, J. B., & Sá, J. C. (2019). The quality management system as a driver of organizational culture: An empirical study in the Portuguese textile industry. *Quality Innovation Prosperity, 23*(1), 1–24.

Ayres, R.U. (1990). Technological transformations and long waves. Part I. *Technological forecasting and social change, 37*(1), 1–37.

Behera, M., Nayak, J., Banerjee, S., Chakrabortty, S., & Tripathy, S. K. (2021). A review on the treatment of textile industry waste effluents towards the development of efficient mitigation strategy: An integrated system design approach. *Journal of Environmental Chemical Engineering, 9*(4), 105277.

Bockmühl, D. P., Schages, J., & Rehberg, L. (2019). Laundry and textile hygiene in healthcare and beyond. *Microbial Cell, 6*(7), 299.

Dangelico, R. M., Pontrandolfo, P., & Pujari, D. (2013). Developing sustainable new products in the textile and upholstered furniture industries: Role of external integrative capabilities. *Journal of Product Innovation Management, 30*(4), 642–658.

de Oliveira Neto, G. C., Correia, J. M. F., Silva, P. C., de Oliveira Sanches, A. G., & Lucato, W. C. (2019). Cleaner Production in the textile industry and its relationship to sustainable development goals. *Journal of Cleaner Production, 228*, 1514–1525.

Denuwara, N., Maijala, J., & Hakovirta, M. (2019). Sustainability benefits of RFID technology in the apparel industry. *Sustainability, 11*(22), 6477.

Ding, W., Chen, D., Ye, J. (2020). Research and practice of whole process product design innovation theory system based on service design thinking. In G. Di Bucchianico, C. Shin, S. Shim, S. Fukuda, G. Montagna, & C. Carvalho (Eds.), *Advances in industrial design. AHFE 2020. Advances in intelligent systems and computing* (Vol. 1202). Springer, Cham.

Dolez, P. I., & Vu-Khanh, T. (2009). Recent developments and needs in materials used for personal protective equipment and their testing. *International Journal of Occupational Safety and Ergonomics, 15*(4), 347–362.

Donges, A., & Selgert, F. (2019). Technology transfer via foreign patents in Germany, 1843–77. *The Economic History Review, 72*(1), 182–208.

Dunne, C. P., Askew, P. D., Papadopoulos, T., Gouveia, I. C., Ahonen, M., Modic, M., & Keinänen-Toivola, M. M. (2020). Antimicrobial coating innovations to prevent infectious disease: A consensus view from the AMiCl COST Action. *Journal of Hospital Infection, 105*(2), 116–118.

Sugg, B. (2022). Circular textiles innovation during COVID-19: not the silver lining some had hoped for. *Journal of Fashion Marketing and Management, ahead-of-print* (ahead-of-print), 1–16

Fu, C., Wang, Z., Gao, Y., Zhao, J., Liu, Y., Zhou, X., & Yang, P. (2023). Sustainable polymer coating for stainproof fabrics. *Nature Sustainability*, 1–11.

Gu, C., Xu, C., Zhou, Q., Shen, C., Ma, C., Liu, S., & Li, F. (2021). Congener-and isomer-specific Perfluorinated compounds in textile wastewater from Southeast China. *Journal of Cleaner Production, 320*, 128897.

Gutarowska, B., Pietrzak, K., Machnowski, W., & Milczarek, J. M. (2017). Historical textiles–a review of microbial deterioration analysis and disinfection methods. *Textile Research Journal, 87*(19), 2388–2406.

Ho, H. T. T., & Watanabe, T. (2020). The dynamics of textile firm responses to the restriction of endocrine disrupting surfactants: A Vietnam case study. *Journal of Risk Research, 23*(4), 504–521.

Ikram, M. (2022). Transition toward green economy: Technological Innovation's role in the fashion industry. *Current Opinion in Green and Sustainable Chemistry, 100657*.

Jiang, H., Liu, J., & Han, W. (2016). The status and developments of leather solid waste treatment: A mini-review. *Waste Management & Research, 34*(5), 399–408.

Kafuku, J. M. (2019). Factors for effective implementation of lean manufacturing practice in selected industries in Tanzania. *Procedia Manufacturing, 33*, 351–358.

Köhler, A. R., & Som, C. (2014). Risk preventative innovation strategies for emerging technologies the cases of nano-textiles and smart textiles. *Technovation, 34*(8), 420–430.

Ledda, A., Palomba, R. (2020). Innovative wearable systems for improving workers' safety. In T. Ahram, R. Taiar, S. Colson, & A. Choplin (Eds.), *Human interaction and emerging technologies. IHIET 2019. Advances in intelligent systems and computing* (Vol. 1018). Springer, Cham.

Levanen, J., Uusitalo, V., Härri, A., Kareinen, E., & Linnanen, L. (2021). Innovative recycling or extended use? Comparing the global warming potential of different ownership and end-of-life scenarios for textiles. *Environmental Research Letters, 16*(5), 054069.

Li, M., & Ma, D. (2016). Research on the officials evaluation system, enterprise pollution discharge and technology innovation under corporate social responsibility base on panel data. *Journal of Computational and Theoretical Nanoscience, 13*(12), 9841–9847.

Lugani, Y., Sooch, B. S., Singh, P., & Kumar, S. (2021). Nanobiotechnology applications in food sector and future innovations. In *Microbial biotechnology in food and health* (pp. 197–225). Academic Press.

Luque, A., & Herrero-García, N. (2019). How corporate social (ir) responsibility in the textile sector is defined, and its impact on ethical sustainability: An analysis of 133 concepts. *Corporate Social Responsibility and Environmental Management, 26*(6), 1285–1306.

Parker, H. (2000). Interfirm collaboration and the new product development process. *Industrial Management & Data Systems, 100*(6), 255–260.

Pfoser, S., Brandner, M., Herman, K., Steinbach, E., Brandtner, P. & Schauer, O. (20210). Sustainable transport packaging: evaluation and feasibility for different use cases. *LOGI—Scientific Journal on Transport and Logistics, 12*(1), 159–170.
Phan, T. T. H., Tran, H. X., Le, T. T., Nguyen, N., Pervan, S., & Tran, M. D. (2020). The relationship between sustainable development practices and financial performance: A case study of textile firms in Vietnam. *Sustainability, 12*(15), 5930.
Romao, M., Simoes, A. (2020). A holistic approach to inclusive mobility under resilience and sustainability goals: a pilot project in Brazil. In G. Di Bucchianico, C. Shin, S. Shim, S. Fukuda, G. Montagna, & C. Carvalho (Eds.), *Advances in Industrial Design. AHFE 2020. Advances in Intelligent Systems and Computing* (Vol. 1202). Springer, Cham.
Rosenberg, M., Ilić, K., Juganson, K., Ivask, A., Ahonen, M., Vrček, I. V., & Kahru, A. (2019). Potential ecotoxicological effects of antimicrobial surface coatings: A literature survey backed up by analysis of market reports. *PeerJ, 7*, e6315.
Seidu, R. K., Eghan, B., & Acquaye, R. (2023). A review of circular fashion and bio-based materials in the fashion industry. *Circular Economy and Sustainability*, 1–23
Shafi, M., Yin, L., & Yuan, Y. (2021). Revival of the traditional handicraft enterprising community in Pakistan. *Journal of Enterprising Communities: People and Places in the Global Economy, 15*(4), 477–507.
Shah, M. A., Pirzada, B. M., Price, G., Shibiru, A. L., & Qurashi, A. (2022). Applications of nanotechnology in smart textile industry: A critical review. *Journal of Advanced Research, 38*, 55–75.
del Solar, M. E. (2019). Highland peasant weavers: Empowered women, heritage keepers and home providers. Incakunaq Ruwaynin project, Cusco, Peru. *TEXTILE, 17*(2), 168–181.
Tsai, W. Y., & Su, C. J. (2022). Digital transformation of business model innovation. *Frontiers in Psychology, 13*, 1017750.
Vanegas-López, J. G., Baena-Rojas, J. J., López-Cadavid, D. A., & Mathew, M. (2021). International market selection: An application of hybrid multi-criteria decision-making technique in the textile sector. *Review of International Business and Strategy, 31*(1), 127–150.
Wang, C., Xia, K., Zhang, Y., & Kaplan, D. L. (2019). Silk-based advanced materials for soft electronics. *Accounts of Chemical Research, 52*(10), 2916–2927.
Wescoat, J. L., Jr., & Rawoot, S. (2021). Blue-green urban infrastructure in Boston and Bombay (Mumbai): A macro-historical geographic comparison. *ZARCH, 15*, 36–51.

António Abreu is currently an Associated Professor with Habilitation of Industrial Engineering at Polytechnic University of Lisbon. He received the Habilitation in Industrial Engineering in 2024 of University of Aveiro and the degree of Ph.D. from Industrial Engineering of New University of Lisbon in 2007. His current research interests are logistics, project management, quality management, open-innovation, risk management, and lean management.

Impact of Effective Leadership Styles to Achieve the Sustainable Growth of Manufacturing Firms

João Lucas

Abstract The article examines the crucial role of leadership in driving sustainable growth within the manufacturing sector, with a specific focus on the textile industry. The aim of this study is to analyse how different leadership styles contribute to achieving sustainability in manufacturing processes, a sector pivotal to global economic development and environmental stewardship. Employing a comprehensive bibliometric analysis using R and Biblioshiny, the study systematically reviews and synthesizes existing literature. This methodological approach facilitates an in-depth exploration of citation patterns, publication trends, and thematic clusters, enabling the identification of influential works and emerging trends in the field of leadership and sustainability within textile manufacturing. Findings reveal that sustainable development is a central theme in the context of manufacturing firms' growth. Leadership styles, particularly transformational and participative, are identified as key drivers in fostering sustainable practices. The study highlights the integration of environmental management and technological advancements as vital components in the pursuit of sustainability in the textile industry. It also uncovers the significant role of human resource management and the necessity for adaptive and ethical leadership competencies to address the industry's unique challenges. This research contributes to the understanding of the dynamic interplay between leadership styles and sustainable manufacturing practices.

J. Lucas (✉)
Instituto Piaget Instituto Superior de Estudos Interculturais e Transdisciplinares de Almada (ISEIT), Avenida Jorge Peixinho, No. 30 Quinta da Arreinela, 2805-059 Almada, Portugal
e-mail: joaolrlucas@gmail.com

Department of Management, University of Évora, Largo dos Colegiais, 2, 7002-554 Évora, Portugal

Lisbon Accounting and Business School, Polytechnic Institute of Lisbon (ed.), *Industry and Innovation: Textile Industry*, SDGs and Textiles,
https://doi.org/10.1007/978-3-031-57804-5_2

1 Introduction

The manufacturing industry, characterized by its reliance on efficient processes, innovation, and adaptability, demands leaders who can not only manage the present but also strategically plan for the future. The type of leadership employed can significantly influence a firm's ability to innovate, maintain competitive advantage, and achieve long-term success. This makes understanding the impact of different leadership styles on the sustainability of manufacturing firms both relevant and essential. Sustainable growth refers to the ability of a business to consistently increase its profits, market share, and stakeholder value without depleting its resources or harming the environment. In manufacturing, this concept takes on added dimensions of resource management, environmental stewardship, and technological innovation. The shift towards sustainable practices is not just a trend but a necessity, driven by regulatory pressures, environmental concerns, and consumer preferences.

The manufacturing sector, especially the textile industry, stands at a crucial juncture where sustainability and growth intersect with leadership. The textile industry, a significant contributor to global economies, faces unprecedented challenges and opportunities in the modern era. This study, through a comprehensive bibliometric analysis, investigates the pivotal role of effective leadership styles in fostering sustainable growth in manufacturing firms, with a keen focus on the textile sector. The textile industry, characterized by its dynamic market demands, rapid technological advancements, and increasing environmental concerns, demands a leadership approach that is not only innovative and adaptive but also responsible and forward-looking. This sector, a significant employer and a major contributor to global trade, has often been at the forefront of industrial revolutions. However, in recent times, it has been grappling with challenges ranging from environmental sustainability to ethical practices in the supply chain.

Leadership styles significantly impact organizational outcomes, including employee motivation, innovation, and adaptability. In the context of the textile industry, the choice of leadership style can determine the success of sustainability initiatives. For instance, transformational leadership can inspire and motivate employees towards a shared vision of sustainability, while transactional leadership may focus on efficiency and short-term goals. The study aims to explore various leadership styles and their implications for sustainable growth in textile manufacturing firms.

This study employs a bibliometric analysis to investigate the relationship between leadership styles and sustainable growth in the textile industry. Bibliometric analysis offers a systematic approach to reviewing and synthesizing existing literature, providing a comprehensive understanding of the research landscape. By analysing citation patterns, publication trends, and thematic clusters, this methodology enables the identification of influential works, emerging trends, and knowledge gaps in the field of leadership and sustainability in textile manufacturing.

The primary aim of this study is to elucidate the impact of different leadership styles on achieving sustainable growth in textile manufacturing firms. The objectives of this bibliometric analysis are to map the existing literature on leadership styles and sustainability in textile manufacturing; analyse the influence of different leadership styles on sustainable practices and outcomes; identify research trends and gaps to guide future studies in this field.

The findings of this study hold relevance for a range of stakeholders in the textile industry. For business leaders and managers, insights into effective leadership styles can inform strategies to foster sustainable growth. Academics and researchers will benefit from a consolidated view of the literature, identifying areas for further exploration. Policymakers can use the findings to develop guidelines and frameworks that encourage sustainable practices in the industry.

This study seeks to bridge the gap in understanding the role of leadership in driving sustainable growth in the textile manufacturing sector. By employing a bibliometric analysis, the study offers a nuanced and evidence-based perspective on how different leadership styles can shape the trajectory towards sustainability in this dynamic and critical industry. The insights garnered from this analysis contribute significantly to both academic discourse and practical applications in the realm of sustainable manufacturing.

2 Systematic Literature Review

The bibliometric analysis presented in Table 1 illustrates a dynamic and impactful field of study, marked by recent and highly cited research, extensive collaboration, and a diversity of sources and approaches. This underscores the importance and complexity of effective leadership styles in driving sustainable growth in manufacturing firms. The analysis encompassed a wide range of sources, including 181 different journals, books, and other publications. This indicates a comprehensive and diverse approach to understanding the subject matter, with a total of 253 documents analysed.

The average age of the documents is approximately 5 years, suggesting that the research is relatively current and reflective of recent trends and developments in the field of effective leadership and sustainable growth in manufacturing. The average number of citations per document, standing at 11.37, highlights the significant impact and relevance of the research in this area. This suggests that the topics covered are of considerable interest and importance within the academic community.

The high number of Keywords Plus (ID) and Author's Keywords (DE), totaling 985 and 781 respectively, demonstrates a rich and varied use of terminology and concepts in the field. This diversity in keywords indicates a broad range of themes and subtopics being explored within the overarching subject.

With 614 authors involved, including 61 authors of single-authored documents, the research field appears to be highly collaborative. The presence of 74 single-authored documents also points to individual contributions that offer unique perspectives.

Table 1 Main information about data

Description	Results
Sources (Journals, Books, etc.)	181
Documents	253
Document average age	4.97
Average citations per doc	11.37
Keywords Plus (ID)	985
Author's keywords (DE)	781
Authors	614
Authors of single-authored docs	61
Single-authored docs	74
Co-authors per doc	2.69
International co-authorships %	27.67
Articles	153
Books	4
Book chapters	15
Book chapter conference reviews	1
Conference papers	57
Conference reviews	11
Notes	1
Short surveys	2

The average number of co-authors per document (2.69) and the percentage of international co-authorships (27.67%) show a healthy level of collaboration and international engagement in the research, fostering a global understanding and approach to the topic. The distribution of document types (e.g., articles, books, conference papers, reviews) demonstrates the multifaceted nature of the research and the different avenues through which knowledge in this area is disseminated.

The co-occurrence network based on studied publications with the key words 'leadership styles', 'sustainability', and 'manufacture' reveals significant insights into the interconnections and relevance of various concepts within this research domain (Fig. 1). The analysis, focusing on key metrics such as betweenness, closeness, and PageRank, uncovers the pivotal role of certain nodes in the network.

The most prominent node is 'sustainable development', which demonstrates the highest betweenness, closeness, and PageRank scores. This indicates its central role in connecting various concepts and its significant influence within the network. The focus on sustainable development reflects the growing emphasis on integrating sustainability into manufacturing practices.

Other important nodes include 'manufacture', 'industrial research', 'environmental management', 'life cycle', and 'environmental impact'. These nodes, though

Fig. 1 Co-occurrence network based on studied publications

scoring lower than 'sustainable development', still play crucial roles in the network, indicating their importance in the study of sustainable manufacturing.

The node 'leadership', placed in a different cluster, also shows significant betweenness and PageRank scores, underscoring the critical role of leadership in driving sustainable practices in manufacturing firms. This is in line with the focus of the article, which explores the impact of leadership styles on sustainable growth in manufacturing.

Interestingly, nodes such as 'manufacturing', 'commerce', 'china', 'innovation', and 'human' in the second cluster, while having lower scores, indicate emerging areas of interest that are gaining traction in the research community.

The co-occurrence network highlights the multifaceted nature of sustainable manufacturing, encompassing a broad spectrum of elements from environmental management to leadership and innovation. This underscores the need for a comprehensive approach in research and practice to achieve sustainable growth in manufacturing firms.

The interplay between leadership styles and the sustainable growth of manufacturing firms, particularly in the textile industry, has garnered considerable attention in contemporary research. This literature review synthesizes findings from various studies, focusing on how different leadership approaches impact the sustainable development of textile manufacturing firms.

A significant body of research has highlighted the role of transformational leadership in enhancing organizational performance and innovation in the textile sector. Studies have shown that transformational leaders, by virtue of their inspirational motivation and intellectual stimulation, foster a conducive environment for creativity and sustainable practices (Awan et al., 2021; Gull et al., 2022). This leadership style, characterized by its focus on visionary thinking, has been linked to improved employee commitment and operational efficiency, which are crucial for the long-term growth of textile firms (Hassan et al., 2020; Nasir et al., 2022).

In contrast, transactional leadership, which relies on structured rewards and penalties, has been associated with short-term efficiency gains in textile manufacturing (Alrowwad et al., 2020; Edwards & Gill, 2012). While effective in certain operational contexts, its limitation lies in potentially stifling innovation, which is essential for sustainability in a rapidly evolving industry (Hossain et al., 2022; Mead et al., 2022). The role of participative leadership, where decision-making is shared with employees, has been emphasized in recent studies. This leadership style has been linked to higher levels of employee engagement and satisfaction, leading to reduced turnover rates and enhanced sustainability of the workforce in the textile sector (Ali et al., 2019; Hayat Bhatti et al., 2019; Li et al., 2019).

The application of situational leadership, which adapts to the context and readiness of employees, has been found to be particularly effective in the dynamic environment of textile manufacturing (Javed et al., 2021; Sharma et al., 2021). It allows leaders to toggle between directive and supportive behaviours, which is essential in an industry characterized by frequent changes in fashion trends and consumer preferences (Ahmad et al., 2020; Paço et al., 2021).

The importance of ethical leadership in driving CSR initiatives has been increasingly recognized. Ethical leaders in the textile industry are seen as instrumental in promoting sustainable practices and ensuring compliance with environmental and social standards, which are critical for the long-term viability of these firms (Luque and Herrero–García, 2019; Ullah et al., 2021).

Innovation is a key driver of sustainable growth in the textile industry. Leadership styles that encourage experimentation and risk-taking have been found to positively influence innovation. This is particularly relevant in the context of adopting sustainable technologies and practices in textile manufacturing (Abbas et al., 2020; Harsanto et al., 2023).

The global nature of the textile industry calls for leaders who can navigate cross-cultural challenges. Studies have highlighted the need for global leadership competencies, including cultural intelligence and adaptability, to effectively manage diverse teams and operations spread across different geographical locations (Kim, 2020; Kim & Kim, 2017).

The role of leadership in enhancing supply chain sustainability in the textile industry has also been explored. Effective leadership is seen as crucial in fostering collaboration and transparency across the supply chain, which is essential for addressing environmental and social issues prevalent in the industry (Chau et al., 2023; Jia et al., 2019).

Despite the recognized importance of effective leadership in achieving sustainable growth, there are challenges in implementation. These include resistance to change, the complexity of measuring the impact of leadership styles, and the need for continuous adaptation to changing market and environmental conditions (Baig et al., 2021; Dhar et al., 2022; Singh et al., 2021).

The literature reveals a consensus on the significance of various leadership styles in driving sustainable growth in the textile manufacturing industry. While transformational and participative leadership styles are generally favoured for their positive impact on innovation, employee engagement, and organizational performance, the

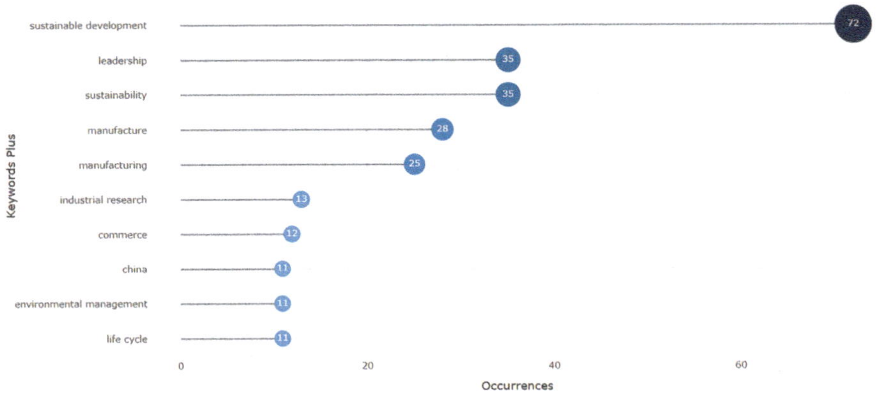

Fig. 2 Most frequently occurring words in the studied documents

importance of situational, ethical, and global leadership competencies cannot be understated. The challenge for textile firms lies in integrating these leadership styles in a balanced manner to address the unique demands of the industry and ensure long-term sustainability.

Figure 2, which illustrates the most frequently occurring words in the reviewed literature, sheds light on the central themes and focal points in the discourse surrounding sustainable growth in manufacturing firms. The prominence of 'sustainable development' with 72 occurrences underscores its criticality as the overarching goal in this sector. This aligns with the growing global emphasis on sustainability, indicating that manufacturing firms are increasingly integrating sustainable practices into their core strategies.

The equal emphasis on 'leadership' and 'sustainability' (35 occurrences each) suggests that effective leadership is considered pivotal in steering manufacturing firms towards sustainable practices. This implies a strong link between leadership styles and the successful implementation of sustainable development strategies in the manufacturing sector. The frequent mention of 'manufacture' and 'manufacturing' (28 and 25 occurrences, respectively) reaffirms the focus of the study, emphasizing the sector's significance in the discourse. Meanwhile, 'industrial research' (13 occurrences) indicates an ongoing investment in exploring new methodologies and technologies for sustainable manufacturing. The inclusion of 'commerce' (12 occurrences) and 'China' (11 occurrences) highlights the influence of global trade dynamics and the role of major manufacturing countries in shaping sustainable manufacturing trends. Finally, the equal frequency of 'environmental management' and 'life cycle' (11 occurrences each) underlines the importance of holistic approaches to environmental stewardship within the manufacturing process, from inception to product end-of-life.

Figure 2 reveals a clear focus on integrating sustainability into manufacturing through effective leadership, research, and global collaboration. It also emphasizes the need for comprehensive environmental management practices that consider the

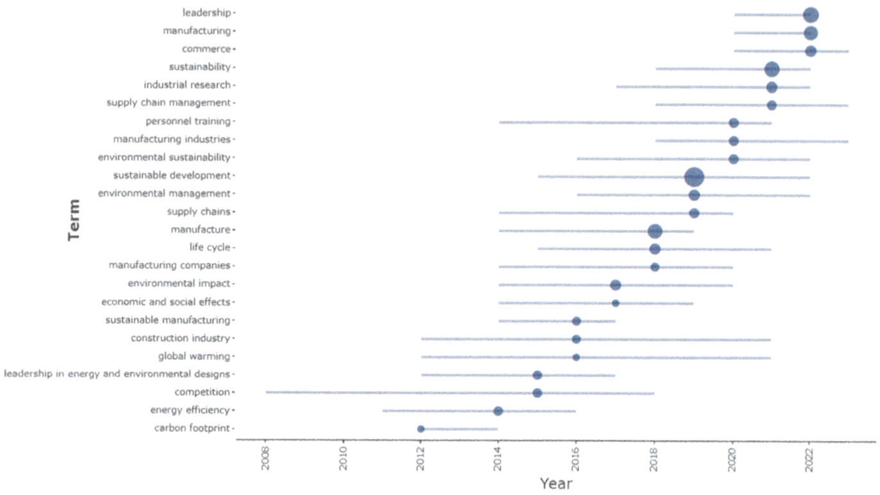

Fig. 3 Trend topics in the studied documents

entire life cycle of manufactured products. This serves as a testament to the evolving landscape of manufacturing, where sustainability and leadership are intertwined in driving forward the industry's sustainable growth.

The most frequent topic is manufacture (Fig. 3), with a significant increase in discussions from 2014 to 2019, indicating a strong focus on the core aspects of manufacturing. This trend suggests that the foundation of sustainable growth in manufacturing firms lies in continuously evolving manufacturing practices. With the highest frequency after 'manufacture', this topic's prominence from 2014 to 2020 reflects the increasing concern and focus on environmental sustainability in manufacturing processes. The leadership in manufacturing firms is evidently prioritizing environmentally friendly practices.

Energy Efficiency and Leadership in Energy and Environmental Designs (LEED): The topics of energy efficiency and leadership in energy and environmental designs (LEED), with a median focus around 2014–2015, imply that there has been a considerable emphasis on energy conservation and sustainable design in manufacturing. The leadership trends in these areas indicate a shift towards more energy-efficient manufacturing processes.

The span from 2012 to 2021 for this topic suggests a long-term, sustained interest, aligning with the broader push towards sustainable practices in this sector. This could be indicative of the construction industry's evolving role in manufacturing and its impact on sustainability goals. Peaking between 2014 and 2017, this topic's prominence indicates that sustainable practices in manufacturing have been a key area of focus, aligning with global trends towards sustainability. These topics, with a consistent presence over the decade, underscore the role of manufacturing leadership in addressing global environmental challenges. The leadership in manufacturing firms seems increasingly aware of their role in mitigating climate change impacts.

The attention to economic and social effects from 2014 to 2019 reflects an understanding of the broader implications of manufacturing practices, beyond just the environmental impact, encompassing socio-economic aspects as well. The increasing frequency of 'competition' from 2008 to 2018 suggests that competitive strategies are being increasingly viewed through the lens of sustainability, indicating a shift in how manufacturing firms position themselves in the market.

The trends identified in Fig. 3 highlight a significant shift in the manufacturing sector towards sustainable practices. This shift is not just limited to environmental aspects but also includes economic and social dimensions. Effective leadership in these firms appears to be increasingly focused on integrating sustainability into all facets of manufacturing, demonstrating a holistic approach towards achieving sustainable growth.

Figure 4 of the thematic map clearly highlights sustainable development as the central theme in the context of manufacturing firms, underlining the critical interplay between environmental management, human and organizational factors, technological advancements, global economic trends, and methodological diversity in achieving sustainable growth in the manufacturing sector.

The most frequent occurrences and central theme revolve around 'sustainable development' with the highest betweenness centrality, closeness centrality, and PageRank centrality scores. This indicates that sustainable development is the core topic in the context of manufacturing firms' growth, emphasizing the importance of environmental and social considerations alongside economic growth. Terms like 'manufacture', 'manufacturing industries', and 'sustainable manufacturing' are prominent, signifying the critical role of manufacturing processes and practices in achieving sustainable development goals. The occurrence of terms such as 'environmental management', 'environmental sustainability', and 'energy efficiency' suggests a strong focus on environmental aspects within the manufacturing sector. This reflects the growing awareness and integration of environmentally friendly practices in manufacturing processes.

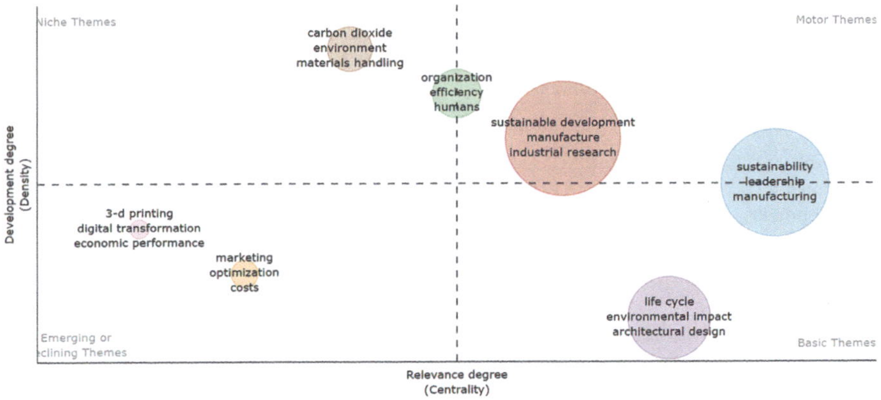

Fig. 4 Thematic map for the studied documents

The presence of terms like ‘human resource management’, ‘personnel training’, and ‘leadership’ underscores the importance of human factors and leadership styles in driving sustainable practices within manufacturing firms. Effective leadership is crucial in steering organizational values and practices towards sustainability. Terms like ‘industrial research’, ‘supply chains’, and ‘Industry 4.0’ highlight the role of technology and research in advancing sustainable manufacturing. This suggests an ongoing trend towards digital transformation and the adoption of advanced technologies to enhance sustainability in manufacturing.

The occurrence of terms such as ‘global leadership’, ‘economic and social effects’, and ‘commerce’ indicates the global scale and economic implications of sustainable manufacturing practices. This emphasizes the need for manufacturing firms to align their strategies with global sustainability standards and economic policies. The presence of diverse methodological terms like ‘factor analysis’, ‘multivariant analysis’, and ‘structural equation modelling’ reflects the complex and multifaceted nature of research in this area, requiring a range of analytical tools to understand the various dimensions of sustainable growth in manufacturing firms.

The cluster analysis, based on Callon’s centrality and density metrics, identifies the relative importance and coherence of various themes (Table 2). The cluster of ‘sustainable development’ emerges as a predominant theme, with the highest frequency of occurrence and strong rankings in both centrality and density. This indicates that sustainable development is not only a widely discussed topic but also holds a central position in the literature, suggesting its pivotal role in driving sustainable growth in manufacturing firms. Similarly, ‘sustainability’ as a cluster also shows a high frequency and strong rankings, although slightly lower in centrality compared to sustainable development. This reinforces the idea that sustainability is a key component in discussions about the manufacturing sector, albeit with a slightly different focus compared to the broader concept of sustainable development.

The ‘organization’ cluster, despite having a lower frequency, exhibits a high density, indicating that while it might not be as widely discussed, the discussions that do happen are highly cohesive and relevant. This suggests that organizational aspects are critical in implementing effective leadership styles for sustainable growth.

Table 2 Cluster analysis according to the thematic map

Cluster	Callon centrality	Callon density	Rank centrality	Rank density	Cluster frequency
Sustainable development	22.6139	117.8308	5	5	414
Sustainability	24.1194	110.0900	7	4	314
Organization	5.1167	128.5042	4	6	32
Life cycle	23.8482	93.4704	6	1	120
Marketing	1.25	93.5185	2	2	14
Carbon dioxide	2.6111	255.5556	3	7	27
3-D printing	1	105	1	3	10

The 'life cycle' cluster, with a high density but lower centrality, points to a specialized, yet significant area within the sustainable manufacturing discourse. This could imply that life cycle considerations are integral to sustainability discussions but are not as central to the overall conversation as sustainable development or sustainability. 'Marketing' and 'carbon dioxide' clusters, although having lower frequencies, show high density and centrality in some aspects. This indicates that these areas, while niche, are significant and well-integrated into the broader conversation about sustainable growth in manufacturing. The '3-D printing' cluster, despite being the least frequent, shows a high density and the highest centrality ranking. This might reflect a growing recognition of the role of innovative technologies like 3-D printing in driving sustainable growth in the manufacturing sector.

The analysis suggests a diverse yet interconnected landscape of themes, where broader concepts like sustainable development and sustainability are central and highly discussed, while specialized areas like life cycle considerations, marketing strategies, carbon dioxide management, and technological innovations like 3-D printing also play significant roles in the discourse on sustainable growth in manufacturing.

2.1 The Common Research Trends and Meta-Analysis

The annual scientific production on effective leadership styles for sustainable growth in manufacturing firms, as depicted in Fig. 5, shows a notable upward trend over the two-decade period from 2003 to 2023. The initial years, from 2003 to 2007, were characterized by a modest output, with an average of less than 2 articles per year, indicating a nascent stage of academic interest in this area. A significant increase in scholarly attention becomes evident from 2008 onwards, with a notable peak in 2009. This surge could be attributed to the increasing recognition of the importance of leadership in driving sustainable practices within the manufacturing sector, possibly influenced by broader environmental and economic challenges of that time. The period from 2010 to 2016 witnessed fluctuations in output, reflecting the evolving nature of research themes and possibly the impact of global economic and industrial trends on academic focus areas.

A remarkable spike is observed from 2017 onwards, peaking in 2022 with 45 articles. This surge can be interpreted as a response to the growing imperative for sustainable practices in manufacturing, driven by global environmental concerns, technological advancements, and shifting market demands. The slightly lower number of articles in 2023, although still high, might suggest a stabilization in research output or a shift in focus to more specific sub-themes within this broad area. It underscores the increasing relevance and urgency of effective leadership in manufacturing firms' journey towards sustainability, reflecting a paradigm shift in academic and industrial circles towards integrating leadership and sustainability as key components of organizational success and resilience.

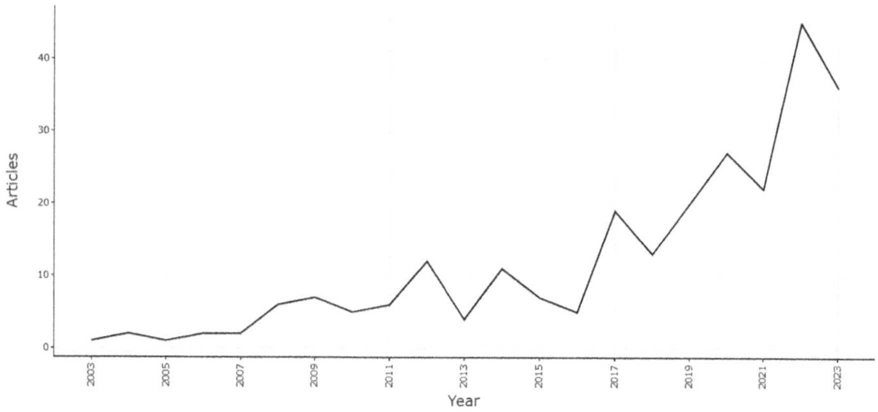

Fig. 5 Annual scientific production on the effective leadership styles for sustainable growth of manufacturing firms

The analysis of average citations per year for the studied documents on the impact of effective leadership styles in manufacturing firms reveals a fluctuating yet insightful trend over the years from 2003 to 2023 (Table 3). In the early years (2003–2007), the citations were relatively low, indicating a nascent interest in this area of research. This trend is exemplified by the minimal average citations per year, with 2005 and 2006 showing particularly low engagement.

There was a notable increase in citations from 2008 onwards, peaking remarkably in 2019. This surge suggests a growing recognition of the importance of leadership styles in the sustainable growth of manufacturing firms, possibly influenced by the evolving challenges and dynamics of the global manufacturing sector.

The subsequent years, particularly 2020–2023, display a mixed pattern in citation averages. While 2021 shows a high average citation per year, akin to the 2019 peak, there is a noticeable drop in 2022 followed by a moderate increase in 2023. This fluctuation could be attributed to various external factors impacting the manufacturing industry and the shifting focus of academic research. The peaks and troughs in the citation patterns over the years also suggest a dynamic field of study, influenced by both internal industry developments and broader economic and social changes. This trend underscores the ongoing relevance and critical nature of this research area, highlighting its importance in both academic and practical realms.

Figure 6 reveals a significant trend in the production of sources over time across various journals. From 2003 to 2013, there was negligible or no activity in the journals 'Sustainability (Switzerland)', 'Journal of Cleaner Production', 'Proceedings of the International Conference on Industrial Engineering and Operations Management', 'Business Strategy and the Environment', and 'Frontiers in Psychology'. This period can be characterized as a latency phase in the field, with minimal academic focus on the intersection of leadership styles and sustainable growth in manufacturing.

Table 3 Average citations per year for the studied documents

Year	Mean TC per art	N	Mean TC per year	Citable years
2003	3	1.00	0.14	21
2004	23.5	2.00	1.18	20
2005	0	1.00	0.00	19
2006	0.5	2.00	0.03	18
2007	1	2.00	0.06	17
2008	14.67	6.00	0.92	16
2009	1.29	7.00	0.09	15
2010	12	5.00	0.86	14
2011	4.83	6.00	0.37	13
2012	8.25	12.00	0.69	12
2013	1	4.00	0.09	11
2014	14.45	11.00	1.44	10
2015	25.57	7.00	2.84	9
2016	11	5.00	1.38	8
2017	8.21	19.00	1.17	7
2018	11.77	13.00	1.96	6
2019	34.75	20.00	6.95	5
2020	15.3	27.00	3.83	4
2021	15.36	22.00	5.12	3
2022	5.69	45.00	2.85	2
2023	3.64	36.00	3.64	1

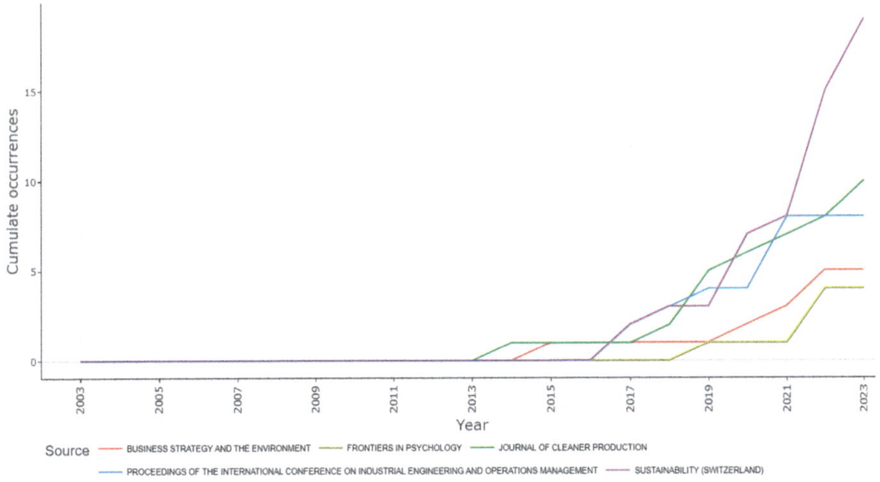

Fig. 6 Sources' production over time

A notable shift occurred from 2014 onwards, with a gradual increase in publications, particularly in 'Sustainability (Switzerland)' and 'Journal of Cleaner Production'. This increase indicates a growing academic interest and recognition of the importance of leadership styles in driving sustainable practices in manufacturing. The year 2022 marked a significant peak, especially for 'Sustainability (Switzerland)', which showed a considerable leap in publications, highlighting its emergence as a leading journal in this area.

The trend reflects a broader academic and industry realization of the critical role of leadership in achieving sustainable growth. The rising number of publications in these journals suggests an expanding body of research and a deepening understanding of how effective leadership can positively influence sustainability and cleaner production practices in the manufacturing sector. This trend is expected to continue, as indicated by the data for 2023, reinforcing the importance of this research area in the current academic and industrial landscape.

The analysis of author production over time in the context of effective leadership styles for the sustainable growth of manufacturing firms reveals a dynamic and evolving landscape of scholarly contributions (Fig. 7). Notably, the frequency of publications and their total citations (TC) per year (TCpY) show a varied pattern among the researchers.

Ahmad N., with publications in 2020 and 2022, shows a significant initial impact with a high TCpY in 2020 that decreases by 2022, indicating a strong initial reception that stabilizes over time. Ahmed's work, spanning from 2021 to 2023, suggests a steady engagement in the field, though the impact as measured by TCpY is modest and shows no citations yet for the most recent year.

Ahmed's trajectory is particularly noteworthy. Starting in 2020 with a high TCpY, this author's most recent publication in 2023 shows a remarkable TCpY of 40, indicating a growing influence in the field. This trend suggests an increasing recognition of Ahmed's work in recent times.

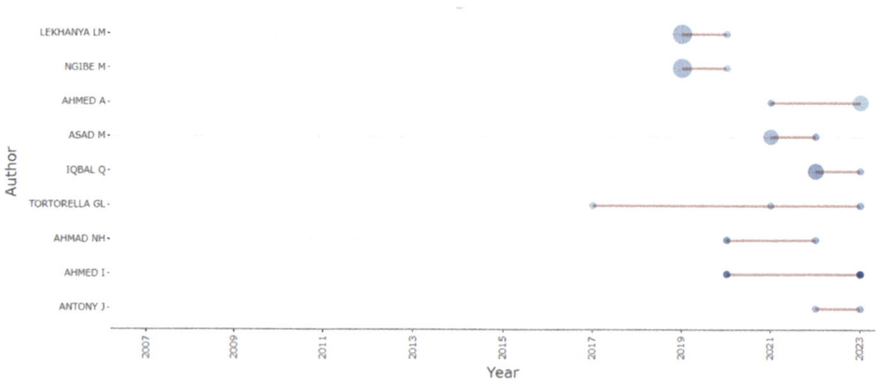

Fig. 7 Authors' production over time

Antony and Asad both demonstrate a consistent presence in the research area across 2022 and 2023, with a gradual increase in TCpY, reflecting a steady accumulation of impact. IQBAL Q's contributions, primarily concentrated in the later years (2022 and 2023), also show a developing influence, though starting from a lower base of citations.

The contributions of Lekhanya and Ngibe, mainly in the earlier phase (2019–2020), had a moderate impact, as indicated by their TCpY, but show a decline in recent years. This could suggest a shift in research focus or the evolving nature of the research topic.

Tortorella's work, spanning from 2017 to 2023, shows a varied pattern, starting with no citations initially, but gaining traction in recent publications, indicating a delayed but growing recognition of their contributions to the field.

The literature on the impact of effective leadership styles on the sustainable growth of manufacturing firms reveals a multifaceted and evolving field (Fig. 8). The importance of leadership in driving sustainable practices in manufacturing has been increasingly recognized, with a focus on how different leadership styles can either facilitate or hinder sustainable growth (Boiral et al., 2015; Raut et al., 2019). The role of transformational leadership in promoting sustainability initiatives has been highlighted, demonstrating that leaders who inspire, intellectually stimulate, and attend to the individual needs of their employees are more likely to foster a culture of sustainability within their organizations (Bhutto et al., 2021).

Additionally, research has explored the influence of transactional leadership on sustainability outcomes. While this leadership style is often contrasted with transformational leadership, findings suggest that its emphasis on clear structures, rewards, and efficiency can also contribute positively to achieving sustainable goals in certain contexts (Graves & Sarkis, 2018). However, the rigidity associated with transactional leadership might limit innovative approaches to sustainability challenges (Khan et al., 2019).

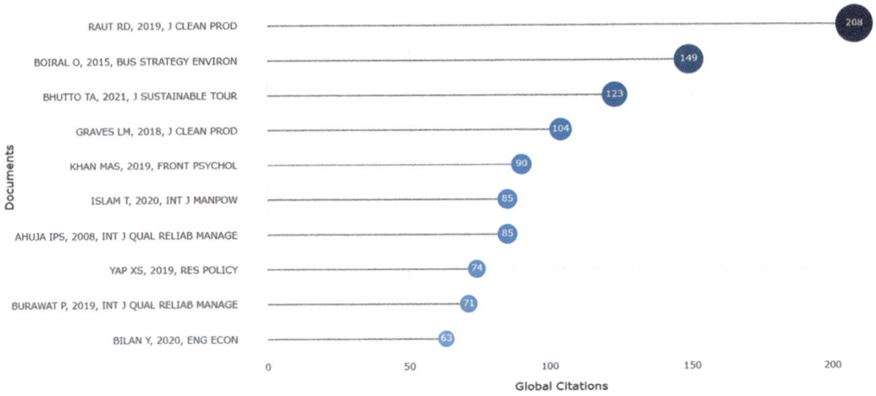

Fig. 8 Most globally cited documents

The participative leadership style, characterized by involving employees in decision-making processes, has been shown to enhance commitment to sustainability initiatives. This approach fosters a sense of ownership and alignment with the organization's sustainability goals among employees (Islam et al., 2021). Furthermore, the role of ethical leadership, which emphasizes honesty, integrity, and fairness, has been found to be crucial in embedding sustainable practices within manufacturing firms. Ethical leaders are seen as role models whose behaviours are emulated by employees, thereby creating a workplace culture that values sustainability (Ahuja & Khamba, 2008).

A few studies have also pointed out the necessity of adaptive leadership in the context of sustainability. As manufacturing firms face rapidly changing environmental standards and market demands, leaders who are flexible and responsive to change are more successful in maintaining sustainable growth (Yap & Truffer, 2019). The impact of leadership on the psychological empowerment of employees has been examined, revealing that empowered employees are more likely to engage in and support sustainability initiatives (Burawat, 2019).

The role of leadership in enhancing corporate social responsibility (CSR) within manufacturing firms has been another focus area. Leaders who prioritize CSR are found to effectively integrate sustainability into their business strategies, leading to long-term benefits for both the organization and society (Bilan et al., 2020).

Most globally cited documents underscore the significance of various leadership styles in promoting sustainable growth in manufacturing firms. While transformational and participative leadership styles are frequently associated with positive sustainability outcomes, the relevance of transactional, ethical, and adaptive leadership styles cannot be overlooked. The effectiveness of these leadership styles appears to be contingent upon the specific context and challenges faced by manufacturing firms. Therefore, a nuanced understanding and strategic application of these leadership styles are essential for fostering sustainable growth in the manufacturing sector.

Figure 9 represents author productivity in the context of manufacturing firm growth, which aligns well with Lotka's law. The vast majority of authors, accounting for 92.2%, have contributed only one document. This highlights a typical pattern in academic publishing where a large number of contributors make singular contributions.

A significantly smaller proportion of authors, 6.7%, have written two documents, further reinforcing Lotka's observation of a steep decline in productivity with increasing output. Those contributing three or four documents represent an even smaller fraction, at 0.7 and 0.3% respectively. Notably, there's an exceptionally rare instance of an author contributing as many as twelve documents, comprising just 0.2% of the total. This distribution is emblematic of the principle that the majority of contributions come from a large pool of less frequent authors, while a minority of authors contribute multiple works. Such a pattern underscores the diverse range of participation and the pivotal role of a small group of highly productive authors in driving research in the field of manufacturing firm growth and leadership.

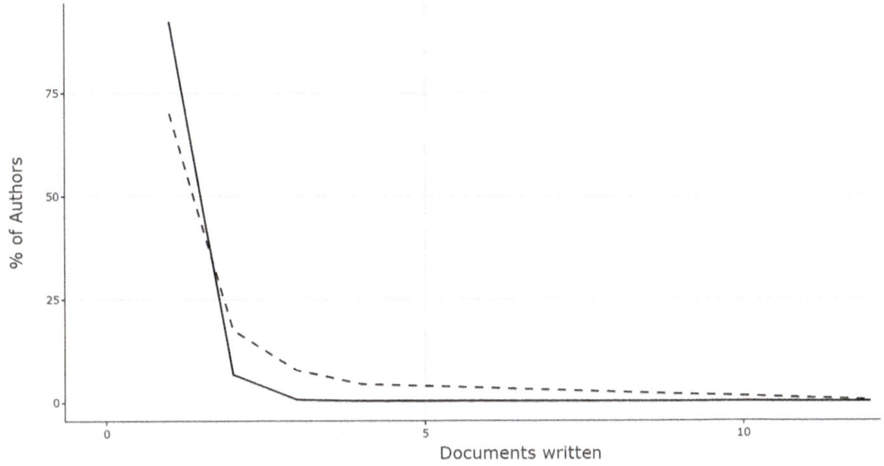

Fig. 9 Author productivity through Lotka's law

Figure 10 highlights significant insights into the research productivity of various academic institutions over time. It is evident that there was a latent phase in the early 2000s, where institutions like the Massachusetts Institute of Technology, Kasetsart University, University of Derby, University of the Punjab, North-West University, and UCSI University showed minimal to no research output in the context of effective leadership styles in manufacturing firms.

This trend underwent a notable change around the mid-2010s. The Massachusetts Institute of Technology, for instance, started contributing to the field in 2016 and demonstrated a steady increase in its research output, reaching a peak of 6 articles

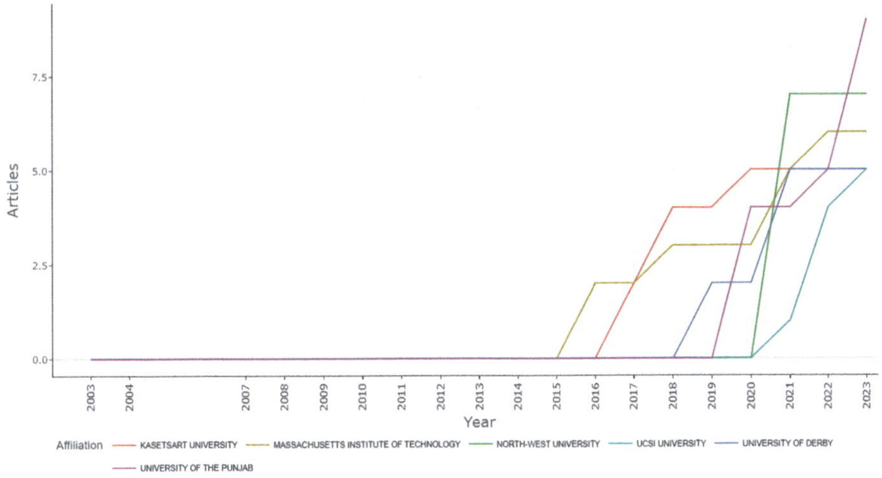

Fig. 10 Affiliations' production over time

in both 2022 and 2023. Similarly, Kasetsart University initiated its contributions in 2017, maintaining a consistent output thereafter. The University of Derby, which had no publications until 2019, showed a rapid increase in its research production from 2019 onwards.

The University of the Punjab and North-West University present unique cases. The former remained inactive in this research area until 2020 but then surged to produce 9 articles by 2023. North-West University, on the other hand, remained inactive until 2021 but then consistently produced 7 articles annually from 2021. UCSI University also displayed a late start with its first article in 2021, followed by a gradual increase in its research contributions. These patterns suggest a growing recognition of the importance of leadership styles in achieving sustainable growth in manufacturing firms. The delayed but eventual surge in research activity may reflect an evolving understanding of leadership's role in this sector, possibly influenced by changing industry dynamics and academic interests. This upward trend in research output from various prestigious institutions underscores the expanding academic and practical relevance of effective leadership in the manufacturing sector.

Figure 11 highlights the global distribution and collaboration patterns among the corresponding authors of the reviewed articles. The data reflects a significant contribution from various countries, with China and the USA leading in terms of the number of articles published, each with 22 articles. However, there's a notable difference in their collaboration patterns. China exhibits an equal distribution between single country publications (SCP) and multiple country publications (MCP), demonstrating a balanced approach between independent research and international collaboration. In contrast, the USA shows a strong inclination towards SCP, indicating a predominance of domestic research activities.

India follows closely with 19 articles, displaying a moderate level of international collaboration. Malaysia, while contributing 10 articles, stands out for its high rate of MCP, suggesting a strong emphasis on international cooperation in research. The

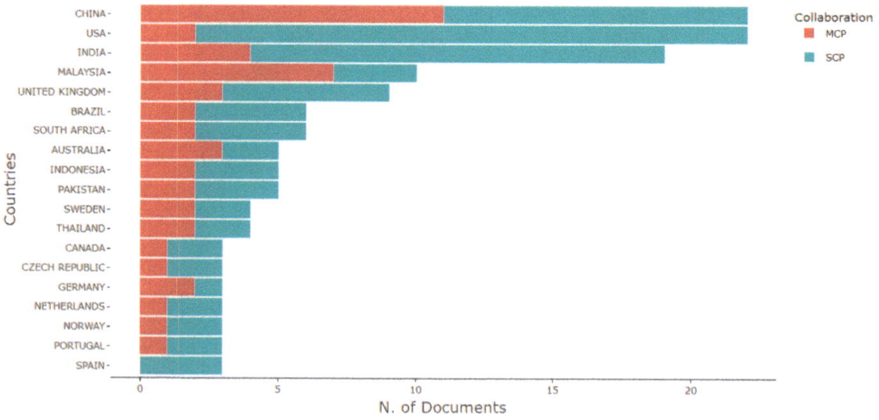

Fig. 11 Corresponding author's countries

United Kingdom, Brazil, and South Africa, each with a relatively smaller number of articles, show a varied approach to collaboration, with the UK leaning slightly towards SCP. Australia and Indonesia, each contributing 5 articles, exhibit different collaboration tendencies. Australia shows a higher inclination towards MCP, whereas Indonesia has a more balanced approach.

It underscores the diverse approaches to research collaboration in the field of effective leadership styles in manufacturing firms. The varying SCP and MCP ratios across different countries reflect the global research community's diverse strategies in pursuing academic excellence and knowledge dissemination in this domain. This diversity not only enriches the research landscape but also offers a multitude of perspectives and insights, essential for the sustainable growth of manufacturing firms worldwide.

The bibliometric analysis revealed a significant concentration of impactful research in certain countries, reflecting their pivotal roles in the study of leadership styles in manufacturing. India leads the list in terms of total citations (TC), indicating a high volume of influential research (Fig. 12). The country's average citations per article also suggest that these works are highly regarded in the academic community. China closely follows, demonstrating a strong research presence, although its average citations per article are slightly lower than India's, suggesting a larger volume of publications with moderate influence each. The USA, while third in total citations, shows a lower average citation count per article, indicating a focus on producing a broader range of research topics rather than concentration on highly cited works.

Norway, despite a lower total citation count, stands out for its extremely high average citations per article. This suggests that while Norway's overall contribution volume is smaller, its research is highly impactful and influential in the field. Similarly, the Czech Republic and the Netherlands, though not leading in total citations, show impressive average citations per article, indicating the high quality and influence of their contributions. Countries like Malaysia, Pakistan, and Thailand, although not at the top in total citations, exhibit significant average citations per

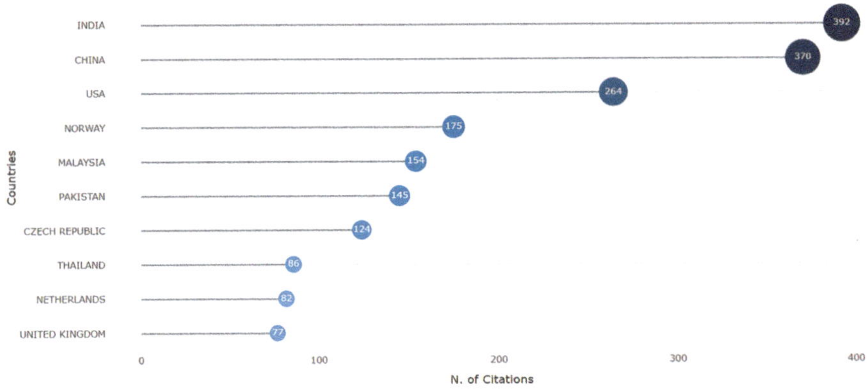

Fig. 12 Most cited countries

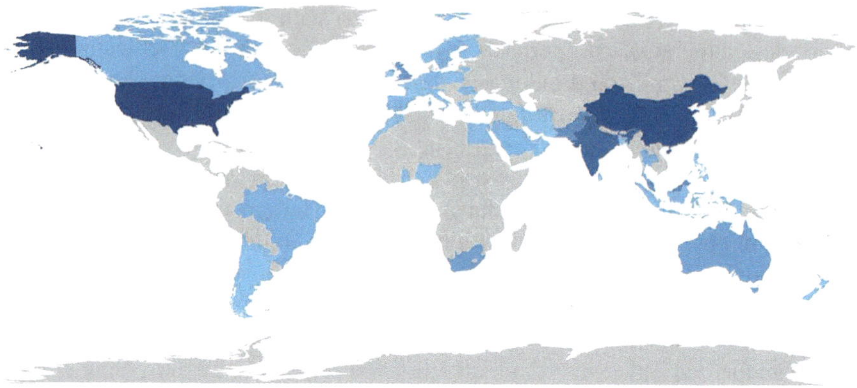

Fig. 13 Countries' scientific production

article, highlighting their important contributions to the field. The United Kingdom, despite its lower position in both total and average citations, still plays a notable role in the global research landscape on this topic.

This bibliometric analysis illustrates a diverse global contribution to research on leadership styles in manufacturing, with certain countries standing out either through the volume of their contributions or the high impact of their individual publications. This diversity highlights the universal relevance and application of effective leadership styles in achieving sustainable growth in manufacturing firms.

Figure 13 presents a compelling overview of the global landscape of scientific production in this field. The United States emerges as the most prolific contributor, with 85 publications, followed closely by China and India, indicating a strong focus on leadership and sustainable growth in manufacturing in these countries. This trend may reflect the economic priorities and industrial advancements in these regions, particularly in the context of the manufacturing sector.

There is a notable presence of countries like Malaysia and Pakistan, alongside traditional research powerhouses like the UK and Australia. This suggests a growing interest and investment in effective leadership within the manufacturing sector across diverse economic landscapes. The representation from countries like South Africa, Thailand, and Brazil further underscores the global recognition of the importance of leadership in driving sustainable manufacturing practices. European countries like Sweden, Germany, and Spain, though having lower publication frequencies, indicate a sustained interest in the area, contributing to a balanced geographical spread of research. This diverse representation highlights the universal relevance of effective leadership in manufacturing, transcending cultural, and economic boundaries. Countries with relatively fewer publications, such as those in the Middle East and Africa, demonstrate an emerging interest in this area of study. This could be indicative of a developing focus on enhancing manufacturing capabilities and implementing

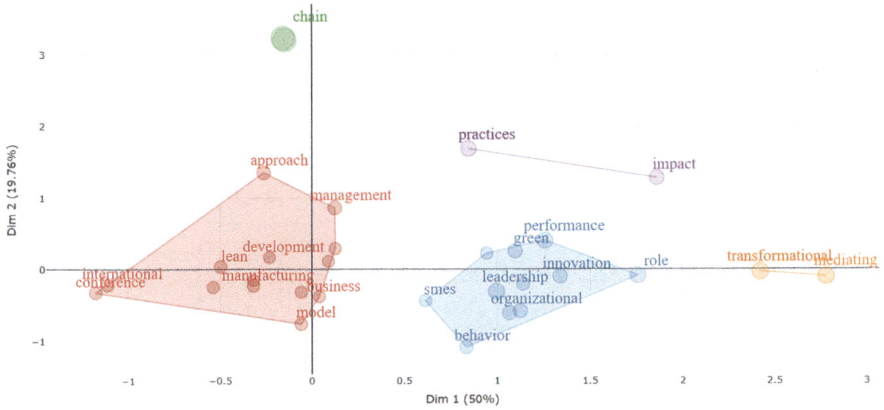

Fig. 14 Word map of the researched publications factorial analysis

sustainable growth strategies in these regions. The distribution of scientific production across these countries reflects a global consensus on the significance of leadership in manufacturing. It also points towards potential areas of collaboration and knowledge exchange that could further enhance the effectiveness and sustainability of leadership practices in the manufacturing sector worldwide.

Figure 14 presents a word map based on a factorial analysis of researched publications. This visual representation offers a nuanced insight into how various concepts and themes are interconnected within the context of manufacturing, sustainability, and leadership.

The word map clusters key terms into distinct groups based on their dimensions and interrelationships. The terms 'manufacturing,' 'sustainability,' 'sustainable,' 'management,' 'industry,' 'lean,' 'development,' 'study,' 'model,' 'approach,' 'business,' and 'factors' cluster around dimensions that suggest a strong focus on the integration of sustainability practices within the manufacturing industry (Fig. 15). This indicates a trend in the literature towards recognizing the importance of sustainable management practices and their role in the development of the manufacturing sector. The emphasis on 'lean' practices within this cluster suggests a convergence of efficiency and environmental considerations.

In contrast, terms like 'leadership,' 'performance,' 'green,' 'environmental,' 'role,' 'innovation,' 'organizational,' 'smes,' 'corporate,' 'behaviour,' and 'product' are clustered differently, highlighting a focus on how leadership styles influence environmental performance and innovation in manufacturing. The prominence of 'green' and 'environmental' in this cluster underscores the growing emphasis on eco-friendly practices. This cluster also suggests that effective leadership is crucial for fostering an organizational culture that prioritizes sustainability and innovation, particularly in small and medium-sized enterprises (SMEs) (Table 4).

The terms 'chain' and 'supply' form a separate cluster, emphasizing the critical role of supply chain management in sustainable manufacturing. This could indicate

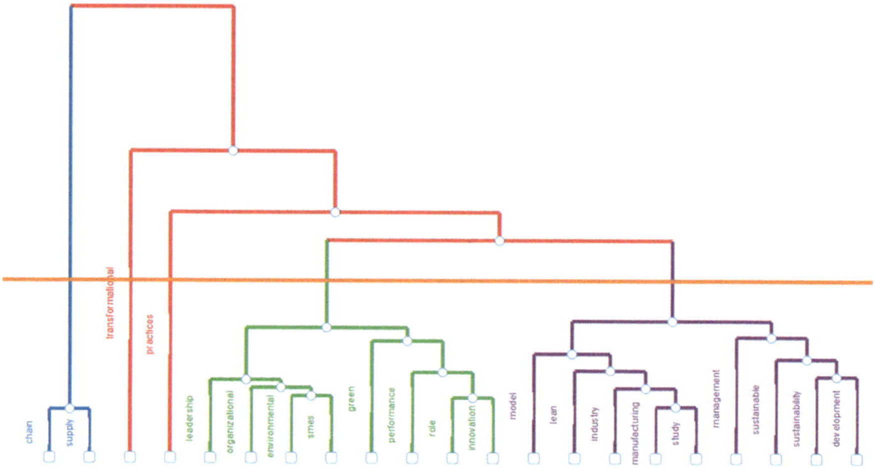

Fig. 15 Factorial analysis dendrogram of the researched publications

an increasing recognition of the broader implications of manufacturing practices beyond the confines of the organization, extending to the entire supply chain.

Interestingly, the terms 'practices' and 'impact' form another cluster, possibly indicating a focus on the tangible impacts of various practices in the field. This might reflect a shift in the literature towards evidence-based studies that assess the real-world implications of sustainability and leadership practices in manufacturing.

The final cluster, including 'transformational' and 'mediating,' suggests a growing interest in transformational leadership styles and their mediating effects on organizational outcomes. This could indicate an acknowledgement of the evolving nature of leadership in the context of sustainability, where leaders are expected to be catalysts for change and innovation.

The word map provides a comprehensive overview of the current state of research in the domain of sustainable manufacturing and leadership. It reveals a multifaceted landscape where the integration of sustainable practices, effective leadership, innovation, supply chain considerations, and tangible impacts are central themes. This suggests that the field is moving towards a more holistic understanding of how leadership styles can drive sustainable growth in manufacturing, recognizing the interplay between organizational culture, environmental stewardship, and business performance.

Table 4 Factorial analysis indicators of the studied documents

Word	Dim.1	Dim.2	cluster
Manufacturing	–0.32	–0.23	1
Sustainability	0.09	0.11	1
Leadership	1	–0.31	2
Sustainable	0.13	0.29	1
Performance	1.26	0.4	2
Green	1.1	0.25	2
Management	0.13	0.87	1
Environmental	1.07	–0.61	2
Industry	–0.32	–0.15	1
Lean	–0.5	0.04	1
Role	1.76	–0.08	2
Innovation	1.34	–0.1	2
Organizational	1.12	–0.58	2
Development	–0.23	0.17	1
Smes	0.61	–0.44	2
Chain	–0.16	3.23	3
Study	–0.06	–0.31	1
Supply	–0.15	3.2	3
Model	–0.06	–0.76	1
Practices	0.85	1.69	4
Transformational	2.42	–0.03	5
Approach	–0.26	1.35	1
Corporate	1.14	–0.21	2
Business	0.04	–0.38	1
Behaviour	0.83	–1.09	2
Factors	–0.54	––0.25	1
Mediating	2.78	––0.1	5
Conference	–1.17	–0.32	1
Impact	1.86	1.28	4
International	–1.11	–0.22	1
Product	0.94	0.22	2

3 Conclusion

The bibliometric analysis offers a panoramic view of the current research landscape, revealing critical insights and identifying future directions in the field. The study underscores the intricate relationship between leadership styles and sustainable growth in manufacturing firms, emphasizing the pivotal role of leadership in steering organizational practices towards sustainability.

A core thematic area that emerges prominently from the analysis is the integration of sustainable development within manufacturing processes. This theme's prevalence indicates a growing consensus on the importance of aligning manufacturing practices with environmental and social sustainability goals. The analysis also highlights the critical role of environmental management, indicating a shift in the manufacturing sector towards more environmentally conscious practices.

Another key area is the influence of different leadership styles on sustainability outcomes in manufacturing firms. The study illuminates the effectiveness of transformational, participative, and ethical leadership styles in fostering a culture of sustainability within organizations. It also suggests that transactional leadership, while often contrasted with transformational styles, can positively contribute to sustainability in certain contexts.

The analysis reveals knowledge gaps and potential areas for future research. One such area is the exploration of the interplay between technological advancements, such as digital transformation and Industry 4.0, and leadership styles in driving sustainable manufacturing practices. Understanding how leadership can leverage technological advancements to enhance sustainability in manufacturing processes represents a promising avenue for future research.

The study suggests the need for more geographically diverse research, especially from regions like Africa, the Middle East, and Latin America, to gain a more comprehensive understanding of how cultural and economic contexts influence the implementation of leadership styles in manufacturing firms.

Another potential area for investigation is the impact of leadership on employee engagement and innovation in the context of sustainability. Further research could explore how different leadership styles influence employee motivation, creativity, and commitment to sustainable practices.

The study also indicates a growing interest in the role of supply chain management in sustainable manufacturing. Future research could delve deeper into how leadership styles affect supply chain decisions and practices, and how these, in turn, impact the overall sustainability of manufacturing firms.

The bibliometric analysis presents a multifaceted picture of the current state of research on leadership styles and sustainable growth in manufacturing firms. It highlights the significance of integrating sustainability into manufacturing practices and the influential role of leadership in this process. The study offers a roadmap for future research, suggesting the need to explore new dimensions of leadership, technological advancements, and cultural contexts to further enhance our understanding of sustainable manufacturing practices.

References

Abbas, J., Zhang, Q., Hussain, I., Akram, S., Afaq, A., & Shad, M. A. (2020). Sustainable innovation in small medium enterprises: The impact of knowledge management on organizational innovation through a mediation analysis by using SEM approach. *Sustainability, 12*(6), 2407.

Ahmad, S., Miskon, S., Alabdan, R., & Tlili, I. (2020). Exploration of influential determinants for the adoption of business intelligence system in the textile and apparel industry. *Sustainability, 12*(18), 7674.

Ahuja, I. P. S., & Khamba, J. S. (2008). An evaluation of TPM initiatives in Indian industry for enhanced manufacturing performance. *International Journal of Quality & Reliability Management, 25*(2), 147–172.

Ali, Z., Sabir, S., & Mehreen, A. (2019). Predicting engagement and performance through firm's internal factors: Evidence from textile sector. *Journal of Advances in Management Research, 16*(5), 763–780.

Alrowwad, A., Abualoush, S. H., & Masa'deh, R. (2020). Innovation and intellectual capital as intermediary variables among transformational leadership, transactional leadership, and organizational performance. *Journal of Management Development, 39*(2), 196–222.

Awan, F. H., Liu, D., Anwar, A., Jamil, K., Qin, G., & Gul, R. F. (2021). Mediating role of innovative climate among leadership and employee performance in textile exporting firms of Pakistan. *Industria Textila, 72*(6), 613–618.

Baig, S. A., Iqbal, S., Abrar, M., Baig, I. A., Amjad, F., Zia-ur-Rehman, M., & Awan, M. U. (2021). Impact of leadership styles on employees' performance with moderating role of positive psychological capital. *Total Quality Management & Business Excellence, 32*(9–10), 1085–1105.

Bhutto, T. A., Farooq, R., Talwar, S., Awan, U., & Dhir, A. (2021). Green inclusive leadership and green creativity in the tourism and hospitality sector: Serial mediation of green psychological climate and work engagement. *Journal of Sustainable Tourism, 29*(10), 1716–1737.

Bilan, Y., Hussain, H. I., Haseeb, M., & Kot, S. (2020). Sustainability and Economic Performance: Role of Organizational Learning and Innovation. *Inzinerine Ekonomika, 31*(1), 93–103.

Boiral, O., Talbot, D., & Paillé, P. (2015). Leading by example: A model of organizational citizenship behavior for the environment. *Business Strategy and the Environment, 24*(6), 532–550.

Burawat, P. (2019). The relationships among transformational leadership, sustainable leadership, lean manufacturing and sustainability performance in Thai SMEs manufacturing industry. *International Journal of Quality & Reliability Management, 36*(6), 1014–1036.

Chau, K.Y., Lin, CH., Altantsetseg, P. *et al.* Green supply chain, green leadership, consumer preferences, sustainability attitudes, and sustainable business advantage: a case of Vietnamese textile industry. *Environ Sci Pollut Res* (2023).

Dhar, B. K., Stasi, A., Döpping, J. O., et al. (2022). Mediating Role of Strategic Flexibility Between Leadership Styles on Strategic Execution: A Study on Bangladeshi Private Enterprises. *Global Journal of Flexible Systems Management, 23*, 409–420.

Edwards, G., & Gill, R. (2012). Transformational leadership across hierarchical levels in UK manufacturing organizations. *Leadership & Organization Development Journal, 33*(1), 25–50.

Graves, L. M., & Sarkis, J. (2018). The role of employees' leadership perceptions, values, and motivation in employees' provenvironmental behaviors. *Journal of Cleaner Production, 196*, 576–587.

Gull, S., Bukhari, S. N. Z., Qamar, U., & Tanvir, A. (2022). Is transformational leadership instrumental to environmental sustainability? A perspective of Pakistani textile sector. *Industria Textila, 73*(4), 411–419.

Harsanto, B., Primiana, I., Sarasi, V., & Satyakti, Y. (2023). Sustainability Innovation in the Textile Industry: A Systematic Review. *Sustainability, 15*(2), 1549.

Hassan, M., Abbas, A., Batool, F., & Ahmad, K. (2020). Role of human resource practices and corporate entrepreneurship: Mediation of transformational leadership and product innovation performance. *Journal of Law & Social Studies (JLSS), 2*(1), 1–9.

Hayat Bhatti, M., Ju, Y., Akram, U., Hasnat Bhatti, M., Akram, Z., & Bilal, M. (2019). Impact of participative leadership on organizational citizenship behavior: Mediating role of trust and moderating role of continuance commitment: Evidence from the Pakistan hotel industry. *Sustainability, 11*(4), 1170.

Hossain, M. B., Nassar, S., Rahman, M. U., Dunay, A., & Illés, C. B. (2022). Exploring the mediating role of knowledge management practices to corporate sustainability. *Journal of Cleaner Production, 374*, 133869.

Islam, T., Khan, M. M., Ahmed, I., & Mahmood, K. (2021). Promoting in-role and extra-role green behavior through ethical leadership: Mediating role of green HRM and moderating role of individual green values. *International Journal of Manpower, 42*(6), 1102–1123.

Javed, B., Fatima, T., Khan, A. K., & Bashir, S. (2021). Impact of inclusive leadership on innovative work behavior: The role of creative self-efficacy. *The Journal of Creative Behavior, 55*(3), 769–782.

Jia, F., Gong, Y., & Brown, S. (2019). Multi-tier sustainable supply chain management: The role of supply chain leadership. *International Journal of Production Economics, 217*, 44–63.

Khan, M. A. S., Jianguo, D., Ali, M., Saleem, S., & Usman, M. (2019). Interrelations between ethical leadership, green psychological climate, and organizational environmental citizenship behavior: A moderated mediation model. *Frontiers in Psychology, 10*, 1977.

Kim, H. T. (2020). Linking managers' emotional intelligence, cognitive ability and firm performance: Insights from Vietnamese firms. *Cogent Business & Management, 7*(1), 1829272.

Kim, H., & Kim, T. (2017). Emotional intelligence and transformational leadership: A review of empirical studies. *Human Resource Development Review, 16*(4), 377–393.

Li, M., Huang, W., Zhang, C., & Yang, Z. (2019). How union participation influences union commitment and job involvement. *Employee Relations, 41*(6), 1348–1363.

Luque, A., & Herrero-García, N. (2019). How corporate social (ir) responsibility in the textile sector is defined, and its impact on ethical sustainability: An analysis of 133 concepts. *Corporate Social Responsibility and Environmental Management, 26*(6), 1285–1306.

Mead, T., Jeanrenaud, S., & Bessant, J. (2022). Sustainability oriented innovation narratives: Learning from nature inspired innovation. *Journal of Cleaner Production, 344*, 130980.

Nasir, A., Zakaria, N., & Zien Yusoff, R. (2022). The influence of transformational leadership on organizational sustainability in the context of industry 4.0: Mediating role of innovative performance. *Cogent Business & Management, 9*(1), 2105575.

Paço, A., Leal Filho, W., Ávila, L. V., & Dennis, K. (2021). Fostering sustainable consumer behavior regarding clothing: Assessing trends on purchases, recycling and disposal. *Textile Research Journal, 91*(3–4), 373–384.

Raut, R. D., Mangla, S. K., Narwane, V. S., Gardas, B. B., Priyadarshinee, P., & Narkhede, B. E. (2019). Linking big data analytics and operational sustainability practices for sustainable business management. *Journal of Cleaner Production, 224*, 10–24.

Sharma, S., Prakash, G., Kumar, A., Mussada, E. K., Antony, J., & Luthra, S. (2021). Analysing the relationship of adaption of green culture, innovation, green performance for achieving sustainability: Mediating role of employee commitment. *Journal of Cleaner Production, 303*, 127039.

Singh, S., Sood, S., & Bala, R. (2021). Passive leadership styles and perceived procrastination in leaders: A PLS-SEM approach. *World Review of Entrepreneurship, Management and Sustainable Development, 17*(1), 20–37.

Ullah, I., Mirza, B., & Jamil, A. (2021). The influence of ethical leadership on innovative performance: Modeling the mediating role of intellectual capital. *Journal of Management Development, 40*(4), 273–292.

Yap, X. S., & Truffer, B. (2019). Shaping selection environments for industrial catch-up and sustainability transitions: A systemic perspective on endogenizing windows of opportunity. *Research Policy, 48*(4), 1030–1047.

João Lucas is currently an Assistant Professor at Piaget Institute, and is a Ph.D. student at the University of Évora. His current research interests are Operations Management, and Leadership Styles.

The Role of the Supply Chain in Developing Innovation Processes in the Textile Industry

João Rita

Abstract This chapter explores the vital interconnection between supply chain management and innovation in the textile manufacturing industry. We delved into how technological advancements, such as digital printing and automation, are reshaping manufacturing processes and necessitating an agile and responsive supply chain. The emergence of smart textiles and eco-friendly materials has further underscored the need for a supply chain capable of sourcing novel materials while maintaining quality and cost-efficiency. A significant focus was placed on sustainability, highlighting how the supply chain is pivotal in implementing ethical and environmentally conscious practices. This includes ensuring transparency in sourcing, adopting sustainable logistics, and aligning with the manufacturers' green initiatives. We also examined how the supply chain must adapt to support energy-efficient practices in manufacturing, reflecting a commitment to reduced carbon footprints. The chapter concluded by emphasizing the dynamic and collaborative nature of the supply chain in the textile industry. It is not just a mechanism for material delivery and product distribution; it is an integral part of the innovation process, facilitating the realization of technological advancements and sustainable practices in textile manufacturing. This synergy between supply chain management and manufacturing innovation is essential for the industry's evolution, ensuring that new developments are conceptualized and effectively brought to market in a sustainable and efficient manner.

1 Introduction

In ancient times, textile production was a craft steeped in tradition and skill. Picture the ancient Egyptians with their flax, spinning and weaving it into linen as early as 5000 BC. The artistry was in creation, and the vibrant dyes were derived from plants,

J. Rita (✉)
Department of Management, Lisbon Accounting and Business School, Av. Miguel Bombarda 20, 1069-035 Lisboa, Portugal
e-mail: jmrita@iscal.ipl.pt

Lisbon Accounting and Business School, Polytechnic Institute of Lisbon (ed.), *Industry and Innovation: Textile Industry*, SDGs and Textiles,
https://doi.org/10.1007/978-3-031-57804-5_3

minerals, and even insects. "In the tapestry of life, we are all connected. Each of us is a gift to those around us, helping each other be who we are, weaving a perfect picture together," Anita Moorjani beautifully put it.

Fast forward to the Middle Ages, and Europe's wool industry flourished. Towns like Bruges and Ghent were bustling with weavers and dyers, transforming raw wool into cloth of astounding quality. This era was a loom on which the foundations of the textile trade were firmly established.

Then came the Industrial Revolution in the eighteenth century—a seismic shift. The invention of machines like the spinning jenny and the power loom revolutionized textile manufacturing. Production shifted from homes to factories, a change as dramatic as turning a page to an entirely new chapter in a novel. Cities like Manchester in England earned the nickname "Cottonopolis" due to their dense concentration of cotton mills.

The nineteenth and twentieth centuries saw further advancements. Synthetic fibers like nylon and polyester emerged, altering the fabric of the industry once again. These innovations paralleled the narrative twists in a complex plot, introducing new characters and possibilities.

Today, the textile industry is a fusion of tradition and innovation, much like a well-crafted story that respects its roots while embracing new ideas. It is a global narrative featuring sustainability and ethical practices as key themes in response to the growing awareness of environmental and social issues.

Much like a pivotal character in a grand novel, the textile industry plays a significant role in the global economy. It is a story of economic impact, international trade, and cultural influence, blending like genres in a well-curated library.

Let us start with its economic impact. The textile industry is a significant player in job creation, particularly in developing countries. It is akin to a character that brings new dynamics into the plot, offering opportunities for growth and development. This industry provides employment and contributes significantly to a country's GDP. Textiles are a cornerstone of economic stability and growth for countries like Bangladesh, India, and China.

Regarding international trade, textiles are like the well-traveled protagonists of adventure novels, crossing borders and connecting countries. This industry is a crucial component in global trade, with a vast network of supply chains. Countries specialize in different stages of textile production—from raw material production to sophisticated fashion design—creating a global tapestry of interdependence.

Moreover, the textile industry is also a catalyst for technological advancement and innovation. Just as a plot twist adds depth to the story, technological advancements in textiles introduce new materials, sustainable practices, and improved production methods, making the industry economically significant and a leader in innovation.

Let us remember the cultural impact. Textiles carry cultural significance, representing traditions, histories, and identities. They are like narrative themes that give a story its depth and color. The global exchange of textile designs and techniques fosters cultural exchange and understanding, enriching the global community like a well-written novel enriches the reader's perspective.

In summary, the economic and cultural significance of the textile industry in the global economy is as intricate and impactful as the most compelling and thought-provoking novels. It is a sector that weaves economic vitality, cultural richness, and technological innovation together in the global narrative.

2 Current State of the Industry, Global Production Hubs

China—As the "protagonist" of the textile world, China plays a leading role in global textile production. With a story as complex and dynamic as a sprawling epic, China excels in the volume and variety of textiles, including cotton, silk, and synthetic fibers. It is a powerhouse in both manufacturing and exporting, with a vast and sophisticated supply chain reminiscent of an intricately plotted novel.

India—India's textile industry is like a classic novel, steeped in history and rich with tradition. Known for its cotton production, India also excels in silk and jute. The country's specialty lies in its blend of traditional handloom fabrics with modern production techniques, creating a historical and contemporary narrative.

Bangladesh—In the narrative of textile production, Bangladesh is like an emerging character that quickly becomes central to the plot. It is one of the world's largest producers of garments, especially in ready-to-wear fashion. The country's textile industry story is marked by rapid growth and increasing sophistication in garment manufacturing.

Vietnam—Vietnam's textile industry is like a fast-paced novella, proliferating and gaining significant attention. Specializing in garment production, Vietnam has become an attractive destination for textile manufacturing, thanks to its skilled workforce and favorable trade agreements.

Italy—Picture Italy as the artist of the textile world, renowned for its luxury fabrics. Known for high-quality wool, silk, and lace, Italy's narrative in textiles is one of artisanal craftsmanship and high-end fashion. It is like the elegant prose in a beautifully written book, appealing to those who appreciate the finer things in literature and fabric.

United States—The U.S. textile industry is a diverse chapter with a story of innovation and advanced technology. Specializing in technical textiles, the U.S. also produces cotton, wool, and synthetic fibers. It is a sector characterized by research and development, much like a science fiction novel that pushes the boundaries of what is possible.

Germany—In the story of textiles, Germany is an innovator, particularly in technical textiles and non-woven fabrics. Its narrative is of precision and high-tech production, akin to a well-researched and brilliantly articulated non-fiction book.

Turkey—Turkey's textile industry is like a bridge between East and West, both geographically and in its production narrative. Specializing in cotton, wool, and silk, Turkey is known for its quality towels, linens, and ready-to-wear garments. The country's textile story is as rich and diverse as a cross-genre novel, blending tradition with modernity.

Japan—In the textile world, Japan is like a finely crafted haiku—small but impactful. Known for its innovative fabrics and technological advancements in textile manufacturing, Japan's industry is like a story of precision, quality, and cutting-edge technology.

South Korea—South Korea's textile industry is like a fast-paced thriller novel, with rapid advancements in technical and synthetic textiles. Known for its innovation and quality, South Korea is a crucial player in producing advanced materials and intelligent textiles.

In these global production hubs, the textile industry weaves a story as diverse and intricate as any literary masterpiece, with each country adding its unique chapter to the narrative. From the traditional looms of India to the high-tech factories of Germany and South Korea, the industry is a tapestry of cultures, technologies, and talents.

3 Market Trends

Like a fast-paced, page-turning thriller, fast fashion is characterized by rapid production and a quick turnover of affordable, trendy clothing. Brands like Zara and H&M are the protagonists in this narrative, constantly updating their collections to reflect the latest styles. However, this trend has complexities and consequences, just like any gripping story.

Environmental Impact—Fast fashion's environmental footprint is like a dark subplot in our tale. The industry is known for its heavy water usage, chemical pollution, and textile waste, much like a narrative that reveals the more challenging aspects of human nature and industry. The production of low-cost, quickly discarded garments significantly contributes to environmental degradation.

Social Impact—The human cost of fast fashion can be likened to a poignant, sometimes tragic, backstory. Workers in this sector often face poor working conditions and low wages, particularly in developing countries where much of the production occurs. This aspect of the narrative brings to light issues of equity and justice, reminiscent of the themes explored in socially conscious literature.

Consumer Behavior—On the brighter side, like a hopeful thread in a complex narrative, there is a growing awareness and shift in consumer behavior. This ongoing story's readers increasingly seek sustainable and ethically produced clothing, turning the pages toward a more conscious and responsible fashion story.

Technology and Innovation—In this evolving plot, technology plays the role of a dynamic character, introducing new possibilities and solutions. Innovations in sustainable materials, recycling of textiles, and digitalization of the supply chain are like plot developments that offer hope and positive change.

Shift to Sustainability—This trend is like a sequel to the fast fashion narrative, where the industry increasingly focuses on sustainable practices. Brands are exploring eco-friendly materials, ethical production methods, and more durable designs. It is a chapter that promises a better future, echoing the themes of redemption and transformation in many great literary works.

Fast fashion's narrative in the textile industry is complex and multifaceted, with its share of challenges and opportunities for change. It is a story that's still being written, with each of us playing a part in shaping its direction.

4 Importance of Supply Chain in the Textile Industry

In the dynamic world of textiles, the supply chain acts as the backbone, orchestrating the flow of materials, information, and finances from the initial raw material to the final product in the hands of the consumer. This comprehensive network is a linear sequence of steps and a complex, interdependent system crucial for the industry's functionality and growth.

The term "supply chain" refers to making and selling commercial goods, from the initial sourcing of materials to the production, distribution, and retail stages. It encompasses a series of steps involving multiple entities, including suppliers, manufacturers, distributors, and retailers. In the textile industry, this translates to a journey from fiber to fashion, involving various phases like fiber production, fabric weaving or knitting, dyeing and finishing, garment manufacturing, and finally, distribution and retail.

Globalized Production Network—The textile industry is a prime example of globalization. Its supply chain often spans multiple continents, with raw materials sourced from one country, processed in another, and sold in markets worldwide. This global network demands meticulous coordination and management to ensure smooth operations and timely delivery of products.

Quality and Cost Control—The supply chain's efficiency directly impacts the quality and cost of textile products. Effective supply chain management ensures high-quality output by maintaining standards at each stage, from raw material selection to final manufacturing. Similarly, optimizing the supply chain can significantly reduce costs, leading to more competitive pricing and higher profit margins.

Responsiveness to Market Demands—In a fast-paced industry like fashion, the ability to quickly respond to changing trends and consumer demands is crucial. A well-managed supply chain allows for rapid adjustments in production and distribution, enabling companies to stay relevant and competitive.

Sustainability and Ethical Manufacturing—The textile industry is increasingly being called upon to address environmental and ethical concerns. A sustainable supply chain focuses on reducing environmental impact through eco-friendly sourcing and manufacturing practices and ensuring fair labor practices across all stages. Transparency in the supply chain has become vital as consumers increasingly demand to know the origins of their garments and the conditions under which they were produced.

Technological Integration—Advancements in technology, such as automation, artificial intelligence, and blockchain, are revolutionizing the textile supply chain.

These technologies enhance efficiency, improve inventory management, and provide greater transparency, allowing for more sustainable and ethical operations.

Navigating Challenges—The textile supply chain has challenges, including fluctuating raw material prices, political and economic instability in manufacturing countries, and evolving regulatory landscapes. Effective supply chain management involves navigating these challenges and anticipating and mitigating potential risks.

In conclusion, the supply chain in the textile industry is a complex yet vital component, influencing everything from production efficiency and product quality to ethical practices and environmental sustainability. As the industry evolves, so must its supply chain strategies, adapting to new technologies, market trends, and consumer expectations.

5 Innovation in Textile Manufacturing

The textile industry's evolution through innovation is not confined to manufacturing units or research labs alone; it is profoundly intertwined with the supply chain, which plays a pivotal role in facilitating and accelerating these advancements. As the industry shifts toward more innovative and sustainable practices, the supply chain evolves as a catalyst and supporter of these changes.

The interplay between manufacturing innovation and supply chain in the textile industry is a story of synergy and mutual dependence. For instance, digital printing technology in textile manufacturing has revolutionized how fabrics are designed and produced. This technological leap demands a supply chain that can promptly deliver the specialized inks and fabrics required for digital printing. Additionally, the ability to produce smaller, customized batches means the supply chain must be agile and responsive, capable of handling diverse, fast-changing orders while ensuring minimal delays.

Similarly, the supply chain's role becomes even more pronounced when it comes to material innovations like smart textiles or eco-friendly fabrics. The development and sourcing of novel materials, such as biodegradable fibers or electronically integrated fabrics, require suppliers and manufacturers to work closely, ensuring the availability and quality of these new materials. This collaboration is crucial in balancing innovation, quality, and cost-effectiveness.

Moreover, sustainability, a driving force in recent textile innovations, highlights the supply chain's role in ensuring ethical and environmentally friendly practices. Sustainable manufacturing is not just about using eco-friendly materials but also involves sustainable sourcing practices. The supply chain must ensure transparency and traceability, confirming that materials are ethically sourced and produced with minimal environmental impact. This extends to logistics and distribution, where eco-friendly practices such as optimized routing and reduced packaging play a vital role.

The supply chain's adaptability is also tested by integrating energy-efficient practices in textile manufacturing. As manufacturers adopt renewable energy and aim for reduced carbon footprints, the supply chain partners must align with these green initiatives, ensuring that their operations reflect these sustainability goals.

In essence, the supply chain in the textile industry is not merely a conveyor belt that delivers materials and distributes products. It is a dynamic network that facilitates innovation, responds to technological advancements, champions sustainability, and adapts to industry's ever-changing demands. The supply chain's efficiency, responsiveness, and ability to evolve are crucial in realizing the full potential of manufacturing innovations in the textile sector.

As the industry continues to innovate, the symbiotic relationship between textile manufacturing and the supply chain will become even more significant. This partnership is vital to ensuring that the innovations not only emerge but are also effectively implemented, reaching the market successfully and sustainably.

6 Sustainable Practices and Circular Economy

Sustainable practices and the concept of a circular economy have become increasingly vital in the textile industry's supply chain, responding to the urgent need for environmental conservation and ethical production methods. This shift represents a fundamental change in how the industry approaches production, consumption, and waste management.

Traditionally, the textile industry has followed a linear model: raw materials are transformed into products, which are used and eventually discarded. This model has led to significant environmental challenges, including resource depletion, pollution, and massive waste. The advent of sustainable practices and the circular economy is an effort to disrupt this pattern, creating a restorative and regenerative system by design.

At the heart of sustainable practices in the textile supply chain is minimizing the environmental impact at every stage of the product life cycle. This involves choosing eco-friendly materials, such as organic cotton or recycled polyester, which require fewer resources and generate less pollution during production. It also includes implementing energy-efficient manufacturing processes, reducing water usage, and minimizing chemical dyes and treatments that can harm ecosystems.

Another critical aspect of sustainability is the social dimension, ensuring fair labor practices and safe working conditions throughout the supply chain. This ethical focus is not just a moral imperative but also responds to increasing consumer demand for ethically produced goods.

The circular economy takes these ideas further, aiming to keep resources in use for as long as possible, extract the maximum value from them while in use, and recover and regenerate products and materials at the end of their life cycle. In the context of the textile industry, this means designing products with longevity in mind, promoting repair and reuse, and facilitating recycling. Brands are increasingly exploring take-back schemes, where consumers can return used garments for recycling or repurposing, thus closing the loop in the product life cycle.

Innovative technologies play a crucial role in enabling the circular economy. For instance, textile recycling technologies are evolving to handle mixed fibers, which has traditionally been challenging. Advances in chemical recycling are making it possible to break down fabrics into their original fibers, which can then be reused to produce new textiles, reducing the need for virgin resources.

Furthermore, the circular economy model encourages a shift from ownership to usership, exemplified by the rise of clothing rental services and second-hand markets. These models extend the life of garments and reduce waste, aligning with the principles of reducing, reusing, and recycling.

In summary, sustainable practices and the circular economy in the textile industry's supply chain represent a transformative approach to addressing environmental and social challenges. By rethinking the way products are designed, used, and recycled, the industry is moving toward a more sustainable future that mitigates its impact on the planet and offers new opportunities for innovation and growth. This shift is a trend and a necessary evolution toward a more responsible and resilient industry.

7 Role of Technology

The role of technology in the supply chain, particularly within textile manufacturing, is multifaceted and profoundly transformative. It revolutionizes how products are made and delivered and reshapes the entire business model and the industry's interaction with its consumers and the environment.

Technology has been a driving force behind significant efficiency improvements and innovation in textile manufacturing. Automation, for instance, has streamlined various manufacturing processes, from spinning and weaving to cutting and sewing. Automated machinery has accelerated production rates and enhanced precision, producing higher quality products with less waste. These advancements are particularly crucial in an industry characterized by high-volume production and a need for consistency and quality control.

Digital technologies, like 3D printing and digital fabric printing, are pushing the boundaries of design and customization. They enable manufacturers to experiment with complex designs and patterns without needing expensive and time-consuming traditional methods. This capability is particularly significant in an industry that thrives on creativity and innovation. Moreover, digital fabric printing is more sustainable, reducing the water and chemical usage typically associated with dyeing fabrics.

Incorporating Artificial Intelligence (AI) and Machine Learning (ML) into supply chain operations marks a leap toward smart and data-driven decision-making. AI algorithms can analyze vast amounts of data to optimize inventory management, forecast trends, and even aid in design aspects by predicting consumer preferences. This predictive capability is invaluable in reducing overproduction and aligning production closely with market demand, thereby reducing waste and increasing efficiency.

The Internet of Things (IoT) brings a new level of connectivity to the supply chain. Sensors and smart devices embedded in machinery and throughout the supply chain provide real-time production processes, inventory levels, and shipment tracking data. This constant flow of information enables visibility and control previously unattainable, allowing for swift responses to production issues or market changes.

Blockchain technology introduces a new dimension of transparency and traceability to the textile supply chain. By securely recording transactions and tracking the movement of materials and products, blockchain can provide undeniable proof of the ethical and sustainable sourcing of materials and adherence to labor standards. This transparency is vital to building trust with consumers and regulators in an industry increasingly scrutinized for its environmental and social impact.

Sustainability is another critical area where technology is making a significant impact. Innovative technologies are enabling the development of sustainable materials and eco-friendly manufacturing processes. Technology is at the forefront of the industry's shift toward a more sustainable future, from recycling technologies that turn textile waste into new fabrics to biotechnological advancements creating sustainable fibers.

In conclusion, technology in the textile industry's supply chain is not just an enabler of efficiency and quality but a catalyst for fundamental change. It is paving the way for more sustainable practices, greater customization, and unprecedented transparency and efficiency. As technology continues to evolve, its role in reshaping the textile supply chain will undoubtedly expand, driving further innovation and sustainability in the industry.

8 Conclusion

In the ever-evolving landscape of the textile industry, the supply chain plays a crucial role in driving and sustaining innovation. While advancements in technology and sustainable practices redefine manufacturing processes, the supply chain acts as the conduit for these innovations, ensuring their successful integration and application in the market.

Facilitating technological advancements

Adoption of New Technologies—For technological innovations like digital printing and automated weaving to be effective, the supply chain must adapt to accommodate these new methods. This includes sourcing advanced machinery, training personnel, and restructuring logistics to handle different production flows.

Ensuring Material Availability—The development of smart textiles and sustainable materials requires the supply chain to source new types of inputs. Finding suppliers for novel materials like organic cotton, recycled polyester, or biodegradable fabrics is integral to bringing these innovations to fruition.

Time to Market—In an industry where trends change rapidly, the supply chain's ability to quickly move products from design to retail is vital. An efficient supply chain allows manufacturers to capitalize on technological advancements, delivering innovative products to consumers faster.

Supporting sustainable practices

Sourcing Eco-friendly Materials—The shift toward sustainability in textile manufacturing heavily relies on the supply chain's ability to source environmentally friendly materials. This involves finding and collaborating with suppliers who adhere to sustainable practices.

Circular Supply Chain Models—Embracing a circular economy requires the supply chain to deliver products and manage the return and recycling of materials. Establishing a system for collecting and repurposing used textiles is essential for sustainable innovation.

Transparency and Traceability—Consumers and regulators increasingly demand transparency in manufacturing processes. The supply chain must provide visibility into the origins of materials and the conditions under which products are made, ensuring ethical and sustainable practices.

Driving efficiency and reducing costs

Optimizing Production—Innovations in manufacturing often aim to increase efficiency and reduce costs. The supply chain supports these goals by optimizing logistics, reducing lead times, and managing inventory effectively, allowing manufacturers to benefit from economies of scale.

Risk Management—Introducing new technologies and materials can involve risks such as supply disruptions or quality issues. A robust supply chain is critical in identifying, assessing, and mitigating these risks.

Weaving innovation into the fabric of the industry

The supply chain in textile manufacturing is not just a series of steps to move products, it is a dynamic ecosystem that plays a vital role in nurturing and implementing innovations. As the textile industry continues to evolve, the strength and adaptability of the supply chain will be key determinants in the successful realization of new technologies and sustainable practices.

9 Challenges and Future Outlook

The supply chain faces challenges as the textile industry strides forward with innovations in manufacturing processes and materials. These challenges stem from integrating new technologies, adapting to sustainable practices, and meeting ever-evolving consumer demands. One of the primary hurdles is the integration of advanced technologies such as AI, IoT, and blockchain into existing supply chain frameworks. While these technologies promise enhanced efficiency and transparency, their implementation requires substantial investment, training, and a shift in traditional operational methodologies.

Sustainability poses another significant challenge. As the industry shifts toward eco-friendly practices, the supply chain must adapt to source sustainable materials, which are often more expensive and less readily available than traditional materials. This shift also demands greater scrutiny and transparency in supplier practices, ensuring that every component of the supply chain adheres to ethical and environmental standards. Moreover, the move toward a circular economy, which emphasizes recycling and reusing materials, necessitates a redesign of the supply chain to accommodate reverse logistics.

Another pressing issue is the supply chain's need for agility and flexibility to respond to fast fashion trends while maintaining quality and ethical standards. This requires a delicate balance between speed, cost, and sustainability, often leading to complex logistical challenges.

Looking toward the future, the textile industry's supply chain stands at a crossroads of opportunity and challenge. Technological advancements, particularly in digitalization and automation, are poised to redefine supply chain operations. These technologies offer solutions to current challenges, enabling faster, more efficient, and transparent supply chains. However, harnessing these technologies requires overcoming initial implementation challenges, including investment costs and training needs.

Sustainability will continue to be a driving force in the evolution of the supply chain. As consumer awareness and regulatory pressures increase, there will be a more significant push toward sustainable and ethical practices. This shift will likely lead to more collaboration across the industry as companies work together to develop sustainable supply networks and shared standards.

While the supply chain in textile manufacturing faces significant challenges, it also stands on the brink of transformative change. By integrating advanced technologies and sustainable practices, the supply chain can become more efficient, transparent, and responsible. As the industry navigates these changes, the supply chain will undoubtedly play a crucial role in shaping the future of textile manufacturing.

References

Jones-Davis, D. (2019). Fashionopolis: The price of fast fashion and the future of clothes. *Science, 365*(6457), 977–977.

Little, T. (2018). *The future of fashion: understanding sustainability in the fashion industry*. New Degree Press.

Rivoli, P. (2005). *The travels of a t-shirt in the global economy*. John Wiley & Sons.

Communication and Market Orientation as Part of an Innovation Strategy in Smart Textile Companies

Jose Agostinho

Abstract Innovation and entrepreneurship are increasingly important in economic development in almost every country and region on the planet, and market competition drives organizations to greater efficiency. In the face of all these global economic transformations and the specificities of the markets, we address the reasons why people are becoming increasingly entrepreneurial in the European textile and clothing industries, and how innovation and communication have become so important in the strategies for entrepreneurship in these industries. The strong economic impact that entrepreneurial companies in the European textile and clothing sector have on economic development and job growth has meant that almost all European governments have created support for entrepreneurship and innovation in the textile and clothing sector.

1 Introduction

The problems of overconsumption and overproduction, as well as the associated problem of environmental pollution, are unfortunately not new phenomena. The EU's textile and clothing strategy for sustainable and circular textiles calls for circularity throughout the entire textile value chain. There are companies applying the following circular business model strategies: reuse and recycling, product as a service, repair and reuse, and material innovation. Closed-loop supply chain companies are not yet prominent, and many textile manufacturing companies are not yet ready for a circular transformation.

The textile and clothing industry is highly complex and is characterized by very long and global supply chains (Pedersen et al., 2015). The European textile and clothing industry is still unable to respond to fundamental challenges related to dominant business models and consumer behaviors. The textile and clothing sector

J. Agostinho (✉)
Department of Management, Lisbon Accounting and Business School, Av. Miguel Bombarda 20, 1069-035 Lisboa, Portugal
e-mail: jeagostinho@iscal.ipl.pt

Lisbon Accounting and Business School, Polytechnic Institute of Lisbon (ed.), *Industry and Innovation: Textile Industry*, SDGs and Textiles,
https://doi.org/10.1007/978-3-031-57804-5_4

in Europe is therefore faced with decisive changes and challenges, because today, digitalization processes and the use of artificial intelligence in industrial processes are essential for the development of the European textile and clothing industry, and for the process of ecological sustainability, toward climate neutrality, thus producing competitive products, with strong support from Industry 4.0. The concept of Industry 4.0 represents the development of industrial automation and the integration of different technologies, such as artificial intelligence, robotics, the Internet of Things and cloud computing, with the aim of promoting the digitization of industrial activities, and thus improving industrial processes and increasing productivity, giving the European textile and clothing sector adaptability to external factors and requirements. Finding sustainable and innovative textile and clothing solutions (product, process, organizational and marketing), including yarns, fabrics and garments, based on natural fibers and recycled fibers from end-of-life garments, as well as energy and resource management solutions for the sector, and intelligent and automated systems are imperative for the European textile and clothing industry.

Innovating with smart textiles means giving renewed potential to the European textile and clothing industry, ensuring competitiveness with the possibility of consolidation and thus providing the European textile and clothing industry with an important green component. As an example, we could point to a new fabric made in Portugal from recycled fabric, incorporating ingredients extracted from agri-food waste, such as that from the meat industry! You don't have to doubt it! The fabric already exists. This innovative product is the result of the Portuguese brand Tenowa, the latest in the Riopele universe. The new brand incorporates the textiles of the future, which is nothing more than clothing made from 100% recycled raw materials. Basically, in this case, it's about creating sustainable fabrics made from textile waste and with functional properties using agri-food waste. Tenowa is the result of the R4 Textiles project, an initiative of the Riopele group, co-financed by Compete 2020 whose aim is to convert waste generated by the textile and agri-food industries into new resources. R4 Textiles aims to add value to Riopele's textile waste, aiming to have an impact on niche markets by introducing novelty/innovation. This project takes a new look at the circular economy, which allows waste from the agri-food industry to be taken into account in order to incorporate the resulting products into the finishing of fabrics and boost the company's sustainability strategy.

A competitive and strategic factor for the European textile and clothing industry, as noted above, are the new opportunities made possible by research into intelligent textiles, in the manufacture of advanced textiles, and a paradigmatic example in which communication plays a key role in market orientation with its potential customers and consumers, as the EU is implementing the digital product passport (DPP), which aims to introduce minimum standards to make almost all products on its market sustainable, durable and environmentally friendly. In order to enter the European market, products must be accompanied by a digital product passport, which provides consumers with all the relevant information to help them make informed decisions. The digital product passport (DPP) will have to provide information on performance, traceability, technical documentation, harmful chemicals, user manuals, information on the environmental impact of the product purchased, as well as how the products can be repaired or recycled. In addition, the new European rules aim to put an end to the destruction of unsold products, as well as promote green public procurement and offer incentives, such as green vouchers or green taxes, to encourage people to buy environmentally friendly products, with communication being a determining factor in the business strategies of innovative organizations.

A key factor in the European textile and clothing industry's strategy is the concept of the circular economy, disseminated globally in the various analog and digital media that exist in the information and communication age, which begins with sustainable production: free of waste, water pollution, low carbon emissions and free of chemicals, replaced by microorganisms, all in addition to textile recycling programs, which give a second life to textile and clothing items. Textiles and clothing constitute a highly complex, diverse and heterogeneous waste stream that covers many end products, from synthetic yarns, wool fabrics, bedding, industrial filters, diapers and clothing. These waste streams can be pre-consumption, i.e. from textile production sites where waste collection is simple, and post-consumption waste that can be dispersed.

Circular economies refer to circular or closed-loop supply chains in which end-of-life products are reintroduced into the value chain through strategies such as recover, recycle, repurpose, remake, renew, repair, reuse, reduce, rethink and delete! It's about promoting manufacturing capacity through fiber and fabric technology and at the same time building the next generation supply chain and manufacturing workforce.

In order to achieve these goals for the European textile and clothing industry, changes must take place on both the consumer and producer sides, which means that individual consumption and value creation patterns must be reconsidered and, if necessary, changed. In this system, waste is reduced to a minimum, since existing materials and products are kept in circulation and in the economy for as long as possible, as this extends the life cycle of these goods. This goal includes a holistic strategy for a fully circular economy that minimizes the use of natural resources, increases efficiency and challenges current production methods. These types of business models aim to promote sustainable development and thus increase or maintain economic prosperity, social equity and environmental quality, although many textile manufacturing companies are not yet ready for circular transformation. In other words, to offset its huge carbon footprint, the European textile and clothing industry is adopting innovation as a strategy throughout the value chain to ensure long-term sustainability (Spaargaren, 2003), and circular business models, in general, are important tools for studying and understanding how a company functions and operates. To this end, a business model can be divided and analyzed into three different domains: (a) Value proposition: this concerns the company's value proposition and corresponds to what value is created and which target group it is aimed at; (b) Value creation and delivery: this is carried out by selling products and services with eco-economic value in return. In sustainable business models, environmental and/or social value is also created; (c) Value capture: corresponds to how the company makes money and what its strategy is, and is related to cost structures, profit streams and strategic objectives for the long-term preservation of the company. These management changes in business focus result in business models, some of which are new and can be considered innovations.

Sustainable business models can be divided into three different levels of innovation: (a) Technological: It includes innovations stemming from a technical innovation, for example in production processes, through innovations in customer offerings that alter customer behavior, and organizational changes in the company, for example changes in responsibilities; (b) Social: It includes notions of sufficiency, functionality and management; (c) Organizational: It results from the expansion of business areas and business purpose. As a result, more and more is being invested in Textiles 2.0, which brings together fibers/textiles with semiconductors and advanced materials to create fibers that are devices and fabrics that work as systems, which means that the fabrics will be able to see, hear, feel, communicate and perform other specific functions designed to solve needs, but also to generate new commercial markets with advanced applications, such as the demand for electronics for clothing, which is more versatile and compatible with the body and which gave rise to the emergence of electronic textiles (e-textiles). By enhancing textiles with electronic functionalities, e-textiles are defining a new frontier for enhancing human capability. The range of applications in market areas is extensive and can include the protection of combatants, in various fields, in the medical field, with the monitoring of wound healing and the monitoring of vital signs, with the possibility of adjusting colors, among many others. Some of the best-known international examples include the MiCoach heart rate monitoring bra from Adidas, the Google jacket from Levi's Jacquard project, which can work with a smartphone because they use a conductive fabric capable of recording interactions that can be mapped, so that the wearer can control their smartphone without having to take it out of their pocket, or clothing from CuteCircuit, such as the Hug Shirt, which is a shirt that allows you to send hugs from a distance, re-creating the sensation of touch and the emotion of a hug for a distant loved one. Sending hugs has become as easy as chatting and it's possible to send hugs while on the move, in the same way and to the same places where you can make phone calls, or the SoundShirt, which is a more specialized garment used for music and games.

As such, it is Communication, in particular through interaction with digital platforms and social networks, which presents a number of potentialities to innovative organizations, through access to a huge capacity for data and influence that plays a key role in business operations, including optimizing product design, targeting consumers and carrying out brand marketing. Consumers exchange shopping experiences via social networks, providing real user feedback for brands, but consumers' interests, life attitudes and consumption habits are also captured internally by organizations for precision marketing. On the other hand, the spontaneous sharing of information by customers and consumers through digital platforms and social networks also facilitates the marketing of products, which not only guarantees the authenticity of the information gathered, but also helps to improve the image and awareness of the brand and reduce marketing costs. Communication, then, is fundamental to core marketing processes, which are considered factors in an organization's market orientation, and which influence business performance (Jaakkola et al., 2016), such as product development management, supply chain management and customer relationship management.

2 Organizational Communication

Communication, which consists of the act of sending a message through different means of communication, which can be verbal or non-verbal, formal or non-formal, as long as it conveys a thought that provokes an idea, a gesture or an action, is considered a learned skill. Most people are born with the physical ability to speak, but we have to learn to speak well and communicate effectively. Speaking and listening, and the ability to understand verbal and non-verbal meanings are skills that we develop in various ways throughout our lives. We start learning the basic meanings of communication skills from an early age by observing other people, and, in this way, we shape our behavior and communication based on what we observe. Communication is therefore the process of transmitting, disseminating or passing on information from one person to another or from one place to another. In other words: Communication is the process of creating, transmitting and interpreting ideas, facts, opinions and feelings; in other words: it is a process that is essentially a process of sharing a mutual process. Furthermore, communication is the result of the exchange/sharing of information between two or more people.

Organizations' interest in communication has received significant attention in recent decades. Previously, communication in organizations was focused on developing advertising campaigns, contacting the press or writing articles about them. Nowadays we can say that organizations have started to pay more attention to the functions that communication performs both inside and outside them, all because the focus on organizations has changed and a new organizational perspective has been adopted when we are faced with the need for organizations to communicate internally and externally, and naturally also in the interest they have in the effects they want to produce in the markets where they operate, particularly in the processes of entrepreneurship and innovation.

Today, managers in the twenty-first century increasingly see communication as a very important tool in the entrepreneurship and innovation policies to be adopted with regard to their organizations, as well as the social environment/markets in which they operate. In this way, organizations are fully aware that they must be permanently attentive to innovation and organizational changes are no exception; on the contrary, they are the order of the day, since demand and supply must be closely harmonized and an organization's communication about products and services must know how to anticipate/innovate what is important to customers and consumers. Therefore, organizations can no longer afford to implement their communication activities in a fragmented way, without having an integral vision of the role that communication plays in an organization, as this will be neither efficient nor effective in a society that is increasingly characterized as an Information and Communication society and where

Sovereign Marketing (SM) reigns. Thus, in order for organizations linked to innovation, such as the new innovative organizations in the European textile and clothing industry, and which pursue a market orientation, to be noticed and perceived in this Information and Communication society of the twenty-first century, a rigorous organization of Communication is necessary; in other words: all organizations depend on Communication for their basic functioning, and even more so those whose focus is on product and service innovation, such as the new innovative organizations in the European textile and clothing industry. Communication is used to transfer information to its audience about the organization's mission and vision, organizational policies and procedures, tasks and duties toward shareholders and clients/consumers, and the various activities necessary for optimal functioning within the company (Farmer et al., 1998). As simple as it is, scientific research has shown that communication can make or break an organization. For this reason, a good communication strategy is essential for the survival of a company, as communication acts as a link between decision-makers and all employees, as well as customers and consumers. When poorly carried out, communication can cause interpersonal conflicts in organizations, preventing them from achieving their strategic business objectives of innovation and market orientation where they operate. In other words, what people hear or understand is largely based on experience and background, because people have preconceptions about what people are going to say, and if these preconceptions don't fit into their mental frame of reference, adjustments must be made until they do (Baskin et al, 1997). Thus, if there is no common understanding resulting from the transmission of symbols (verbal or non-verbal), there is no Communication (Donnelly et al, 1984). Therefore, effective communication is strategic for achieving organizational goals aimed at entrepreneurship and innovation, because ineffective communication in an organization can result in uncertainty, apprehension and dissatisfaction, which translates into low productivity and loss of entrepreneurial focus on the intended innovation. It is therefore necessary for managers to communicate effectively with employees; in other words: the extent to which a manager can achieve the organization's objectives/innovation depends on their ability to communicate effectively (Herich, 2008). The reality is that many managers, still formatted in assigning a higher value to their professional skills in the area of performance, have yet to understand what communication is and its organizational value, due to a cultural deficit and a lack of openness to the world of communication that surrounds us and soaks us, and the importance of its role in the success of an innovative company (Nnamseh, 2009).

Communication improves organizational relationships by favoring innovation and correct market orientation, and it has been proven, for example, that communication can minimize organizational strikes and blockages, because organizational objectives and goals are sometimes defeated when communication is not effective, i.e. waste/inefficiency and costly mistakes are made in organizations due to communication failures. The absence of Communication is one of the main reasons that can lead to confusion and poor planning in many innovation-driven organizations, such as the new innovative organizations in the European textile and clothing industry, and this is because some managers have not yet learned that Communication is the most important source of information used by managers in making decisions that affect the organization's performance.

Therefore, a manager's communication skills are vital not only for making decisions, but also for conveying the results and intent of the decisions made to other people involved in the whole process of business management and innovation.

2.1 Organizations' Concern with Communication

Communication is fundamental to the development and participation of organizations. We create, change and dissolve organizations through our discursive communication practices. Therefore, organizational communication is much more than the communication that takes place in organizations.

Organizations' concern with communication is an asset, in other words: it is of great value for the organization. Its aim is to promote a favorable climate for the organization, as a precondition for its functioning, because concern for Communication makes it possible to establish a social basis for transactions with groups of the public in the organization's environment, and at its core is the image of the organization as a whole. Normally, the first step organizations take is to carry out an analysis of the organization's mission, the direction the organization is taking and how competitors are positioning themselves. Target orientation means, in the case of Concerned Communication, taking into account how the public perceives the organization and interprets existing trends.

An organization aiming for success and innovation needs to be well informed about all these extremely important variables, which can be achieved through the communication process. In other words, internally, all the participants in an organization need to interact, the strategies defined need to be maintained, the how-to policies need to be formulated, the strategies for achieving the objectives need to be developed and the programs need to be planned, executed and evaluated, and the employees need to be well paid and motivated. In fact, without an effective means of communication, there will be no management function.

2.2 Communication in Organizations

In today's society there are countless social, financial, economic, political, legal, environmental, cultural and technological developments that justify greater attention and a change in the focus that communication represents for organizations, that is: there is a greater dependence on a healthier environment, society is increasingly faced with an increase in the scale of its demographics and the complexity of relationships in society is increasing, as is the pace at which change occurs. Today, the need for transparency on the part of organizations has become a requirement that requires constant visibility with the world around them, as well as a connection to that world in various forms, since the market for organizations, particularly innovative ones, is no longer a sellers' market, but a buyers' market. Today, it's not the sellers who dominate the market, but the buyers, as customers and consumers have become more aware of their consumption, among other factors, as noted by the formation of consumer associations, which means that organizations have to take into account the aspirations of these same customers or consumers when marketing their products. Governments, clients/consumers, investors, employees and other public groups are increasingly critical of how organizations operate. We can see, then, that organizations have taken the initiative to communicate their values more clearly to their target audiences, and in many more advanced organizations Communication has become a basic discipline in the strategic planning process, i.e. Communication skills are not just used to carry out activities such as media contacts, but have been given an improved position through their close link with the organization's general policies and innovation objectives.

Communication, as a management function, is the process of creating, communicating and interpreting ideas, facts, opinions and feelings about work performance, organizational effectiveness and efficiency, as well as the organization's objectives. A manager, in addition to other professional skills, must be an effective communicator and no organization can succeed, progress or build a reputation without effective communication skills. We all know, although some managers with cultural deficits are still resilient in this regard, that a poor communication system can result in bad management and poor business results. So, the success of any business lies in effective Communication and effective Communication is essential for the survival and progress of an organization, even more so for those that want to be innovative, such as the new innovative organizations in the European textile and clothing industry. Communication skills therefore need to be developed continuously, especially in a turbulent business environment such as the one we live in. The stage of development of the communication skills area differs from organization to organization, depending on the degree to which communication professionals have access to decision-makers and are involved in decision-making processes, and the degree to which they support top management.

In general, we can summarize how the communication process develops in organizations:

(a) Focus on the recipient and the communication process.
(b) Recognition that communication is important for the whole organization.
(c) Attention to the social environment in which the organization carries out its activities.
(d) A strategic approach to Communication with a view to the future.

2.3 *Integrated Communication in Organizations*

The concept of integrated communication refers to the organization's strategy, which aims to promote effective communication by improving various organizational coherences in various ways, that is: providing coordination between the organization's internal and external audiences; it is not a function that only concerns the communications department, but rather a transversal attitude throughout the organization. Integrated communication is a broader concept than integrated marketing communication, because integrated marketing communication presupposes the use of different marketing communication tools, and its driving force is a reorientation from transaction-based marketing to relationship marketing. So, in addition to the traditional inside-out approach, attention should be paid to an outside-in approach, i.e. thinking about customers first and building relationships. Therefore, integrated communication is a broader concept that not only addresses customers and consumers, but also investors, employees and other audiences of organizations. It requires a holistic vision, linking the communication strategy to the strategy and objectives of the organization/innovation.

The concept of integrated communication has undergone some changes over time and initially, its focus was on the media or activities within the scope of multimedia campaigns and projects. Later, the focus shifted to harmonizing the various areas of communication policy, such as internal and external communication, marketing communication and corporate communication, as well as the communication policy of the various business units within a corporate group. Subsequently, the concept expanded, and its focus shifted from Communication policy to Organizational policy. Therefore, an integrated approach to communication can reinforce organizational management policies. As such, integrated communication can be conceptualized at the following levels:

(a) Integrated communication at the (micro) level of operational activities.
(b) Integrated communication at the (meso) level of communication policy.
(c) Integrated communication at the (macro) level of organizational policy.

These three levels reinforce each other, and integrated communication can be seen as a harmonization of these three levels, that is: communication is internally consistent in these three areas, there is a policy with a common denominator and the communication policy is based on organizational policies. All this is accompanied by attention to the social environment, and only in this way is it possible to maintain a coherent image of the organization and its brands. Therefore, integrated communication must be achieved through the participation of various protagonists, working in different fields of communication, who collaborate on projects, formulating common starting points and promoting horizontal mobility. In addition, top managers must be aware of the roles that communication can play in their organizations and, as such, communication specialists must actively participate in shaping the set of reflections on the positioning of the organization and its products and services on the market, and they must be properly aware of the organization's social environment.

Integrated communication is an approach that aims not only to ensure the cohesion of communication activities and modalities, but also represents a strong link to company policy. In short: each organization needs to have its own vision of the contribution Communication can make to solving the problems the organization faces. These ideas can be fleshed out in the form of starting points for communication policy, including decisions on the identity structures of organizations. Measures can then be taken to standardize working methods, as well as ways of consulting and coordinating communication activities.

2.4 The Importance of Communication for a Manager

The most effective and innovation-minded managers are those who understand communication and its use in the context of the organization. Communication is the vehicle that enables managers to carry out every management function. In order to plan and innovate successfully, managers must be able to effectively communicate their business management vision to the rest of the organization (Radovic Markovic & Omolaja, 2009). To organize successfully, managers must enable and encourage fluid communication, both up and down the hierarchy, as well as between departments and colleagues. To lead successfully, managers must be able to clearly communicate organizational objectives and, through this communication, inspire employees to trust their leadership and perform at the highest level. To control management successfully, managers must communicate effectively with employees to monitor progress, to emphasize organizational objectives and to correct any deviant processes that are underway. Consequently, communication is more than simply speaking, writing, reading and listening (Radovic Markovic & Omolaja, 2009). Effective communication is the key to successful and innovative management, because communication allows managers to share the objectives and processes of innovation with stakeholders, both inside and outside the organization. It allows managers to stimulate behavioral changes in workers and suppliers. It allows managers to inspire loyalty in workers, customers and consumers. In this way, managers can convince workers and unions to abandon counterproductive practices. It also allows managers to persuade leaders to grant funding, and it also allows managers to appease dissatisfied customers and impress new ones. So, managers must be effective communicators in order to succeed in their roles in organizations, but what can make managers successful communicators?

First, they need to understand what communication is. Next, they need to understand how communication works, both on an interpersonal and organizational level and, finally, they need to understand what barriers can prevent communication, so that they can overcome these constraints and improve communication throughout the organization.

Communication, at its basic levels, involves three basic elements or components, which are the encoder, the channel and the decoder. The source or origin of the information is known as the encoder. In other words, the encoder is the originator of the information to be communicated to the other party. The channel is the medium chosen or used to communicate the message or information to the other party. The decoder is the person to whom the message or information is being sent and is the recipient of the message. However, they may or may not be the user of that information. However, the Communication process in modern organizations follows the pattern below:

1. The sender has an idea.
2. The idea is transformed into a message. It is important to put the message into a form that the receiver understands (called the encoding process).
3. The message is transmitted to the receiver.
4. The receiver receives the message and interprets it, a process known as decoding.
5. The receiver gives feedback to the sender on the effectiveness of the information that has been communicated.

Therefore, the coding process involves determining how the message should be written or spoken in order to communicate with another person. The use of words and understanding can, however, alter the meaning of a message. Facial expression, tone of voice, emphasis and gestures all play a role in the coding process when conversation is used. In addition, of course, decoding is the process of achieving understanding of the message. We can see, then, that the encoding process involves determining how the message should be written or spoken in order to be able to communicate with another person. Varying the use of words and understanding can, however, alter the meaning of a message. Facial expression, voice, emphasis and gestures all play a role in the encoding process when conversation is used. In addition, of course, decoding is the process of achieving understanding from the message. And as we know, different people extract different meanings from the same message, influenced by their experience, attitudes and value systems.

2.5 Leadership Communication

According to Nelissen et al. (2007), organizational culture is the result of organizational communication, both consciously and unconsciously. Good Communication is a complex, idiosyncratic and sometimes subtle process and determines the success or failure of the organization, and is a reflection of the organization's reputation. In short, the Communication factor determines whether others follow the leadership or not (Hensley & Burmeister, 2004).

The term entrepreneurial leadership means the ability to lead by knowing how to provide market orientation, motivation and inspiration to others in an organization, in order to achieve organizational goals and promote innovation that a leader establishes with careful business objectives, keeping an eye on the rapid changes in the business environment (Barret, 2006). Business leadership thus encompasses the ability to promote, support and sustain a business organization, taking risks if necessary. Leaders of business organizations perform a variety of functions, including knowing how to design the models of action in the markets in which they operate, so that the business organization has a clear vision of the objectives and innovation they are pursuing, with intellectual stimulation and creativity being critical elements for business leadership (Gupta et al, 2004).

The entrepreneurial leadership of an organization has to take certain risks at crucial moments in its activity, either by seizing opportunities/innovation, making decisions or promoting changes in its direction of progress, thus adding value to the goals it pursues. Thus, the entrepreneurial leadership of an organization is a fundamental task, and its purpose is to assume an effective leadership style (Cohen, 2004; Femald et al, 2005). In order to achieve this goal, there are two fundamental skills: the ability to lead and the ability to take risks (Kuhn & Weinberger, 2005; Kumar, 2012). However, it is quite clear how the communicative capacity, which forms the core of every organizational competence, continues to be ignored in this whole process. The assumption is that the effectiveness and success of organizations' business leadership depend first and foremost on the ability to communicate effectively. Axelrod (2002) observation that leadership effectiveness is primarily a "communication issue" serves as a starting point for conceptualizing the notion of leadership in organizations. Leadership in organizations is based on winning the affection of followers who share a common vision and mission. The leader then seeks to inspire each member of his team in some way, and thus motivate them to give the best of their abilities in order to achieve the desired goals. Ireland and Hitt (1999) emphasized five components of organizational/business leadership:

(a) Determining the vision of the organization/company.
(b) Developing human capital.
(c) Sustaining an effective organizational culture.
(d) Establishing balanced organizational control.
(e) Placing special emphasis on ethical practices.

In any business venture, the manager's leadership skills play a vital role in achieving the objectives that are intended to promote the business, thereby ensuring that employees are committed to their work with a proactive, independent approach and quality work (Haleblian & Finkelstein, 1993; Hambrick, 1994; Finkelstein and Hambrick, 1997). So, to say that business leadership is unimaginable without Communication would not be an overstatement. In fact, it is extremely crucial and therefore pertinent to any conceptualization of business leadership in organizations.

In an organization, communication takes the form of an organizational process because it involves various people and group activities. As such, in an organization, the success of corporate vision, market orientation and innovation depends on how well things are communicated, and also on how well the leader is able to communicate and share with his team, followers or co-workers. Mintzberg (1973) rightly stated that managers spend most of their day co-communicating, since three-quarters of their time in a day is spent on it. Similarly, in an organization, the leader's communication with his co-workers plays a significant role and therefore has to be persistent, recurrent and reliable. Being critical in the Communication process means remaining focused and incessantly analytical about what is being said, along with its implicit, inferential and suggested implications. The prerequisite for critical communication is understanding, generosity, frankness and firmness in its articulation. To be critical in Communication is also to be open to a situation of reciprocity, since we also have to allow others to be critical. The biggest challenge in critical communication is to be able to convey what you want without ever letting the content of what you say go unnoticed/misinterpreted. Thus, the communicative business leader is the one who actively and incessantly engages employees/collaborators in dialogue, and shares his vision with them through serious attention to their feedback, encouraging stakeholder participation in decision-making, remaining genuinely open and involved at all times. According to Booth (1986), systems theory is best suited to communication, as it recognizes the role that communication plays in helping the various components of organizations to function efficiently. In systems theory, the whole determines the character and functions of the parts (Waweru, 2002). This means that the main systems are designed into subsystems, making up the system, which functions in a wider environment. Systems theory works from the primary principles of interrelationship and interdependence. Every organization has a hierarchical structure with the various sections functioning as an overall structure and all staff working toward a common goal. Therefore, the business leader must convey the message effectively and have it understood; this is a prerequisite for progress and innovation. There is no point in moving forward if the basic message is lost or if there are barriers in the process. A manager/leader must be an effective communicator and no organization can succeed or progress, be innovative or build a reputation without effective communication skills.

2.6 *Effective Communication in an Organization*

Heron (1942) identified certain elements necessary for effective communication. His work outlined objectives, attitudes and criteria for effective communication. In the existing literature on organizational communication, he is credited with having outlined two-way communication between employees and executive management in organizational communication. However, Pigors (1949) pointed out that effective communication is only possible if management and employees collaborate and work together. The results of his study showed that employees are significantly very important to the success of internal communication and that internal communication played a very positive role in shaping the performance of employees in the organization. Therefore, effective communication can encourage an organization's employees to improve their working environment through positive contributions. The measure of the relationship between effective communication and work productivity, and other explanatory variables, results from the duration of work in a company, academic qualifications, timely information about changes at work, cooperation at work and adequate training provided to employees. In other words: to achieve the desired level of productivity, managers must confirm by communicating that instructions are clear or understandable, provide sufficient training for employees, ensure that there is cooperation at work through incentives and, finally, develop a good Communication plan to provide timely information about changes that affect work.

2.7 Communication and Organizational Performance

Organizations carry out various activities to achieve their organizational goals, and it is these repeatable activities, which use processes to make the organization successful, that must be quantified to determine the level of performance so that management can make informed decisions about where, if necessary, within the processes to initiate actions to improve performance and innovation. Communication must then take place in an open communication environment in which all members of the organization feel comfortable sharing feedback, ideas and even criticism at all levels. Organizations with a secretive communication environment exclude people, which can stifle involvement and result in the loss of innovative ideas and opportunities. Companies whose managers are able to engage employees in conversations about their work, their ideas and their perspectives on issues related to products/innovation, services/innovation, customers and the business environment develop a culture of inclusion that generates positive results. Organizations need to think about what employees need to know in order to do their jobs effectively, to interact with customers effectively and to serve as ambassadors outside the organization. Therefore, organizations should try to incorporate effective communication strategies that will help improve organizational performance. The lack of adequate communication is a challenge for organizational performance. Organizational performance can be defined as the evaluation of constituents trying to assess an organization's capacity and ability to achieve constituents' aspiration levels, using criteria of efficiency, effectiveness or social reference.

2.8 Challenges and Obstacles in Organizational Communication

There are several problems with communication during business negotiations: negotiators don't communicate with each other in order to understand and solve a problem more easily, or to reach a solution with the right support. Negotiators often listen only to themselves and pay attention only to their actions, instead of listening to those they represent and, on whose behalf, they are negotiating. Finally, there is often a misunderstanding, when one party can misinterpret what the other is saying and thinking. Communication barriers include, but are not limited to, everything that prevents a message from being received and understood. Barriers are mostly of human origin and arise through group opinions, conflicts of values and beliefs, stereotypes, language and slang. In addition, there are psychological barriers that relate to the individual and their mental and emotional state. Often, people only hear what they want to hear and only see what they want to see. Entrepreneurs often rely on experience and tradition to process information and make decisions. So, organizational barriers are related to organizational structure, culture, work model and communication channels/flows. The most common barriers are information overload, which creates a feeling that a person is overloaded with work tasks; message competition: important messages are separated from irrelevant ones; message filtering: managers filter messages to choose the communications they should pay special attention to or to reproduce them in a simpler and more effective way. So, all the obstacles that arise can be overcome if employees develop and improve their communication skills. The communication relationship depends on a multiplicity of information being transmitted through different communication channels (important and irrelevant, formal and non-formal, correct and incorrect).

3 Entrepreneurship

In the European textile and clothing industry, we have seen a gradual increase in the emergence of new companies that undertake innovation. As such, we can define entrepreneurship as a mechanism by which society converts technical information into products and services, and, as Christensen et al. state, the area of entrepreneurship is defined as how, by whom and with what effects opportunities to create future goods and services are discovered, evaluated and exploited. Today, new entrepreneurs in the European textile and clothing industry are creating greater recognition for their innovative brands through analog and digital communication, whether in the cinema, on television, in specialist magazines, sponsoring sporting activities and music, in digital and social networks and have built successful textile and clothing companies, because with the Internet and its possibilities for communication/generating business, new European textile and clothing companies are increasingly manufacturing locally and selling globally, and the will to undertake through innovation seems to be a factor in global success. Digital social networks such as Facebook, MySpace, LinkedIn, Instagram and others have become part of the daily routines of most citizens and professionals, who seek to communicate and establish connections with each other through digital communication, and we are also seeing an ever-increasing growth in digital commentators and influencers, as well as e-commerce. All these technological, economic, financial and social developments in twenty-first century society have made social and professional relationships more complicated. Let's say, then, that relationships in the world of business organizations, particularly those linked to entrepreneurship and innovation, are also more complicated than they were years ago.

So, we say that communication in organizations is more important today than it has ever been, partly because business and market conditions are more complex. Communication today has become the core business of entrepreneurial/innovative organizations and has come to be the most important of all business competencies, because an organization's ability to convey information helps both customers/consumers and employees/collaborators to communicate with the company and ultimately trust it. The development of a strategic Communication strategy and its implementation in the new innovative organizations in the European textile and clothing industry can provide a number of benefits, such as keeping employees/collaborators motivated and engaged, and it also allows clear and coherent messages to be shared with employees/collaborators in a timely manner, which in turn contributes to the productivity and success of the organization.

4 Innovation Strategy

Since the dawn of mankind, the desire to explore something new or to try to build has become a defining characteristic of human beings, and this desire has evolved over time. Studies on innovation take us back to the early work of Schumpeter (1934), who attributed to innovation the role of driving economic progress through more evolved alternatives resulting from technological development. Every day, new ideas are generated for the purpose of innovation, which seek to improve efficiency and production, but only when they provide some value to the market can they be considered innovations, i.e. they accompany and anticipate the needs of customers/consumers, so that organizations can serve up ever-changing value propositions. However, innovation is not only linked to the generation of new ideas and inventions that can be implemented in organizations, because it only happens when investments and the mobilization of various organizational resources come together (Innoskills, 2014), it is also a global process of the organization that extends over time. Innovations can be translated into the type of change they cause in society and drive the market, motivating organizations to define strategies that give them competitive advantages, offering differentiated products and services that make them unique in the market. There are several ways in which the new European textile and clothing industry can acquire competitive advantages through innovation, against the backdrop of the importance of Communication and its development in organizations: through the efficient communication of a new product or service offering; of a new process that offers something in a way that the competition cannot imitate and in a faster, cheaper and more personalized way; through complexity: informing about something offered that other organizations have difficulty mastering; through the legal protection of intellectual property: subject to the payment of a licensing fee; expansion of competitive factors aimed at competition: communicating changes in the price of the product by referring to price and quality, or price, quality, variety; communicating the opportunity and timing to the markets: the advantage of being the first to enter the market; communicating the development of a new platform (components, methods or technologies that serve as the basis for building products or services): offering a product that is the basis on which other variations and generations can be built; rewriting the market rules: adopting a different way of doing things; reconfiguring parts of the process: building more efficient distribution networks, outsourcing and extending their activities through the virtual medium; transferring different contexts of application: recombining the elements already known in different markets (Tidd et al, 2008).

Freeman and Soete (1997) typify innovation strategies as.

(a) Offensive strategies (organizations seeking technological leadership in certain industry segments).
(b) Defensive strategies (organizations that don't want to take the risk of being the first to innovate, but also don't want to miss out on being present in technological terms).
(c) Imitation strategy (organizations that just want to make a mark on the market by offering a product similar to existing ones).
(d) Dependent strategy (organizations that do not take the initiative to promote technical changes in their products or processes except through the explicit demand of their clients/consumers).
(e) Traditional strategy (does not have the technical capacity to initiate changes, but can develop new innovations and make minor changes to product design).
(f) Opportunistic strategy (exploiting market niches or temporary opportunities).

We can therefore define innovation strategy as the organization of resources, processes and people that are directly related to carrying out short- and long-term innovation actions, which in turn enable the company's strategy and vision. Strategy consists of a model or plan that integrates objectives, policies and the sequence of actions into a coherent whole (Quinn, 1980); in other words: the set of decisions and actions relating to the choice of means and the articulation of resources with a view to achieving an objective (Thietart, 1984). In an innovation strategy, particularly in new innovative organizations in the European textile and clothing industry, it is necessary to have a precise description of future-oriented innovation, thus determining in which business areas and with which innovative products and services the organization wants to integrate the market (Bullinger, 2008), integrating into this innovation strategy the vision, values and mission statement, which describes the character and corporate values of the innovative organization (Bullinger, 2008). In order to achieve this goal, information/communication and knowledge management are key factors, as the organization must have the ability to build and adapt the information/communication flows relating to its activity, as a way of boosting the organization's overall performance (Rasmussen, 2011). As an initial strategic development strategy, in order to achieve innovative results in the new innovative organizations in the European textile and clothing industry, organizations must first identify the sources of knowledge/communication in their organizations, which are roughly divided into two types: systematic—systems-oriented, or informal, experience-based—human resource-oriented (Rasmussen, 2011). They must then be able to recognize the different information/communication needs in each sector of the organization, with regard to employees and their roles (Rasmussen, 2011), so that these needs can be transformed into learning and decision-making models in the management of individual and collective knowledge/communication, as a competitive factor in the market (Rasmussen, 2011). The managers of the new innovative organizations in the European textile and clothing industry must begin by developing the innovation strategy by formulating and describing the innovation strategy, aiming to clearly and precisely define the vision, i.e. the intention as to what its future position in the competitive environment will be; the mission: the central objective of its existence, i.e. its reason for being and existing and where it wants to go in relation to a business area in which the company operates or intends to operate; and the values of the innovative organization: a set of corporate beliefs and principles that guide the behavior of the organization's members. This requires strategic definitions of how to get there, using the means, resources, capabilities and competencies available or to be acquired. Exploiting threats and opportunities to achieve these objectives, bearing in mind that there is a difference between the intended strategy and the realized strategy, since, due to the difficulty of forecasting and changes in the environment, the realized strategy is not always exactly the same as the intended strategy (Mintzberg et al, 2000). In this sense, in this process of strategic definition of new innovative organizations in the European textile and clothing industry, several management tools are used to analyze how they can best define and implement, in various dimensions, the innovation strategy, such as PESTEL analysis (Gupta, 2013) and SWOT (Gupta et al, 2016), that is: before developing specific business strategies, one must analyze and understand, at the time of strategic diagnosis, the competitive dynamics of the segment of the market for new innovative organizations in the European textile and clothing industry, and the available internal resources and capabilities (Dyson, 2004; Simons, 2000), in order to identify critical success factors. Porter's (1992) 5 forces analysis will also be one of the management tools that will allow new innovative organizations in the European textile and clothing industry to carry out a sectoral analysis of the main competitors, actors and forces that interact within the clothing market, and new innovative organizations in the European textile and clothing industry. After this previous phase, new innovative organizations in the European textile and clothing industry must seek to define a strategic positioning strategy (Kotler, 2000), in which new innovative organizations in the European textile and clothing industry will have the opportunity, through Communication, to determine what image they want to be created in the mind of the client/consumer (Ries & Trout, 2002). It will also be important to know how to use resource-based strategies (RBV—Resources Based View of the Firm) (Fleury & Fluery, 2003), which serve to optimize the process of planning and scheduling the resources and activities of new innovative organizations in the European textile industry, and clothing: a resource can include everything from equipment and funds to labor and technological tools, which contribute to the completion of the innovation project, as the organization's resources are transformed into skills and capabilities, which create and profitably exploit a potential for latent differentiation in the textile and clothing innovation markets, through a systematic process of learning and innovation. It will also be important, in strategic terms of innovation, to use the blue ocean strategy (Kim & Mauborgne, 2016), through which it will be possible for new innovative organizations in the European textile and clothing industry to redraw the boundaries of the industry's innovation market. Textile sector in which they wish to operate, as in blue oceans/new markets, demand is created rather than competed for and there are ample growth opportunities for new innovative organizations in the European textile and clothing industry, which are at the same time profitable and fast. In blue ocean markets, competition is irrelevant because business rules are waiting to be defined. Emergent strategies (Mintzberg, 1987) can also be used as management tools: those that emerge from the organization as a response to an opportunity in the environment, generally associated with the difficulty of accurately predicting the behavior and interrelationships of agents in the environment, and the consequent response to this change. As a next step, managers of the new innovative European textile and clothing industry must seek to know how to plan the innovation strategy, that is: managers must design the strategic innovation plan through the strategic definition of objectives embodied in metrics, goals/targets, initiatives and budget that allow guiding the allocation of resources, with management instruments: strategic maps and use of the balanced scorecard. New innovative organizations in the European textile and clothing industry must also know how to align the organization according to the innovation business strategy which means executives of new innovative organizations in the European textile and clothing industry must know how to link the innovation strategy of the organization to the strategies of the business units and functional units of the company, through a multifunctional system of clear understanding/communication of these innovation strategies. Highlighting in this particular the tools to be used: communication with employees of the business units and business support units, seeking to know how to level corporate strategies and service levels. The managers of the new innovative organizations in the European textile and clothing industry must also, in terms of innovation strategy, know how to plan operations by defining the priorities of management processes and allocating the resources that will provide the innovation strategy, such as improvements in key processes, sales planning, budgeting, resource capacity planning. They will also be able to use the lean management system which consists of increasing efficiency and reducing waste in order to improve results, through the use of six sigma production systems, which consists of a philosophy of continuous qualitative and quantitative improvement of what is of value to the customer and what is necessary to define the scope of an improvement project, thus guaranteeing a high level of confidence (6 sigma level) in the customer/consumer's processes, i.e., going through its five (5) phases: process definition, process measurement, process analysis, process improvement and process control. The use of dashboards is fundamental throughout this innovation strategy process of new innovative organizations in the European textile and clothing industry: dashboards are ways of presenting various types of visual data in a single place. Roughly speaking, a dashboard is intended to convey different but related information in an easy-to-understand way, including things like key performance indicators (KPI) or other important business metrics for the innovative organization to see and understand. New innovative organizations in the European textile and clothing industry, in strategic terms of innovation, must also know how to provide proper monitoring of operations and initiatives that should lead to the execution of the innovative business strategy, through the promotion/communication of meetings periodic and frequent, in order to examine the performance of the different departments in achieving the proposed objectives and goals, reviewing them when necessary, in order to monitor the strategic indicators of innovation for the business. Finally, organizations in the European textile and clothing industry must have as a fundamental concern trying to test and adapt the innovation strategy: managers must be able to assess whether the organization is executing its strategy or whether they need to change or review procedures to its operationalization.

5 Market Orientation

The academic world considers market orientation as a business philosophy that focuses on collecting information from customers and competitors and that takes advantage of the synergy of shared efforts to generate value for customers and the company as a whole. Characteristics of market orientation include customer orientation, competition, product development and functional coordination. The market orientation technique has four phases: initiation, reconstitution, institutionalization and maintenance. The concept of market orientation then results in a culture that effectively and efficiently creates superior value for customers through customer orientation, competitor orientation, as well as interfunctional coordination (Narver & Slater, 1990). New innovative organizations in the European textile and clothing industry increasingly follow a market orientation, which can be defined as the method used by companies to create superior performance, adopting in their management the behaviors necessary to improve the company's performance, in which communication plays a prominent role, that is: the company understands the needs, desires and demand of the market, assuming in its management an orientation toward the client and consumer, always taking into maximum account the actions of its competitors and internally carrying out increasingly greater interfunctional coordination. Therefore, we can state that market orientation is determined through the congruence between the functional level and the underlying organizational culture, that is: when the customer is the first priority of the company's top management and when the marketing strategy is a direct extension of business strategy: the exploration of organizational factors will facilitate leadership in organizations in the design and implementation of business practices and processes, aiming to make them more market-oriented (Jaworski & Kohli, 1993; Slater & Narver, 1994).

Within this organizational dynamic, it is also important that innovative companies in the new European textile and clothing industry have the capacity to develop differentiated, high-standard products that help satisfy the customer in the long term and are thus able to generate the best deals. Customers and consumers are the best judges of innovation and, therefore, have influence on the extent to which the innovation strategy and the way it is communicated will influence the company's performance, as ethics and value morals are increasingly important to customers/consumers, affecting their purchasing decisions. Issues such as the environment, sustainability, animal welfare, production and work practices are elements that are increasingly considered when purchasing a textile or clothing product, particularly among younger European generations, who admit to taking different actions to reduce its environmental impact, including changing textile and clothing purchasing habits.

5.1 Communication and Market Orientation

We have, then, that the competitive advantage of the new European textile and clothing industry is defined with the aggregation of several items, which differentiates it from its global competitors, thus providing a unique and superior position in the textile and clothing market, and which basically consists of innovation, differentiation, low-cost leadership and the way it communicates. The degree to which an organization has a market orientation has an impact on communication within or between companies and the markets in which it operates, as companies that create strategic alliances depend on market information regarding the current and future needs of customers/consumers to be successful (Jaworski & Kohli, 2000). In this way, the market orientation of the new European textile and clothing industry is predominantly focused on constantly creating excellent value for customers and, consequently, obtaining a sustainable competitive advantage. Thus, the new European textile and clothing industry is able to achieve good business performance through its communication and operational capacity to satisfy the desires of the company's main shareholders. Roughly speaking, the dimensions that measure business performance consist of the performance of markets, the performance provided by suppliers, the performance of production processes, the performance of people in the organization and the performance of companies' relationships with customers and consumers. Thus, the market orientation of the new European textile and clothing industry is predominantly focused on constantly creating excellent value for customers and, consequently, achieving a sustainable competitive advantage, and there is a significant relationship between communication, market orientation and business performance.

5.2 Product, Process, Management and Marketing Innovation

In a world in constant evolution, innovation appears as a necessity, even taking into account the difficulty in promoting it and the risks, as its results are subject to numerous random factors. Innovating is not just about discovering and placing a product on the market, as the concept of innovation involves the discovery of new market segments, not least increasing existing ones. An innovative product must meet an undeniable condition, which consists of satisfying a latent need that had not been satisfied until then, be it technical, commercial, organizational or socio-institutional. To achieve this goal, it is necessary to create appropriate structures in organizations, even though we know that there is no single solution. In general, there are 4 (four) variants regarding the implementation of structures in organizations favorable to innovation: (a) product development committees (production, finance, marketing, research): especially responsible for product development of new ones, where information/communication circulates well and the confrontation of ideas is beneficial; (b) the new products department: when the size of the organization and its power of innovation allow it, it may be advantageous to create a department whose objective is specifically to deal with new products; (c) the product manager: organizations that use a matrix internal organization replace departments with functions (product managers), whose function is to develop new products, managing the product throughout its life; (d) specialized groups: these are working groups, which, although with a limited duration in time, have the specific tasks of promoting a product. In this variant, the group identifies itself with the specific product or products and accompanies it until serial launch, being then dissolved and one of the members being able to eventually become the manager of the product in question.

During the innovation process, the creativity that generates innovation is not the result of chance and, broadly speaking, innovation results from a culmination of reflection, from a creative and patiently constructed process, whether individual or collective. In this innovation process, the organization's internal sources are used: (a) the research department as the first source of ideas for new products; (b) production departments: through ideas for new features and improvements; (c) the sales department: through knowledge of the desires expressed by customers, competitors' offerings; (d) managers: due to the type of functions they perform and their openness to the surrounding environment, they will be able to discover new ideas at exhibitions, at fairs and through their personal contacts. Organizations' external sources consist of (a) customers, who know the problems caused by the use of a product, generating ideas for the development of new products; (b) competitors, as they pursue the same objectives as the organization and a careful examination of the ranges they propose is of interest, in order to create new products; (c) specialized documentation, inter-professional research centers, inventors and study offices, all of which can give rise to new ideas.

Innovative organizations' main characteristics are a shared vision, leadership and desire to innovate, translated into an articulated and shared purpose, with a long-term vision for the business, that is, without being focused on the short-term return on investment. In this sense, they establish an organizational design that allows and encourages creativity, learning and interaction through multidisciplinary teams internally and between organizations. In innovative organizations, there is involvement and participation of the entire organization in the processes of continuous improvement and innovation, through the valorization of creative ideas and the implementation of motivation/reward systems. Innovation then consists of a process that aims to transform an opportunity into a marketable idea, generally represented by an advance in science or technology and with a practical and economic application capable of producing innovations in products or services, which can generate acceptance, widespread in the product or service market. Product innovation consists of improving the product or service. Process innovation consists of an improvement in the way of manufacturing or service and is a strategic option for technological innovation, which is adopted to improve organizational efficiency. Management innovation refers to something that changes the way a management task is carried out or that significantly modifies usual organizational forms and brings advances in organizational objectives. Managing this process requires a clear innovation strategy, an organizational culture that supports innovation and processes that ensure that innovative ideas are developed appropriately. Marketing innovation is associated with the implementation of new methods involving improvements in product design, packaging, price, distribution and promotion. The objective is to increase sales by better satisfying market needs, changing positioning or opening new markets.

Innovations can be, depending on their degree of novelty, revolutionary/radical, which implies a significant improvement, incremental, with small improvements, or disruptive, generally associated with a process in which a technology, product or service is transformed or replaced by a superior innovative solution. We have, then, that innovation has several components associated with its concept, such as knowledge, novelty, usefulness and the creation or preservation of value. Let's say that innovation allows us to win new business in markets with advantageous conditions, value the brand, sell existing products at a better price, strengthen connections with existing customers and consumers and be ahead of competitors in the main competitiveness factors; it also allows improving relationships with business partners, enables the creation of better conditions for the development of new businesses, improves work processes and reduces organizational problems and conflicts. Innovation, on the other hand, also allows organizations to ensure that there is no decrease in the value of products to customers, no decrease in revenues inherent to the sale of these products, no decrease in the profitability of the organization and its business, also avoiding the loss of business opportunities and competitiveness in fundamental markets, avoiding the loss of positions and market share, avoiding the devaluation of the image of the organization and the product, avoiding the aging of the product life cycle and also avoiding the obsolescence of processes with increased costs. Process innovation can increase customers' appreciation of the organization and its products, increasing the value of the product, its quality and reducing delivery times. We can therefore say that the innovations implemented in the business and process management models of the European textile and clothing industry began to encompass concepts such as the reduction of organic raw materials, the reuse, the restoration of garments and other textile products, the recycling and sharing, which has transformed the way the European textile and clothing business is carried out and value is generated, thus limiting as much as possible the resources and inputs of materials needed in the value chain of the textile and clothing industry and thus minimizing the ecological impact of its activity, promoting greater acceptance of its products among customers and consumers. In this way, the European textile and clothing industry has focused its efforts on reducing the environmental impacts of textiles at all stages of their life cycle. New innovative organizations in the European textile and clothing industry have also been adopting standards and introducing codes and conduct to better manage the social and environmental dimensions of the supply chain. In addition, social aspects are also being improved, such as the institution of fairer working conditions, with the definition of new social standards, the establishment of minimum wages, the guarantee of safety at work and the imposition of a ban on child labor and of forced labor.

The new model of business management and innovative process of the new innovative organizations of the European textile and clothing industry follows the principles of sufficiency and circular economy, since if in the first concept/sufficiency it pursues reductionist strategies, of producing/consuming less, the concept of circular economy aims at closed-loop business management models, such as the concepts of reuse and sharing. A circular economy is an economy designed to be capable of regenerating itself, using two types of materials: organic or renewable materials, designed to be reused and re-entered at the end of their life cycle in the biosphere, and technical or non-renewable materials, designed to cyclically shift from production to consumption with minimal loss of quality or value (Dryzek, 1997).

6 Conclusion

Currently, more and more innovative organizations with a strong market orientation, particularly the European textile and clothing industry, are realizing that innovative strategies that include new methods and new management technologies are necessary to survive in the modern economy. Organizations are created and organized through Communication processes, and are maintained by managers who communicate with each other.

In developed environments, the connection between organizations and the environment in which they operate is increasing. Successful organizations have realized that they must adapt their activity to the local environments in which they operate, as this is a fundamental parameter of the company's reputation and communication.

Communication is one of the most important factors in establishing the best innovative strategies in the European textile and clothing industry, proposing innovation with the most appropriate market orientation in organizations. The most successful managers in the European textile and clothing industry are aware that a large part of their success is due to their good Communication skills, which is why they are constantly developing and improving them. Taking care of the capabilities of Communication subjects is a necessary condition to establish successful Communication feedback. With better information and education of customers and consumers, they can make sustainable purchasing decisions and thus influence better collaboration along the entire value chain, as the European textile and clothing industry must invest strategically in networks, partnerships and collaborations in financial ecosystems, in order to overcome obstacles to transformation, and in this way implement new forms of collaboration between manufacturers, retailers and customers/consumers, ensuring that materials are kept in use for as long as possible.

The importance of Communication in innovative and market-oriented European textile and clothing industry organizations is most recognized by those who have missed important business opportunities, or caused problems for themselves and others due to a lack of this skill. That is why Communication is the basis of business strategy, which enables innovation and excellent market orientation of the new European textile and clothing industry.

References

Agrawal, V. (2012). Managing the diversified team: challenges and strategies for improving performance. *Team Performance Management, 18*, 384–400.

Anderson, N., De Dreu, C. K., & Nijstad, B. A. (2004). The routinization of innovation research: A constructively critical review of the state-of-the science. *Journal of Organizational Behavior, 25*(2), 147–173.

Anderson, N., Potocnik, K., & Zhou, J. (2014). Innovation and creativity in organizations: A state of the science review, prospective, commentary and guiding framework. *Journal of Management, 40*(5), 1297–1333.

Ansoff, H. I., & Stewart, J. M. (1967). Strategies for a technology-based business. *Harvard Business Review, 45*(November–December), 71–83.

Antony, A. (2013). *Impact of effective communication on labour productivity in civil engineering projects: a case study of Kampala Central Division.*

Awamleh, R., & Gardner, W. L. (1999). Perceptions of leader charisma and effectiveness: The effects of vision content, delivery, and organizational performance. *The Leadership Quarterly, 10*(3), 345–373.

Axelrod, A. (2002). *Elizabeth I CEO: Strategic lessons from the leader who built an empire.* Penguin.

Bagheri, A., & Pihie, Z. A. L. (2011). Entrepreneurial leadership: Towards a model for learning and development. *Human Resource Development International, 14*(4), 447–463.

Baker, W. E., & Sinkula, J. M. (1999). Learning orientation, market orientation and innovation: Integrating and extending models of organizational performance. *Journal of Market Focused Management, 4*, 295–308.

Bala, M., & Verma, D. (2018). A critical review of digital marketing. *International Journal of Management, IT & Engineering, 8*(10), 321–339.

Bandura, A. (1986). *Social cognitive theory.* Lawrence Erlbaum Associates.

Bandura, A. (1970). *Principles of behavior modification.* Holt, Rinehart& Winston.

Barber, E. J. W. (1991). *Prehistoric textiles: the development of cloth in the neolithic and bronze ages with special reference to the Aegean.* Princeton University Press.

Baron, R. A. (2007). Behavioral and cognitive factors in entrepreneurship: Entrepreneurs as the active element in new venture creation. *Strategic Entrepreneurship Journal, 1*(1–2), 167–182.

Barrett, D. J. (2006). Leadership communication: A communication approach for senior-level managers. *Handbook of Business Strategy, 7*(1), 385–390.

Bartoloni, E., & Baussola, M. (2016). Does technological innovation undertaken alone have a real pivotal role? product and marketing innovation in manufacturing firms. *Economics of Innovation and New Technology, 25*, 91–113.

Baskin, O., Aronoff, C., & Lattimore, D. (1997). *Public relations—The profession and the practice* (4th ed.). McGraw-Hill.

Booth, A. (1986). *The communication audits.* Gower Publishing Company Ltd.

Bullinger, A. C. (2008). Innovation and ontologies: structuring the early stages of innovation management (1st ed.). Gabler, GWV Fachverlage GmbH, Wiesbaden.

Burke, S. (2013). *Fashion entrepreneur: Starting your own fashion business.* Burke Publishing.

Cacciolatti, L., & Lee, S. H. (2016). Revisiting the relationship between marketing capabilities and firm performance: the moderating role of market orientation, marketing strategy and organisational power. *Journal of Business Research, 69*, 5597–5610.
Cameron, S., & Price, D. (2009). *Business research methods: a practical approach*. CIPD.
Charles, K. (1998). *Peak performance the art of communication* (pp. 117–121). Black Enterprise.
Chase, R. (2016). We need to expand the definition of disruptive innovation. *Harvard Business Review, T14*(00), 45Z.
Christensen, C. M., & Overdorf, M. (2000). Meeting the challenge of disruptive change. *Harvard Business Review, T05*(00), 00Z.
Christensen, C. M., Hall, T., Dillon, K., & Duncan, D. S. (2016). Know your customers' jobs to be done. *Harvard Business Review, T04*(00), 00Z.
Christensen, C. M. (2001). *O dilema da inovação*. São Paulo [etc.]: MAKRON Books.
Christensen, C. M. (2010). Editorial managing innovation in emerging economies: an introduction to the special issue. *IEEE Transactions on Engineering Management, 57*(1), 4.
Christiansen, J. A. (2000). *Competitive innovation management*. McMillan Press Ltd.
Cohen, A. R. (2004). Building a company of leaders. *Leader to Leader, 2004*(34), 16–20.
Cooren, F., Kuhn, T., Cornelissen, J. P., & Clark, T. (2011). Communication, organizing and organization: An overview and introduction to the special issue. *Organization Studies, 32*(9), 1149–1170.
Cooren, Taylor, J. R., & Van Emery, E. J. (2006). *Communication as organizing (1–18)*. Lawrence Erlbaum.
Damanpour, F. (2006). Research on innovation in organizations: Distinguishing innovati-on-generating from innovation-adopting organizations. *Journal of Engineering and Te-Chnology Management, 23*(4), 269.
De Fries, R. E., Bakker-Pieper, A., & Oostenveld, W. (2010). Leadership = communication? The relations of leaders' communication styles with leadership styles, knowledge sharing and leadership outcomes. *Journal of Business and Psychology, 25*(3), 367–380.
Deakins, D., & Freel, M. (1998). Entrepreneurial learning and the growth process in SMEs. *The Learning Organization, 5*(3), 144–155.
Deakins, D., & Freel, M. S. (1999). *Entrepreneurship and small firms*. McGraw-Hill.
Dobni, C. B. 2010. Achieving synergy between strategy and innovation: The key to value creation. *International Journal of Business Science and Applied Management 5*(1), 49.
Donnelly, I. H., Gibson, I., & Vancerich, I. M. (1984). *Fundamentals of management*. Business Publication Inc.
Dryzek, J. S. (1997). The politics of the earth: environmental discourses. *Oxford University Press, 1997*, 220.
Dyer, B., & Michael Song, X. (1998). Innovation strategy and sanctioned conflict: A new edge in innovation? *Journal of Product Innovation Management, 15*(6), 505–519.
Dyer, J. H., Gregersen, H. B., & Christensen, C. (2008). Entrepreneur behaviors, opportunity recognition and the origins of innovative ventures. *Strategic Entrepreneurship Journal, 2*(4), 317–338.
Dyson, R. G. (2004). Strategic development and SWOT analysis at the University of Warwick. *European Journal of Operational Research, 152*, 631–640.
E-Textiles Market: Global Opportunity Analysis and Industry Forecast 2021–2031. (2023). Allied Market Research.
Ezezue, B. O. (2008). Effective communication. An essential tool for organization growth and sustainability. *Nigeria Journal of Management Research, 3*.
Farmer, B. A., Slater, J. W., & Wright, K. S. (1998). The role of communication in achieving shared vision under new organizational leadership. *Journal of Public Relations Research, 10*, 219–235.
Feisel, E., Hartmann, E., & Giunipero, L. (2011). The importance of the human aspect in the supply function: Strategies for developing PSM proficiency. *Journal of Purchasing and Supply Management 17*(1), 54–67.

Femald, L. W., Solomon, G. T., & Tarabishy, A. (2005). A new paradigm: Entrepreneurial leadership. *Southern Business Review, 30*(2), 1–10.
Finkelstein, S., & Hambrick, D. C. (1997). Review: strategic leadership: top executives and their effects on organizations. *Australian Journal of Management*
Fleishman, E. A. (1953). The description of supervisory behavior. *Journal of Applied Psychology, 37*(1), 1.
Fleishman, E. A. (1957). Factor structure in relation to task difficulty in psychomotor performance. *Sage Journal, 17*(4).
Fletcher, M. (1999). The effects of internal communication, leadership and team performance on successful service quality implementation: A South African perspective. *Team Performance Management, 5*, 150–163.
Fleury, A. C. C., & Fleury, M. T. L. (2003). Estratégias competitivas e competências essências: Perspectivas para a internacionalização da industria no Brasil. *Gestão Da Produção, 10*(2), 129–144.
Forti, V., Balde, C. P., Kuehr, R., & Bel, G. (2020). *The global e-waste monitor 2020: quantities, flows and the circular economy potential.* UNU/UNITAR, ITU and ISWA.
Freeman, C., & Soete, L. (1997). *The Economics of industrial innovation.* MIT Press.
Gajanand, G., & Rajesh, M. P. (2016). A SWOT analysis of reliability centered maintenance framework. *Journal of Quality in Maintenance Engineering, 22*(2), 130–145.
Garrido-Moreno, A., Lockett, N., & Garcia-Morales, V. (2015). Exploring the role of knowledge management practices in fostering customer relationship management as a catalyst of marketing innovation. *Baltic Journal of Management, 10*, 393–412.
Gerbner, G. (1985). Field definitions: Communication theory. In *1984–85 U.S. Directory of Graduate Programs International Development Research Centre*
Gloy, Y. S. (2020). *Industrie 4.0 in Der Textilproduktion.* Springer Vieweg
Grant, R. M. (2010). *Contemporary strategy analysis. 7th* (ed). John Wiley.
Greenberg, J., & Baron, R. A. (2008). *Behavior in organizations* (9th ed.). Pearson Education.
Gupta, A. (2013). Environmental and pest analysis: An approach to external business environment. *Merit Research Journal of Art, Social Sciences and Humanities, 1*(2), 13–17.
Gupta, S., Malhotra, N. K., Czinkota, M., & Foroudi, P. (2016). Marketing innovation: a consequence of competitiveness. *Journal of Business Research, 69*, 5671–5681.
Gupta, V., MacMillan, I. C., & Surie, G. (2004). Entrepreneurial leadership: Developing and measuring a cross-cultural construct. *Journal of Business Venturing, 19*(2), 241–260.
Haleblian, J., & Finkelstein, S. (1993). Top management team size, CEO dominance, and firm performance: The moderating roles of environmental turbulence and discretion. *Academy of Management Journal, 36*(4), 844–863.
Hambrick, D. C. (1994). Top management groups: A conceptual integration and reconsideration of the "team ". In L. L. Cummings & B. M. Staw (Eds.), *Research in Organizational Behavior* (pp. 171–214). CTJAI Press.
Hammel, G., & Prahalad, C. K. (1995). *Competindo Pelo Futuro, Editora Campos.*
Hannan, M. T., & Freeman, J. H. (1977). The population ecology of organizations. *American Journal of Sociology, 82*, 929–964.
Hayward, J. (2020). *E-textiles and smart clothing 2020–2030: Technologies, markets and players.* IDTechEx.
Hensley, P. A. & Burmeister, L. (2004).*The artistry of communication leadership* (pp. 30–33).
Herich, L. B. (2008). *Effective communication.* Purkus Publishing Company.
Heron, A. R. (1942). *Sharing information with employees.* Stanford University Press.
Howells, J. (2003). The sourcing of technological knowledge: Distributed innovation pro-cesses and dynamic change. *R & D Management, 33*(4), 395.
Innoskills. (2014). *Competências de Inovação para PMEs.* Características e Tipos de Inovação.
Ireland, R. D., Hitt, M. A., & Sirmon, D. G. (2003). A model of strategic entrepreneurship: The construct and its dimensions. *Journal of management, 29*(6), 963–989.

Ireland, Duane, R., Hoskisson, R. E., & Hitt, M. A. (2011). *The Management of strategy: Concepts, International Edition* (9th ed.). South-Western Cengage Learning
Ireland, R. D., & Hitt, M. A. (1999). Achieving and maintaining strategic competitiveness in the 21st century: The role of strategic leadership. *The Academy of Management Executive, 13*(1), 43–57.
Jaakkola, M., Frösén, J., Tikkanen, H., Aspara, J., Vassinen, A., & Parvinen, P. (2016). Is more capability always beneficial for firm performance? market orientation, core business process capabilities and business environment. *Journal of Marketing Management, 32*, 1359–1385.
Jansen, K. M. (2019). How to shape the future of smart clothing. In *Adjunct Proc. 2019 ACM International Joint Conference on Pervasive and Ubiquitous Computing and Proc. 2019, ACM International Symposium on Wearable Computers* (pp. 1037–1039). ACM.
Jaworski, B. J., & Kohli, A. K. (2000). Market-driven versus driving markets. *Sage Journal, 28*(1).
Jaworski, B. J., & Kohli, A. K. (1993). Market orientation: Antecedents and consequences. *Journal of Marketing, 57*(3), 53–70.
Johansson, C., Miller, V. D., & Hamrin, S. (2014). Conceptualizing communicative leadership: A framework for analyzing and developing leaders' communication competence. *Corporate Communications: An International Journal, 19*(2), 147–165.
Kaleka, A., & Morgan, N. A. (2019). How marketing capabilities and current performance drive strategic intentions in international markets. *Industrial Marketing Management, 78*, 108–121.
Kansikas, J., Laakkonen, A., Sarpo, V., & Kontinen, T. (2012). Entrepreneurial leadership and familiness as resources for strategic entrepreneurship. *International Journal of Entrepreneurial Behavior & Research, 18*(2), 141–158.
Kaplan, R. S., & Norton, D. P. (1996). *The balanced scorecard*. Harvard Business School Press.
Kaplan, R. S., & Norton, D. P. (2008). *The execution premium linking strategy to operations for competitive advantage*. Harvard Business Press.
Kaplan, R. S., & Norton, D. P. (2004). *Strategy maps : Converting intangible assets into tangible outcomes*.
Kast, F. E., & Rosenzweig, J. E. (1972). General systems theory: applications for organization and management. *Academy of Management Journal, 15*, 447–465.
Kim, W.C., & Mauborgne, R. (2016). *A Estratégia Oceano Azul Como criar mais mercado e tornar a concorrência irrelevante*. Actual Editora.
Kohli, A. K., & Jaworski, B. J. (1990). Market orientation: the construct, research propositions, and managerial implications. *Journal of Marketing, 54*, 1–18.
Kotler, P., & Keller, K. L. (2006). *Marketing management*. Pearson Prentice Hall.
Kotler, P. (2000). *Marketing management: The millennium edition*. Person Prentice Hall, Upper Saddle River.
Kube, C. W. (2014). Effects of communication strategies on organizational performance: a case study of Kenya Ports authority. *European Journal of Business and Management, 6*, 6–10.
Kuhn, P., & Weinberger, C. (2005). Leadership skills and wages. *Journal of Labor Economics, 23*(3), 395–436.
Kumar, A. (2012). *Entrepreneurship: creating and leading an entrepreneurial organization*. Pearson Education India.
Kuratko, D. F. (2007). Entrepreneurial leadership in the 21st century. *Journal of Leadership and Organizational Studies, 13*(4), 1.
Kuria, K. L. (2008). *Effective communication on organizational performance: a case study of National Health Insurance Fund Nairobi Branch*. University of Nairobi East African Collection.
Wing Yan Man, T. (2006). Exploring the behavioral patterns of entrepreneurial learning: a competency approach. *Education & Training, 48*(5), 309–321
Markaki, E., Sakas, D., & Chadjipantelis, T. (2013). Communication management in business. *The Latent Power for Career Development, Procedia—Social and Behavioral Sciences, 73*(2013), 319–326.
McCarthy, D. J., Puffer, S. M., & Darda, S. V. (2010). Convergence in entrepreneurial leadership style: Evidence from Russia. *California Management Review, 52*(4), 48–72.

McShane, S. L., & Von Glinow, M. A. Y. (2001). *Organizational behavior* (2nd ed.). Tata McGraw Hill.

Mintzberg, H. (1973). *The nature of managerial work*. Harper and Row Publishers Inc.

Mintzberg, H. (1990b). The design school: Reconsidering the basic premises of strategic management. *Strategic Management Journal, 11*, 171–195.

Mintzberg, H., & Quinn, J. B. (2001). *O processo da Estratégia*. Bookman.

Mintzberg, H., Ahlstrand, B., & Lampel, J. (2000). *Safari de Estratégia: Um roteiro pela selva do planejamento estratégico*. Bookman.

Mintzberg. H. (1987). *Stratégie et artisanat* (pp. 94–104). Harvard l'Expansion, no. 47. Hiver.

Mintzberg. H. (1988). Strategy-making in three modes. In Quinn, J. B., Mintzberg, H., & R. M. James (Ed.), *The strategic process-concepts, contexts and cases*. Prentice-Hall Inc.

Mintzberg, H. (1988a). Opening up the definition of strategy. In J. B. Quinn, H. Mintzberg, & R.M. James (Ed.), *The strategic process—concepts, contexts and cases*, Prentice-Hall Inc.

Mintzberg. H. (1990a). Strategy formation: Schools of thought. In J. W. Fredrickson (Ed.), *Perspectives on strategic management*. Harper Business

Morand, D. A. (2001). The emotional intelligence of managers: assessing the construct validity of a nonverbal measure of people skills. *Journal of Business and Psychology, 16*(1), 21–33.

Mustonen-Ollila, E., & Lyytinen, K. (2003). Why organizations adopt information system process innovations: A longitudinal study using Diffusion of Innovation theory. *Information Systems Journal, 13*(3), 275–297.

Narver, J. C., & Slater, S. F. (1990). The effect of a market orientation on business profitability. *Journal of Marketing, 1990*(54), 20.

Nelissen, R. M. A., Dijker, A. J. M., & de Vries, N. K. (2007). Emotions and goals: Assessing relations between values and emotions. *Cognition and Emotion, 21*(4), 902–911.

Nnamseh, M. (2009). The role of communication in business success. *Nigeria Journal of Management Research, 1*.

Pedersen, E. R., Gwozdz, W., & Hvass, K. K. (2018). Exploring the relationship between business model innovation, corporate sustainability, and organizational values within the fashion industry. *Journal of Business Ethics, 149*(2), 267–284.

Pedersen, E. R. G, Andersen, K. R. (2015). Sustainability innovators and anchor draggers: a global expert study on sustainable fashion. *Journal of Fashion Marketing and Management*.

Petrick, J. A., Scherer, R. E., Brodzinski, J. D., Quinn, J. F., & Ainina, M. F. (1999). Global leadership skills and reputational capital: Intangible resources for sustainable competitive advantage. *The Academy of Management Executive, 13*(1), 58–69.

Pfeffer, J., & Salancik, G. R. (1978). *The external control of organizations*. Harper & Row.

Pigors, P. J. (1949). *Effective communication in industry: what is its basis?* National Association of Manufacturers.

Porter, M. E. (1986). *Estratégia competitiva: Técnicas para análise de indústrias e da concorrência*. Campus.

Porter, M. E. (1992). *Vantagem competitiva: Criando e sustentado um desempenho superior*. Editora Campus.

Porter, M. E. (1999). *Competição: Estratégias competitivas essenciais*. Campos.

Puth, G. (2002). *The communicating leader, the key to strategic alignment* (2nd ed.). Van Schaik.

Quinn, J. B. (1980). *Strategies for change: logical incrementalism*. Irwin Inc.

Quinn, J. B. (1988). Strategic change-logical incrementalism. In J. B. Quinn, H. Mintzberg, & R. M. James (Eds.), *The strategic process—concepts, contexts and cases* (pp. 94–104). Prentice-Hall Inc.

Radovic Markovic, M., & Omolaja, M. (2009). *Information management-concepts*. Himalaya Publishing, Delji, India.

Radovic Markovic, M. (2011).*Organisation behaviour and culture: globalization and the changing environmemt of organizations*. VDM Verllag Dr Muller, p. 348.

Ramirez, J., Francisco, G.-R., Ruiz-Ortega, M. J., & Garcia-Villaverde, P. M. (2018). From external information to marketing innovation: the mediating role of product and organizational innovation. *Journal of Business & Industrial Marketing, 33*, 693–705.

Rasmussen, P. (2011). Knowledge management in the firm: Concepts and issues. *Inter-National Journal of Manpower, 32*(5/6), 479.

Richard, et al. (2009). Measuring organizational performance: towards methodological best practice. *Journal of Management, 35*, 718–804.

Ries, A. l., Trout, J. (1993). *As 22 consagradas leis do marketing*. Makron Books.

Ries, A. I., & Ries, L. (2000). *As 22 consagradas leis de marcas*. Makron Books.

Ries, A. I., & Trout, J. (2002). *Posicionamento: A batalha por sua mente* (20a ed.). Makron Books.

Robert, K. M., George, C. R., & Kendall, P. (1957). *Social theory and social structure* (pp. 287–293). Harvard University Press.

Rogers, E. M. (2003). *The diffusion of innovation* (5th ed.). Free Press.

Ruão, T. (2016). *A organização comunicativa: teoria e prática em Comunicação Organizacional*. CECS.

Schumpeter, J. A. (1934). *Theory economic development*. Harvard University Press.

Scott, W. G. (1974). Organization theory: a reassessment. *Academy of Management Journal, 17*, 242–254.

Shergill, G., & Nargundkar, R. (2005). Market orientation, marketing innovation as performance drivers: extending the paradigm. *Journal of Global Marketing, 19*(1), 27–47.

Simons, R. (2000). *Performance measurement and control systems for implementing strategy: text and cases*. Prentice-Hall. Inc.

Skinner, T, Michael, P. (2008). *Using media intelligence tools to drive communications success-the benefits of engaging traditional media and social media*. CNW Group

Slater, S. F., & Narver, J. C. (1994). Does competitive environment moderate the market orientation-performance relationship? *Journal of Marketing, 58*, 46–55.

Spaargaren, G. (2003). Sustainable consumption: a theoretical and environmental policy perspective. *Society & Natural Resources, 16*(8), 687–701.

Supply–Demand Mismatch Looms Large For Smart Clothing, Technical Textile. (2023). https://www.technicaltextile.net/articles/supply-demand-mismatch-looms-large-for-smart-clothing-9134

Survey: Consumer Sentiment on Sustainability in Fashion, McKinsey & Company. (2023). https://www.mckinsey.com/industries/retail/our-insights/survey-consumer-sentiment-on-sustainability-in-fashion

Thietart, R. A. (1984). *La Stratégie d'entreprise*. McGraw-Hill.

Tidd, J., Bessant, J., Pavitt, K. (2008). *Gestão da inovação* (3rd ed.). Book-man.

Trout, J. (1995). *The new positioning: The latest on world's #1 business strategy*. McGraw Hill.

Van de Ven, A. H., Angle, H. L., & Poole M. S. (2002). *Research on the management of innovation 1989*. Ballinger.

Waweru, S. N. (2002). *Secondary school administration*. Kenyatta University Press.

The Influence of the Marketing Orientation of Textile Companies in Increasing Their Competitiveness

Muhammad Farrukh Shahzad, Iqra Javed, and Imran Zahid

Abstract In today's globalized and fiercely competitive business environment, textile companies constantly seek strategies to enhance their competitiveness. Adopting a marketing orientation is a critical factor that has emerged as a key driver of success in the textile industry. The study's main aim is to examine the influence of the marketing orientation on the competitive advantage of textile firms in terms of market innovation and global strategy. It helps to explain how market turbulence moderates the market orientation, strategy, and competitive advantage relationships. In this study, data was collected from 320 employees from industrial firms listed on the Pakistan Stock Exchange (PSE). This study found that market orientation correlates positively with competitive advantage. Findings of all moderation and mediated variables show significant correlation. The study found that market turbulence moderates the association between marketing orientation, strategy, and competitive advantage. Additionally, this study underscores the significance of marketing orientation as a potent tool for textile companies seeking sustainable growth and success in a highly competitive industry.

Keywords Marketing orientation · Market innovation · Global strategy · Competitive advantages · Market turbulence

M. F. Shahzad (✉)
College of Economics and Management, Beijing University of Technology, Beijing 100124, PR China
e-mail: farrukhshahzad207@gmail.com

I. Javed
Institute of Business and Management, University of Engineering and Technology, Lahore 54000, Pakistan
e-mail: iqra.javed@uet.edu.pk

I. Zahid
Department of Mechanical Engineering and Technology, Government College University, Faisalabad, Pakistan
e-mail: imranzahid@gcuf.edu.pk

Lisbon Accounting and Business School, Polytechnic Institute of Lisbon (ed.), *Industry and Innovation: Textile Industry*, SDGs and Textiles,
https://doi.org/10.1007/978-3-031-57804-5_5

1 Introduction

In today's dynamic business environment, organizations constantly seek ways to establish a competitive advantage and secure long-term success (Alghamdi & Agag, 2024). Understanding the complex interplay between market orientation and competitive advantage is critical to achieving such benefits in the Pakistani textile sector. Market orientation, defined as a keen emphasis on customer wants and dynamic market conditions, has emerged as an important predictor of an organization's competitiveness (Dayan et al., 2023). Meanwhile, competitive advantage refers to the characteristics that enable an organization to outperform its competitors. Competitive advantages can come from various factors, including cost structure, product offerings, brand reputation, and customer service. The crucial point is that these advantages are long-term and difficult for competitors to replicate. The concept of global strategy is considered a comprehensive plan created by a company to operate in foreign markets. A global strategy evaluates how to efficiently manage and execute company activities in multiple locations efficiently, considering local market conditions, cultural variations, and international rules (Arici & Gok, 2023).

Market innovation refers to developing and executing new ideas, products, services, or procedures for the marketplace. Along with global strategy, the idea of market turbulence describes the rate and extent of change in the market environment (Rokkan, 2023). This study uses market turbulence to moderate the relationship among market innovation and competitive advantage. Furthermore, its position as a mediator in the relationship between global strategy and competitive advantage is largely unexplored in previous research (Suandi et al., 2023). However, technological improvements, alterations in customer tastes, regulatory changes, and economic swings can all contribute to market volatility. High market turbulence means a fast-changing environment in which enterprises must adapt to remain competitive.

There is a significant gap in how these processes interact in Pakistan's textile sector. Existing research frequently takes a broad or worldwide view, with little focus on the interplay of market orientation, global strategy, market innovation, and market volatility in the context of Pakistan's textile industry (Ed-Dafali et al., 2023). This study gap is critical because Pakistan's textile industry faces unique problems and opportunities created by its cultural, economic, and geopolitical environment. Understanding these interrelationships more clearly could provide significant insights for Pakistani textile companies looking to improve their domestic and international competitiveness.

Moreover, this study is well-contributed by the theoretical underpinning of the VIRN framework, which is derived from the resource-based view of the firm (RBV) theory (Malik et al., 2020). According to this framework, how internal resources and capabilities (Valuable), such as market orientation, global strategy, market innovativeness, and adaptability to market turbulence, contribute to sustained competitive advantages in Pakistan's textile sector. Hence, the current study used the existing measures from the research to evaluate the firm's competitive advantage and market orientation. This study helps answer the following questions: (1) Does market orientation affect the competitive advantage of firms in Pakistan's textile sector? (2) Does the Pakistan textile sector's market orientation and competitive advantage affect the global strategy and market innovation? (3) How much the market turbulence affect the innovation, strategy and firm's competitive advantage?

2 Literature Review and Supporting Theory

2.1 Supporting Theory

This study is best comprehended with the VIRN (valuable, imitate, rare, and non-substitutable) framework derived from the resource-based view (RBV) of the firm theory, which is defined as evaluating a company's competitive advantage by determining whether its resources or capabilities are valuable, rare, unique, and non-substitutable (Larbi-Siaw et al., 2022). The resources that match these criteria will likely provide a long-term competitive advantage because they are distinctive, difficult to copy, and lack substitutable alternatives. In the context of Pakistan's textile business, the VRIN framework provides insights into potential sources of sustained competitive advantage resulting from market orientation, global strategy, market innovativeness, and the ability to navigate market turbulence. A strong market orientation is valuable because it allows businesses to effectively understand and respond to client needs, resulting in a unique and differentiated strategy within the sector. This method may be rare because not many organizations prioritize customer-centric strategy to the same amount and its uniqueness stems from competitors' difficulty when reproducing a deeply embedded market orientation (Ed-Dafali et al., 2023).

Moreover, the non-substitutability of a strong market orientation is clear, as the unique insights gained from such a strategy are difficult to replicate with other approaches (Bano et al., 2016). Similarly, applying the VRIN paradigm to the global strategy dimension demonstrates its usefulness in exploiting economies of scale and knowledge transfer across borders. A well-executed global strategy distinguishes organizations and its uniqueness stems from the complex combination of talents and market knowledge required for its implementation (Wagner & Schaltegger, 2004). A global strategy cannot be substituted because tailored global approaches are difficult to replace with standard industry procedures. Examining market innovativeness under the VRIN framework demonstrates its importance in promoting competitiveness through continual innovation. The uniqueness of market-driven innovativeness distinguishes organizations that prioritize such approaches and its inimitable quality stems from the unique knowledge, skills, and technology required for effective innovation (O'Connor et al., 2011). The non-substitutability of market innovativeness is clear, as distinctive and market-driven ideas are difficult to replace with standard industry methods.

While navigating market turbulence, the VRIN framework reveals the value of adaptability and resilience. Effectively managing and adapting to market turbulence is considered valuable, and its rarity lies in the differentiation of firms that can successfully navigate such challenges (Pal et al., 2014). The inimitability of adaptability to turbulence is rooted in unique processes or organizational cultures that competitors find challenging to replicate. The non-substitutability of effective strategies in turbulent markets is evident, as the combination of unique capabilities required for success is not easily substituted with standard approaches (Chen et al., 2016). In short, the VRIN framework provides a comprehensive understanding of how internal resources and capabilities, such as market orientation, global strategy, market innovativeness, and adaptability to market turbulence, contribute to sustained competitive advantages in Pakistan's textile sector.

2.2 *Marketing Orientation and Competitive Advantages*

According to (Freixanet, 2012), market orientation entails creating and distributing information within an organization on customer demands and preferences and competitive data. When an organization adopts this perspective, its performance improves. This is because senior management and employees understand what customers expressly and intuitively demand and have insights into competitors' strengths. Researchers (Yayla et al., 2018) highlighted that this knowledge fosters a strong dedication to delivering great client satisfaction. Such capabilities can develop into a competitive advantage when a firm is unique in having this knowledge and uses it efficiently and effectively as part of its operating process. (Kumlu, 2014) defines competitive advantage as a firm's ability to outperform its rivals. This elaborates on this with the resource-based approach, which assigns competitive advantage to specific resources and capabilities. Several studies in Pakistan's textile sector have found a positive association between marketing orientation and business performance. A past study (Tjahjadi et al., 2022) discovered that companies with a stronger market orientation had higher levels of customer satisfaction and attained more competitive benefits.

Innovation in the Pakistani textile business was a crucial mediator between market orientation and competitive advantage. They contended that knowing market trends and client needs fosters innovative methods. Market orientation is critical for Pakistani textile enterprises in foreign markets, as the flexibility to global customer preferences considerably improves competitive positioning (Baierle et al., 2020). The government emphasized the need for incentives and assistance for export-oriented marketing tactics. Government policies significantly impact the competitive dynamics of the textile sector (Osorio Tinoco et al., 2020).

Market orientation theory emphasizes the necessity of gathering, analyzing, and responding to market information, which ideally aligns with the textile industry's need to adapt to continuously changing market needs and client preferences. Simultaneously, the RBV emphasizes using distinctive organizational resources and competencies, such as consumer insights and new processes, to maintain a competitive advantage (Gyedu et al., 2021). This theoretical fusion demonstrates how Pakistani textile manufacturers can achieve and sustain market superiority by leveraging market-oriented tactics based on their unique. It is argued from the above literature that adopting a marketing approach for practitioners in Pakistan's textile business entails not just focusing on customers but also analyzing competition strategies and using the firm's particular advantages. Strategic alignment with market requirements and trends is critical to maintaining competitive advantages in both domestic and foreign markets. Hence, it is hypothesized that:

Hypothesis 1: Marketing orientation is positively related to the competitive advantages of pakistan's textile sector.

2.2.1 Mediating Role of Market Innovation

Market innovation refers to creating new market strategies and channels or major enhancements to existing ones to gain a competitive advantage. It includes product, process, and administrative improvements (Dhiman et al., 2020). Marketing orientation, which includes customer, competitor, and inter-functional orientation. This provides a favorable atmosphere for innovation. Market-oriented organizations excel at discovering and capitalizing on new opportunities. Marketing orientation, which incorporates client, competitor, and inter-functional orientations (Dayan et al., 2023), fosters innovation. Market orientation leads to a better awareness of client wants and developing market trends, allowing for more innovative solutions. These market-oriented businesses thrive at identifying and capitalizing on new opportunities.

Market orientation has been found to have a favorable impact on innovation, which improves corporate performance. It has been demonstrated that market orientation helps new products succeed, particularly through innovation (Arici & Gok, 2023). Empirical research conducted in China discovered a significant positive association between market orientation, innovation capability, and firm performance (Rokkan, 2023). Studies such as (Shahzad et al., 2021; Taçoğlu et al., 2019) have found comparable tendencies in other industries and countries, confirming the mediating function of innovation in the marketing orientation-competitive advantage link. Additional elements strengthen market innovation's critical significance as a bridge between marketing orientation and competitive advantages. Organizational learning, market flexibility, proactive technological integration, and strategic resource allocation all help to facilitate the mediation process.

A past study (Nguyen & Khoa, 2020) investigated the positive impact of market orientation on innovation and organizational learning, revealing a direct relationship with company success. From the perspective of Pakistan's textile sector, (Suandi et al., 2023) found that Pakistani textile companies that embraced market orientation were likelier to adopt innovative methods, resulting in increased competitive advantage. This mediating function of market innovation is consistent with Schumpeter's view of innovation as a critical driver of competitive advantage. It is consistent with the resource-based view (RBV), which holds that market innovation is a valuable, uncommon, and inimitable resource. Thus, prior studies argue that market innovation is a key mediator between marketing orientation and competitive advantages (Udriyah et al., 2019). This is especially important in areas such as textiles in Pakistan, where market dynamics are constantly changing, and innovation can provide a major competitive advantage in domestic and international markets. Hence, it is hypothesized that:

Hypothesis 2: Market innovation positively mediated the relationship between marketing orientation and competitive advantages.

2.2.2 Mediating Role of Global Strategy

Global strategy involves aligning an organization's resources and competences with opportunities in the global market. A global strategy enables firms to leverage their marketing orientation to capture international competitive advantages (Ed-Dafali et al., 2023). The marketing orientation approach goes beyond home markets, emphasizing the significance of understanding global client demands and tailoring strategies appropriately. A strong marketing orientation encouraged the adoption of a global strategy, resulting in increased competitive advantages (Risitano et al., 2022). They found that a global marketing approach improved a firm's international competitive position. Global marketing perspective was positively connected with implementing a worldwide strategy in technology enterprises, resulting in superior competitive advantages. A global strategy serves as an independent competitive strategy, cultivating differentiation and cost leadership worldwide (Saleh et al., 2021). This strategic orientation is critical for organizations seeking to properly use their marketing orientation, providing a long-term competitive advantage that extends beyond the boundaries of local markets.

A framework for market-based organizational learning connects values, knowledge, and behavior. The research, which focuses on market-based organizational learning, investigates how organizations might improve their learning capacity by combining marketing orientation and global strategies (Naidoo, 2010). Moreover, the mediating role of a global strategy is consistent with the resource-based view (Kraus et al., 2020), which suggests that strategic resources and competencies used internationally can contribute to long-term competitive advantages. A country like Pakistan needs a global strategy to bridge the gap between marketing focus and competitive advantages in a culturally varied environment (Farrukh et al., 2023a, 2023b).

Past studies (Chan et al., 2012; Singh et al., 2021) discovered in Asian markets that a global strategy bridges marketing orientation and competitive advantages, showing the relationship's resilience across varied cultural settings. In Europe's, manufacturing firms have a strong marketing orientation to acquire a global competitive advantage through strategic global efforts (Osorio Tinoco et al., 2020). In a Middle Eastern environment, (Tjahjadi et al., 2022) found that a global strategy makes translating marketing orientation into a competitive advantage easier, particularly in service-oriented businesses. Hence, from the above arguments, it is claimed that a global strategy is a beneficial bridge between marketing orientation and competitive advantages. This emphasizes the significance of expanding marketing focus beyond home borders, strategically synchronizing global initiatives to capitalize on different market opportunities and achieve long-term competitive success. Based on arguments, it is hypothesized that:

Hypothesis 3: Global strategy positively mediated the relationship between marketing orientation and competitive advantages.

2.2.3 Moderating Role of Market Turbulence

Market turbulence is instability, uncertainty, and rapid change in a market environment. It reflects the market's dynamic and unpredictable nature, which is defined by swings in demand, competitive forces, technology breakthroughs, regulatory changes, and other external variables (Bamfo & Kraa, 2019). High market turbulence means frequent and severe alterations in market conditions, making it difficult for enterprises to forecast and respond to changes successfully. Organizations functioning in volatile marketplaces may face increasing competition, shifting client preferences, and greater uncertainty about future trends (Gugler & Shi, 2009). Adapting strategy, developing agility, and remaining current on market dynamics are critical for organizations navigating moments of market upheaval. A past study (O'Connor et al., 2011) has underlined the favorable influence of market innovation on acquiring a competitive edge. Innovations in products, processes, and marketing tactics help to improve market positioning and competitiveness. Market volatility can modify the link between innovation and performance (Pal et al., 2014). In volatile markets, the impact of innovation on competitive advantages may be greater. (Freixanet, 2012) investigated the relationship among market innovation and competitive advantages in high-tech businesses. According to the findings, market turbulence positively moderates this link, boosting the beneficial benefits of innovation on competitive advantages.

A past study (Tjahjadi et al., 2022) scrutinized the moderating role of environmental turbulence on the innovation-performance connection. The results found that in volatile markets, the favorable impact of innovation on competitive advantages is more significant. Within the RBV paradigm, the unique, valuable, and non-substitutable resources created by market innovation are significant determinants of long-term competitiveness (Dhiman et al., 2020). The review also incorporates ideas from contingency theory, which suggests that market turbulence has a moderating role in changing the dynamics of the link between market innovation and competitive advantage (Osorio Tinoco et al., 2020). According to contingency theory, market innovation has a greater influence and effectiveness in bestowing competitive advantages in high-turbulence situations. This theoretical integration sheds light on the intricate interaction of internal capabilities, external environmental conditions, and organizational results, providing a comprehensive knowledge of how businesses can strategically navigate dynamic marketplaces. Therefore, the literature consistently supports the notion that market turbulence improves the link between market innovation and competitive advantage (Wong & Ngai, 2023). Understanding this dynamic is crucial for companies seeking a competitive advantage in quickly changing business settings. It is hypothesized that:

Hypothesis 4a: Market turbulence positively moderated the relationship between market innovation and competitive advantages.

Global strategy can lead to competitive advantages by capitalizing on economies of scale, scope, and knowledge transfer across borders. The RBV theory emphasizes that global strategies donate to competitive advantages by evolving and utilizing distinctive, valuable, and difficult-to-copy resources on a global scale (Nguyen & Khoa, 2020). In volatile markets, a well-executed global strategy can influence competitive advantages more. Market turbulence modulates this relationship positively, amplifying the favorable benefits of a global strategy on competitive advantages (Udriyah et al., 2019). However, the findings revealed that in volatile markets, a global approach had a more favorable influence on competitive advantages (Risitano et al., 2022). This paper combines insights from the resource-based view framework, emphasizing the role of global strategy as a basis of competitive advantage, with contingency theory, which proposes that market volatility works as a moderator determining the strength of this link.

Further, prior study argued (Udriyah et al., 2019) that dynamic global strategies, including international diversification and cross-border coordination, contribute significantly to enterprises' worldwide competitive advantages. Global strategies are necessary components of a firm's dynamic capacities. Adjusting and configuring resources internationally improves a company's competitive advantage. Global strategies act as adaptive processes in volatile markets, allowing enterprises to navigate uncertainty efficiently (Saleh et al., 2021). This adaptive capacity provides a competitive advantage. Industrial organizations that strategically leverage global activities in volatile situations earn greater competitive advantages (Tjahjadi et al., 2022). This study concluded that the benefits of a global strategy on competitive advantages are more obvious in highly volatile markets. Hence, the research suggested that market turbulence improves the link between global strategy and competitive advantages. Firms that use dynamic global strategies in unstable contexts can capitalize on uncertainties, transforming them into strategic opportunities to maintain and improve global competitiveness (Dhiman et al., 2020). Hence, it is hypothesized that:

Hypothesis 4b: Market turbulence positively moderated the relationship between global strategy and competitive advantages.

Moreover, Fig. 1 displays the theoretical framework.

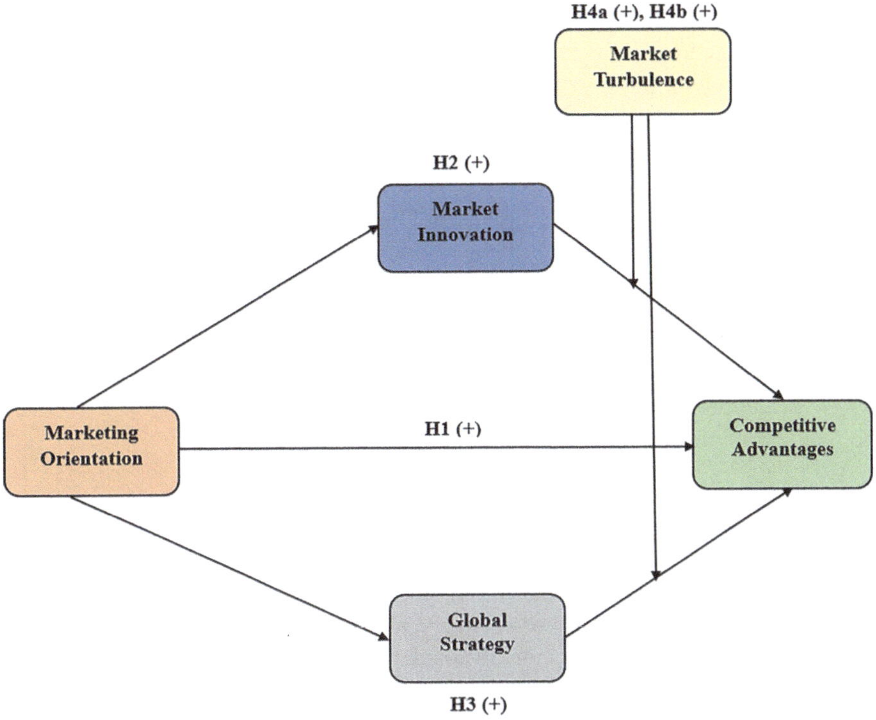

Fig. 1 Theoretical framework

3 Research Methodology

This study concentrates on the marketing orientation impacts toward competitive advantages of textile industries through market innovation, global strategy, and market turbulence. Professionals who are working in industrial sector firms listed on the Pakistan Stock Exchange (PSE) make up the target market. Pakistan's industrial sector was the focus of this study due to its recent economic crisis and the need for improvements in the industrial zone. The manufacturing sector, which includes Pakistan's leading textile industries, was the primary target sector. The textile industry was selected because it is important to Pakistan's economy, generating employment, GDP growth, and export revenue. This study may be helpful for the industrial sector as it contributes to the nation's business and GDP. The industrial sector of Pakistan implements these market orientation strategies in their firm to improve competitiveness in the market. 8.5% of Pakistan's yearly GDP is generated by its manufacturing industries (Farrukh et al., 2023a, 2023b).

We decided to check the impact of market orientation on increasing their competence in the textile industry because manufacturing has much potential to spur economic growth in developing countries like Pakistan. The researchers' selection of manufacturing companies listed on the PSE indicates that they are interested in researching publicly traded businesses subject to reporting and regulatory requirements. A popular statistical method in social sciences studies for examining the connections between latent variables is partial least squares (PLS-SEM). The SmartPLS 4.0 software is employed per earlier research (Farrukh et al., 2023a, 2023b; Shahzad et al., 2023a, 2023b), offering a suitable context for investigating this study variable. The study looks into usual technique bias using Harman's single-factor check (Podsakoff et al., 2003). If the total variation for a single element is greater than 50%, there is no general bias problem in the research; nevertheless, if it is less than 50%, there is a bias issue in the data (Howard & Henderson, 2023). Nevertheless, our analysis found (16%) no bias issue in the study data related to the suggestions made by a single-factor check.

3.1 Participants and Procedure

To choose a suitable sample size for our study, we considered the number of latent variables, the complexity of our model, and the resources that were accessible. PLS-SEM has become known for its adaptability and capacity to manage models with numerous variables or intricate associations, irregular and non-linear data, and small sample sizes (Shahzad et al., 2021). Because of the large population size and complexity of reaching every participant, a convenient sampling technique was used as an efficient medium in this study (Martins et al., 2023a, 2023b, 2023c). Participants in convenience sampling are selected according to their availability and desire to participate. Convenience sampling is widely used in online surveys since it's easy to reach respondents by email, social media, or other online methods. Emails and e-messages were used to distribute and reconcile the questionnaires during the survey administration process. We used both online and offline structured questionnaires to get information from the management of various production companies. They were asked to use a five-point Likert scale, ranging from 1 ("strongly disagree") to 5 ("strongly agree"), to answer and express their thoughts on the efficacy of the factors that had been evaluated in their organization. In the survey, middle-level, senior, and junior officials from manufacturing (textile) companies were asked to discuss how market orientation influences the increase of their competitiveness. They were also asked to discuss how market innovation, global strategy, and turbulence influence market orientation's competitiveness.

The process of collecting data for this purpose began in the last quarter of 2023. Initially, participants were given 410 questionnaires; of those, 320 responses were obtained; 90 of them were missing. Therefore, this indicates a 78% response rate. The sample size was adequate to conclude because it complied with the research's guideline (Martins et al., 2023a, 2023b, 2023c) that the SmartPLS 4.0 analysis technique requires at least 150 sample sizes (Roscoe et al., 1975). PLS-SEM is a high-level modelling approach that uses t-values to examine the link between the inner

Table 1 Demographic details

Demographics	Distribution	n = 320
Gender	Male	189 (59%)
	Female	131 (41%)
Age	18–26 years	109 (34%)
	27–35 years	138 (43%)
	More than 35 years	73 (23%)
Qualification	Diploma	41 (13%)
	Bachelor	129 (40%)
	Masters	112 (35%)
	PhD	38 (12%)
Position	Higher authorities	81 (25%)
	Middle authorities	125 (39%)
	Junior authorities	114 (36%)
Experience	5 or less years	124 (39%)
	6–15 years	156 (49%)
	16–25 years	40 (12%)
Textile Mills	Fateh Textile Mills	44 (14%)
	Husein Industries Limited	51 (16%)
	Nishat Mills Limited	54 (17%)
	Interloop Limited	61 (19%)
	Soorty Enterprises (Pvt) Ltd.	48 (15%)
	Kohinoor Mills Limited	62 (19%)

model and the PLS method. 41% of the 320 respondents are women, and 59% of the respondents are men. Further, demographic details are provided in Table 1.

3.2 *Survey Development and Measures*

All of the variables considered in this analysis were modified from previous studies to meet the standards set by the literature on information sharing. Initially, marketing orientation, market innovation, global strategy, competitive advantages, and market turbulence criteria were researched and constructed based on existing literature. Marketing orientation five items were taken from past study (Genc et al., 2019). Eight items of competitive advantages were taken from past studies (Naidoo, 2010; Udriyah et al., 2019). Market innovation six items were assessed from the prior study (Kumlu, 2014). Six items of global strategy were taken from past study (Nguyen & Khoa, 2020). Market turbulence five items were taken from the prior study (Chen et al., 2016). This study examined both observable and unobserved factors and indicators using the PLS-SEM statistical technique.

4 Analysis and Results

The proposed model is measured in the current study using SmartPLS 4.0 statistical software using PLS-SEM. The PLS-SEM consists of two major techniques. First, the outer model is the term used to refer to the measurement model. The second model is the structural model, often known as the inner model (Hair et al., 2011). A causal forecasting system assesses statistical models to characterize the relationships between variables. PLS-SEM determines the independent regression analysis for measurement and structural models, which allows it to handle small sample sizes, many variables, and related problems well (Shahzad et al., 2024b).

4.1 Measurement Model

The validity and reliability tests are looked at first to establish if a measurement tool is acceptable. Validity refers to how variables of a particular idea are represented, whereas reliability is the degree of consistency in the results evaluated by the instrument (Shahzad et al., 2023a, 2023b). Using a variety of techniques, we computed the validity and reliability of successfully applying the constructs. Table 2 presents the results of evaluating the internal consistency of all the variables using Cronbach's alpha and composite reliability. These findings reveal that the outer model's regular factors loadings were over the minimum requirement of 0.70. It shows the degree to which constructions are related to one another. Every item in Table 2 meets the range when the threshold value is 0.70. Each construct's convergent validity was tested by the AVE. At least 0.50 was the lowest cutoff value (Sarstedt et al., 2022). Every item's AVE fits inside Table 2's range. We investigated the likelihood of multicollinearity through the examination of the variance inflation factor (VIF) and acceptance. It is implied that the readings are accurate by the low threshold number. Less than five is required for the VIF cutoff value (Shahzad et al., 2022). The multicollinearity of the dataset was significant, as Table 2 of VIF shows. Additionally, the correctness and convergence of our results were demonstrated by the fact that all constructs' Cronbach's alpha, composite reliability, and AVE values were higher than the threshold levels.

To ascertain if the employed constructs were distinct from one another, discriminant validity was investigated (Fornell & Larcker, 1981). Different methods were employed to estimate the discriminant validity. We looked at the relationship between the share variance of constructs and the extracted average variance (AVE). Table 3 shows that the root of the AVE was greater than the correlational values for every construct. Furthermore, the Heterotrait-Monotrait (HTMT) ratio was computed (Sarstedt et al., 2022), which is derived using a conventional analysis method. The HTMT cutoff number shouldn't be higher than 1.

Table 2 Factor analysis

Latent constructs	Items	Factor loadings	α	CR	AVE	VIF
Marketing orientation (MO)			0.895	0.922	0.704	
	MO1	0.822				2.451
	MO2	0.835				2.267
	MO3	0.850				2.726
	MO4	0.843				2.393
	MO5	0.843				2.388
Competitive advantages (CA)			0.942	0.952	0.713	
	CA1	0.871				2.603
	CA2	0.872				2.691
	CA3	0.878				2.147
	CA4	0.878				2.904
	CA5	0.865				1.709
	CA6	0.770				2.506
	CA7	0.834				2.011
	CA8	0.777				1.684
Market innovation (MI)			0.924	0.940	0.724	
	MI1	0.799				2.347
	MI2	0.841				2.516
	MI3	0.843				2.892
	MI4	0.848				2.626
	MI5	0.894				1.433
	MI6	0.879				2.305
Global strategy (GS)			0.917	0.935	0.706	
	GS1	0.842				1.332
	GS2	0.795				2.421
	GS3	0.842				1.137
	GS4	0.853				2.223
	GS5	0.855				2.943
	GS6	0.853				2.701
Market turbulence (MT)			0.878	0.911	0.671	
	MT1	0.821				1.545
	MT2	0.807				2.458
	MT3	0.873				1.692
	MT4	0.802				1.177
	MT5	0.790				2.963

Note α = Cronbach's alpha value, CR = Composite reliability, AVE = Average variance extracted, VIF = Variance inflation factors

Table 3 Discriminant validity

Constructs	1	2	3	4	5
1. Competitive advantages	0.844				
2. Market innovation	0.280	0.851			
3. Marketing orientation	0.122	0.831	0.839		
4. Global strategy	0.470	0.223	0.122	0.840	
5. Market turbulence	0.496	0.273	0.150	0.537	0.819

4.2 *Structure Model*

For the structural model, SEM was employed to assess the hypotheses and address pertinent research issues. The bootstrapping method was used to examine the hypothesis' relevance (Martins et al., 2023a, 2023b, 2023c). Initially, the predictive power of the study model assessment was ascertained by evaluating the path coefficient values and R^2 values. In accordance with this, R^2 analysis computes the coefficient of determination based on the dependent variable's total variance and variation ratio on independent constructs (Zhang et al., 2024). The R^2 value for the current study model as market innovation = 69%, global strategy = 1.5% and competitive advantages = 35%. A research methodology from a previous study (Hair et al., 2011) was used to determine the goodness of fit (GOF). The GOF values show that the model fits the data well overall. For illustration, Chi-Square = 7702.002, SRMR = 0.078, and NFI = 0.465.

We used SEM analysis with the PLS bootstrapping technique to test our research hypotheses. The results were assessed using t-statistics, standardized path (Beta coefficient) values, and p-statistics (Hair et al., 2011). These findings show whether the specific variable has a positive or negative effect on other factors. It proved how some independent constructs significantly impact the dependent constructs. The structural model comprises the following constructs: marketing orientation, market innovation, global strategy, competitive advantages, and market turbulence. The p-value criterion ($p < 0.05$) was used to calculate the variables' importance. Table 4 provides a summary of the findings. The hypothesis test results show that marketing orientation significantly influences competitive advantages ($\beta = 0.225$, $t = 3.754$, and $p = 0.000$). Similarly, marketing orientation significantly influences market innovation ($\beta = 0.831$, $t = 15.976$, and $p = 0.000$). Market innovation significantly influences competitive advantages ($\beta = 0.360$, $t = 5.581$, and $p = 0.000$). Marketing orientation significantly influences global strategy ($\beta = 0.122$, $t = 2.145$, and $p = 0.032$). Global strategy significantly influences competitive advantages ($\beta = 0.212$, $t = 3.37$, and $p = 0.001$). Market turbulence significantly influences competitive advantages ($\beta = 0.268$, $t = 4.244$, and $p = 0.000$).

For mediation analysis, market innovation significantly mediates the relationship between marketing orientation and competitive advantages ($\beta = 0.299$, $t = 5.224$, and $p = 0.000$). Global strategy significantly mediates the relationship between

Table 4 Hypothesis testing

Constructs	β-values	STDEV	t-statistics	p-values	Decision
Direct effect					
MO -> CA	0.225	0.060	3.754	0.000	Supported
MO -> MI	0.831	0.052	15.976	0.000	Supported
MI -> CA	0.360	0.065	5.581	0.000	Supported
MO -> GS	0.122	0.057	2.145	0.032	Supported
GS -> CA	0.212	0.063	3.337	0.001	Supported
MT -> CA	0.268	0.063	4.244	0.000	Supported
Mediation effect					
MO -> MI -> CA	0.299	0.057	5.224	0.000	Supported
MO -> GS -> CA	0.031	0.015	2.037	0.040	Supported
Moderation effect					
MI -> MT*CA	0.094	0.041	2.308	0.021	Supported
GS -> MT*CA	0.098	0.040	2.446	0.015	Supported

Note MO = Marketing orientation, MI = Market innovation, GS = Global strategy, CA = Competitive advantages, MT = Market turbulence

marketing orientation and competitive advantages (β = 0.031, t = 2.037, and p = 0.040).

For moderation analysis, market turbulence significantly moderates the relationship between marketing orientation and competitive advantages (β = 0.094, t = 2.308, and p = 0.021). Similarly, market turbulence significantly moderates the relationship between global strategy and competitive advantages (β = 0.098, t = 2.446, and p = 0.015). Therefore, all hypotheses are accepted, as shown in Table 4.

5 Discussion

The main objective of this study is to examine the influence of marketing orientation on competitive advantages with the mediating role of market innovation and global strategy in Pakistan's industrial sector. It further reviews the contribution of market turbulence as a moderating variable among marketing orientation and competitive advantages. The researchers developed all hypotheses to examine the relationship between variables. Data was gathered by combining offline and online usage survey forms filled out by 320 employees of industrial companies listed in the PSE market. There are four noteworthy research findings from the study. First, the study's conclusion indicates that marketing orientation significantly influences competitive advantages in Pakistan's industrial sector. A past study (Gugler & Shi, 2009) focused on marketing orientation involves deeply understanding customer needs and preferences. When industries in Pakistan adopt this approach, they are more likely to align their products and services with what the market demands, leading to the development of products that are better suited to the local market, ultimately resulting in a competitive advantage. Identifying specific customer segments can be crucial in a diverse and rapidly changing market like Pakistan's industrial sector (Baierle et al., 2020). Companies that invest in market research to understand these segments can tailor their strategies and offerings to meet the unique needs of each group, which can lead to a competitive edge over those that do not (Kumlu, 2014).

In Pakistan's industrial sector, where market dynamics can be unpredictable, companies that continuously monitor and respond to changing customer preferences and market conditions are more likely to maintain a competitive edge. Industrial companies that invest in building and promoting their brand in Pakistan's industrial sector tend to command higher prices and enjoy customer loyalty (Osorio Tinoco et al., 2020). Over time, this can result in a sustainable competitive advantage as customers are likelier to choose a trusted and recognized brand over lesser-known competitors. In a competitive industrial sector, businesses that invest in research and development and are open to customer feedback and market trends are more likely to introduce innovative solutions that can set them apart from rivals (Dayan et al., 2023).

Second, the findings of this study indicate that market innovation significantly mediates the relationship among marketing orientation and competitive advantages. Marketing orientation is a strategic approach that places customer needs and market information at the center of a company's decision-making processes (Wong & Ngai, 2023). It involves gathering market intelligence, understanding customer preferences, and aligning products or services accordingly. Competitive advantage states to a company's ability to outstrip its rivals regarding profitability, market share, or other key performance metrics. It is the business strategy that can be achieved through leadership, product and niche markets (Fraj-Andrés et al., 2009). Market innovation bridges the gap between understanding customer needs and effectively meeting those needs in a way that outpaces competitors. Market innovation encompasses developing and introducing new products, services, marketing strategies, and business models that resonate with the target market. It can involve technological advancements, process improvements, or even novel ways of communicating and engaging with customers (De Clercq et al., 2018).

Marketing orientation underscores that strong marketing orientation is a foundational element, but it needs to be complemented by innovative practices to realize its full potential. Similarly, market innovation can lead to sustained competitive advantages. Companies that continuously innovate in response to changing market conditions are better positioned to adapt and thrive in the long run (Ed-Dafali et al., 2023). The combination of marketing orientation and market innovation places a company in a better position to cater to evolving customer preferences, thereby building stronger customer relationships and loyalty. Recognizing the mediating role of market innovation highlights that in highly competitive markets, where product lifecycles are short and customer demands are ever-changing, being innovative can be the key differentiator (Battaglia et al., 2014).

Third, this study's findings reveal that global strategy significantly mediates the relationship among marketing orientation and competitive advantages. Past research highlighted that a company's approach to international markets and global expansion significantly impacts how effectively it can leverage its marketing efforts to gain a competitive edge. Strategic alignment underscores the importance of aligning marketing orientation with global strategy (Milewska, 2022). A strong marketing-oriented company focuses on customer needs and market dynamics. This study suggests that businesses can achieve a greater competitive advantage when their global strategy complements this orientation, allowing them to serve diverse international markets better (Gyedu et al., 2021).

An effective global strategy enables a firm to respond more to various international markets' unique demands and preferences. This responsiveness can be seen as a direct result of the mediation between marketing orientation and competitive advantages. Past study findings emphasize that businesses that successfully mediate their marketing orientation through a global strategy are better positioned to achieve and sustain competitive advantages (Nguyen & Khoa, 2020). These advantages could manifest in various forms, such as cost leadership, product differentiation, or customer loyalty. It underscores that a global strategy tailored to specific markets can provide a sustainable competitive advantage. Companies must allocate resources effectively to maximize the benefits of this mediating relationship. This means investing in research and development, market research, and talent acquisition to support their marketing orientation and global strategy. Resource allocation should be aligned with the strategic priorities that emerge from the mediation process (Naidoo, 2010).

Fourth, this study's findings indicate that market turbulence significantly moderates the relationship between market innovation, global strategy, and competitive advantages. Market turbulence refers to the rapid changes and fluctuations in market conditions, including shifts in consumer preferences, technological advancements, regulatory changes, or economic instability (Inga-Ávila, 2022). The fact that it moderates the relationship between market innovation, global strategy, and competitive advantages implies that the impact of these variables on a company's success is not consistent across all market environments. Market innovation, which can include developing new products, services, or business models, is often considered a driver of competitive advantage (Naidoo, 2010). In a stable market, a company's innovation ability may directly translate into a sustainable competitive advantage. However, in a turbulent market, the relationship might be more complex. It suggests that during times of market turbulence, simply being innovative may not be enough; businesses may need to adapt their innovation strategies to navigate and thrive in uncertain conditions (Bamfo & Kraa, 2019).

Global strategies involve expanding operations and reaching customers in international markets. Traditionally, going global has been seen as a way to diversify and reduce risk. However, the study's findings indicate that market turbulence can also influence the effectiveness of global strategies (Wong & Ngai, 2023). In turbulent markets, companies may need to be more agile and flexible in their global strategies, adapting to changing conditions and tailoring their approaches to specific markets. Competitive advantages are typically considered a source of sustained superior performance (De Clercq et al., 2018). The study's findings imply that the nature of competitive advantage may change in turbulent markets. Companies may need to focus on developing dynamic capabilities that allow them to quickly adjust to changing circumstances and exploit opportunities as they arise rather than relying solely on static advantages (Nuvriasari et al., 2020).

5.1 Study Implications

This study has several implications. Textile companies in Pakistan should prioritize and invest in developing a strong marketing orientation. This includes adopting a customer-centric approach, focusing on market research, and understanding customer needs and preferences. Companies should also prioritize building strong customer relationships to gain a competitive advantage. This study suggests that textile companies should actively seek and embrace market innovations. This could involve adopting new technologies, production methods, or design processes to stay ahead of competitors. Investing in research and development (R&D) to create innovative products and services tailored to changing market demands can significantly enhance competitiveness. Pakistani textile companies should consider expanding their reach beyond domestic markets. Implementing global strategies, such as exporting products, forming international partnerships, or establishing overseas operations, can help them tap into new markets and diversify their revenue streams. This can reduce dependence on local markets and increase competitiveness. Companies should recognize the importance of effectively managing market turbulence. This includes monitoring market dynamics, responding swiftly to changes, and having a flexible business strategy. Developing contingency plans and risk management strategies can help textile companies navigate uncertainties, such as economic downturns, supply chain disruptions, or changes in consumer preferences.

Textile companies need to foster a culture of continuous learning and adaptation. They should encourage their employees to stay updated with industry trends, technologies, and marketing strategies. This adaptability will enable companies to stay competitive in dynamic markets. Collaboration within the textile industry and networking with other stakeholders in the value chain (e.g., suppliers, manufacturers, distributors) can provide knowledge sharing and resource pooling opportunities. This collaborative approach can lead to economies of scale, reduced costs, and improved competitiveness. The findings of this study can also be used to advocate for supportive government policies that facilitate the growth of the textile industry in Pakistan. Policies that encourage innovation, ease of doing business, and international trade can positively impact the competitiveness of textile companies. Investing in industry-specific training and employee education can help textile companies develop the skills and knowledge required to effectively implement marketing orientation, market innovation, and global strategies. This could include specialized training programs, workshops, and partnerships with educational institutions. Textile companies can benefit from benchmarking their marketing orientation, innovation strategies, and global initiatives against industry leaders and best practices worldwide. Identifying gaps and areas for improvement can guide companies in enhancing their competitiveness. In addition to short-term competitiveness, textile companies should also focus on long-term sustainability. This includes responsible and sustainable materials sourcing, ethical manufacturing practices, and environmentally friendly processes. Such practices can align with global trends and consumer preferences, enhancing the company's image and competitiveness in the long run.

6 Conclusion

The textile industry is a crucial sector in the global economy, with a history dating back centuries. This industry has changed significantly over time due to globalization, shifting customer preferences, and technological improvements. Textile companies face numerous challenges and opportunities in today's highly competitive and dynamic business environment. One of the key factors that can significantly impact their success is their marketing orientation. The current research study explores the role of marketing orientation toward competitive advantages with the mediating role of market innovation and global strategy. Moreover, the moderating role of market turbulence is also explored, contributing to the literature. This research study has a three-fold contribution; first, the study's conclusion indicates that marketing orientation significantly influences competitive advantages in Pakistan's industrial sector. Second, the findings of this study indicate that market innovation significantly mediates the relationship between marketing orientation and competitive advantages. Third, this study's findings reveal that global strategy significantly mediates the relationship between marketing orientation and competitive advantages. Fourth, this study's findings indicate market turbulence significantly moderates the relationship between market innovation, global strategy, and competitive advantages. Incorporating these concepts into a holistic business strategy can help companies navigate the complexities of the modern marketplace and position themselves for long-term success. Organizations must remain agile, innovative, and customer-focused to thrive in an ever-evolving global economy.

6.1 Limitations and Future Research

There are a few restrictions to this study. First, this study focuses on variables such as marketing orientation, market innovation, global strategy, and market turbulence, which may be limited to textile companies in Pakistan. Future researchers may face challenges in obtaining up-to-date and relevant information, which could affect the quality of the study's findings. Second, his study relies on a specific sample of textile companies, and this sample may not fully represent the entire industry in

Pakistan. There might be biases in the selection of companies or limitations in the representativeness of the chosen sample, which could impact the generalizability of the results. Future researchers would increase the sample size to test the variables. Third, this study focuses on textile companies in Pakistan, and the findings may not apply to other industries or regions. For future researchers, generalizing the results to different sectors or countries should be done cautiously. Fourth, this study may utilize cross-sectional data, which provides a snapshot of the relationship between marketing orientation, market innovation, global strategy, and market turbulence at a single point in time. Future research can benefit from longitudinal studies that track the performance and strategies of textile companies in Pakistan over an extended period. Last, this study may find associations between marketing orientation, market innovation, global strategy, and competitiveness, it may not establish causality or the direction of causality. Future studies can explore the mediating and moderating effects of other variables that may influence the relationship between marketing orientation, market innovation, global strategy, and competitiveness.

References

Alghamdi, O. A., & Agag, G. (2024). Competitive advantage: A longitudinal analysis of the roles of data-driven innovation capabilities, marketing agility, and market turbulence. *Journal of Retailing and Consumer Services, 76*(August 2023), 103547.

Arici, T., & Gok, M. S. (2023). Examining environmental turbulence intensity: A strategic agility and innovativeness approach on firm performance in environmental turbulence situations. *Sustainability, 15*(6), 5364.

Baierle, I. C., Benitez, G. B., Nara, E. O. B., Schaefer, J. L., & Sellitto, M. A. (2020). Influence of open innovation variables on the competitive edge of small and medium enterprises. *Journal of Open Innovation: Technology, Market, and Complexity, 6*(4), 1–17.

Bamfo, B. A., & Kraa, J. J. (2019). Market orientation and performance of small and medium enterprises in Ghana: The mediating role of innovation. *Cogent Business and Management, 6*(1).

Bano, A., Salleh, N., Mendes, E., Grundy, J., Burch, G., & Nordin, A. (2016). The effect of software engineers ' personality traits on team climate and performance : A systematic literature review. *Information and Software Technology, 73*, 52–65.

Battaglia, M., Testa, F., Bianchi, L., Iraldo, F., & Frey, M. (2014). Corporate social responsibility and competitiveness within SMEs of the fashion industry: Evidence from Italy and France. *Sustainability (switzerland), 6*(2), 872–893.

Chan, R. Y. K., He, H., Chan, H. K., & Wang, W. Y. C. (2012). Environmental orientation and corporate performance: The mediation mechanism of green supply chain management and moderating effect of competitive intensity. *Industrial Marketing Management, 41*(4), 621–630.

Chen, K. H., Wang, C. H., Huang, S. Z., & Shen, G. C. (2016). Service innovation and new product performance: The influence of market-linking capabilities and market turbulence. *International Journal of Production Economics, 172*, 54–64.

Dayan, M., Ng, P. Y., Husain, Z., & Zacca, R. (2023). Effects of constructive politics and market turbulence on entrepreneurial orientation–performance relationship: A moderated mediation model. *European Management Journal, 41*(3), 385–394.

De Clercq, D., Thongpapanl, N., & Voronov, M. (2018). Sustainability in the face of institutional adversity: Market turbulence, network embeddedness, and innovative orientation. *Journal of Business Ethics, 148*(2), 437–455.

Dhiman, R., Kumar, V., & Rana, S. (2020). Why export competitiveness differs within Indian textile industry? Determinants and empirical evidence. *Review of International Business and Strategy, 30*(3), 375–397.

Ed-Dafali, S., Al-Azad, M. S., Mohiuddin, M., & Reza, M. N. H. (2023). Strategic orientations, organizational ambidexterity, and sustainable competitive advantage: Mediating role of industry 4.0 readiness in emerging markets. *Journal of Cleaner Production, 401*(March), 136765.

Farrukh, M., Xu, S., Baheer, R., & Ahmad, W. (2023a). Unveiling the role of supply chain parameters approved by blockchain technology towards firm performance through trust : The moderating role of government support. *Heliyon, 9*(11), e21831.

Farrukh, M., Xu, S., Naveed, W., & Nusrat, S. (2023b). Investigating the impact of artificial intelligence on human resource functions in the health sector of China : A mediated moderation model. *Heliyon, 9*(11), e21818.

Farrukh, M., Xu, S., An, X., & Javed, I. (2024a). Assessing the impact of AI-chatbot service quality on user e-brand loyalty through chatbot user trust , experience and electronic word of mouth. *Journal of Retailing and Consumer Services, 79*(March), 103867. https://doi.org/10.1016/j.jretconser.2024.103867

Farrukh, M., Xu, S., Marc, W., & Yang, X. (2024b). Artificial intelligence and social media on academic performance and mental well-being : Student perceptions of positive impact in the age of smart learning. *Heliyon, 10*(8), e29523. https://doi.org/10.1016/j.heliyon.2024.e29523

Fornell, C., & Larcker, D. F. (1981). Evaluating structural equation models with unobservable variables and measurement error. *Journal of Marketing Research, 18*(1), 39.

Fraj-Andrés, E., Martinez-Salinas, E., & Matute-Vallejo, J. (2009). A multidimensional approach to the influence of environmental marketing and orientation on the firm's organizational performance. *Journal of Business Ethics, 88*(2), 263–286.

Freixanet, J. (2012). Export promotion programs: Their impact on companies' internationalization performance and competitiveness. *International Business Review, 21*(6), 1065–1086.

Genc, E., Dayan, M., & Genc, O. F. (2019). The impact of SME internationalization on innovation: The mediating role of market and entrepreneurial orientation. *Industrial Marketing Management, 82*(January), 253–264.

Gugler, P., & Shi, J. Y. J. (2009). Corporate social responsibility for developing country multinational corporations: Lost war in pertaining global competitiveness? *Journal of Business Ethics, 87*(SUPPL. 1), 3–24.

Gyedu, S., Tang, H., Ntarmah, A. H., & Manu, E. K. (2021). The moderating effect of environmental turbulence on the relationship between innovation capability and business performance. *International Journal of Innovation Science, 13*(4), 456–476.

Hair, J. F., Ringle, C. M., & Sarstedt, M. (2011). PLS-SEM: Indeed a silver bullet. *Journal of Marketing Theory and Practice, 19*(2), 139–152.

Howard, M. C., & Henderson, J. (2023). A review of exploratory factor analysis in tourism and hospitality research: Identifying current practices and avenues for improvement. *Journal of Business Research, 154*(October 2022), 113328.

Inga-Ávila, M. F. (2022). Dynamics of the behavior of competitiveness factors in the textile sector. *Uncertain Supply Chain Management, 10*(3), 877–886.

Kraus, S., Rehman, S. U., & García, F. J. S. (2020). Corporate social responsibility and environmental performance: The mediating role of environmental strategy and green innovation. *Technological Forecasting and Social Change, 160*(August), 120262.

Kumlu, Ö. (2014). The effect of intangible resources and competitive strategies on the export performance of small and medium sized enterprises. *Procedia - Social and Behavioral Sciences, 150*, 24–34.

Larbi-Siaw, O., Xuhua, H., Owusu, E., Owusu-Agyeman, A., Fulgence, B. E., & Frimpong, S. A. (2022). Eco-innovation, sustainable business performance and market turbulence moderation in emerging economies. *Technology in Society, 68*(October 2021), 101899.

Malik, S. Y., Cao, Y., Mughal, Y. H., Kundi, G. M., Mughal, M. H., & Ramayah, T. (2020). Pathways towards sustainability in organizations: Empirical evidence on the role of green human resource management practices and green intellectual capital. *Sustainability (switzerland), 12*(8), 1–24.

Martins, J. M., Muhammad, F. S., & Shuo, X. (2023a). Examining the factors influencing entrepreneurial intention to initiate new ventures: Focusing on knowledge of entrepreneurial skills, ability to take risk and entrepreneurial innovativeness in open innovation business model. *Research Square*, 1125–

Martins, J. M., Shahzad, M. F., & Javed, I. (2023b). Assessing the impact of workplace harassment on turnover intention : Evidence from the banking industry. *7*(5), 1699–1722.

Martins, J. M., Shahzad, M. F., & Xu, S. (2023c). Factors influencing entrepreneurial intention to initiate new ventures : evidence from university students. *Journal of Innovation and Entrepreneurship.*

Milewska, B. (2022). The impact of instability in the business environment on the competitiveness of enterprises using the example of the Apparel industry. *Sustainability (Switzerland), 14*(22).

Naidoo, V. (2010). Firm survival through a crisis: The influence of market orientation, marketing innovation and business strategy. *Industrial Marketing Management, 39*(8), 1311–1320.

Nguyen, M. T., & Khoa, B. T. (2020). Improving the competitiveness of exporting enterprises: A case of Kien Giang Province in Vietnam. *Journal of Asian Finance, Economics and Business, 7*(6), 495–508.

Nuvriasari, A., Ishak, A., Hidayat, A., Mustafa, Z., Haryono, S., Islam, U., Yogyakarta, I., & Yogyakarta, U. M. (2020). The effect of market and entrepreneurship orientation on SME's business performance: The role of entrepreneurial marketing in Indonesian Batik industries. *European Journal of Business and Management, 12*(5), 29–37.

O'Connor, N. G., Vera-Muñoz, S. C., & Chan, F. (2011). Competitive forces and the importance of management control systems in emerging-economy firms: The moderating effect of international market orientation. *Accounting, Organizations and Society, 36*(4–5), 246–266.

Osorio Tinoco, F. F., Hernández-Espallardo, M., & Rodriguez-Orejuela, A. (2020). Nonlinear and complementary effects of responsive and proactive market orientation on firms' competitive advantage. *Asia Pacific Journal of Marketing and Logistics, 32*(4), 841–859.

Pal, R., Torstensson, H., & Mattila, H. (2014). Antecedents of organizational resilience in economic crises - An empirical study of Swedish textile and clothing SMEs. *International Journal of Production Economics, 147*(PART B), 410–428.

Podsakoff, P. M., MacKenzie, S. B., Lee, J. Y., & Podsakoff, N. P. (2003). Common method biases in behavioral research: A critical review of the literature and recommended remedies. *Journal of Applied Psychology, 88*(5), 879–903.

Risitano, M., Romano, R., Rusciano, V., Civero, G., & Scarpato, D. (2022). The impact of sustainability on marketing strategy and business performance: The case of Italian fisheries. *Business Strategy and the Environment, 31*(4), 1538–1551.

Rokkan, A. I. (2023). Market orientation (once again): Challenges and a suggested solution. *AMS Review, 13*(1–2), 71–91.

Roscoe, A. M., Lang, D., & Sheth, J. N. (1975). Follow-Up methods, questionnaire length, and market differences in mail surveys. *Journal of Marketing, 39*(2), 20.

Saleh, M. H. M., Azmin, A. A., & Saraih, U. N. (2021). The effects of market orientation, product innovation and marketing ethics on firm performance: A theoretical framework. *International Journal of Innovation and Industrial Revolution, 3*(7), 31–47.

Sarstedt, M., Ringle, C. M., & Hair, J. F. (2022). Partial least squares structural equation modeling. In *Handbook of market research* (pp. 587–632). Springer.

Shahzad, F., Shahzad, M. F., Dilanchiev, A., & Irfan, M. (2022). Modeling the influence of paternalistic leadership and personality characteristics on alienation and organizational culture in the aviation industry of Pakistan: The mediating role of cohesiveness. *Sustainability (Switzerland), 14*(22).

Shahzad, M. F., Khan, K. I., Saleem, S., & Rashid, T. (2021). *What factors affect the entrepreneurial intention to start-ups? The role of entrepreneurial skills, propensity to take risks, and innovativeness in open business models.*

Shahzad, M. F., Xu, S., Khan, K. I., & Hasnain, M. F. (2023a). Effect of social influence, environmental awareness, and safety affordance on actual use of 5G technologies among Chinese students. *Scientific Reports, 0123456789*, 1–16.

Shahzad, M. F., Xu, S., Rehman, O., & Javed, I. (2023b). Impact of gamification on green consumption behavior integrating technological awareness, motivation , enjoyment and virtual CSR. *Scientific Reports*, 1–18.

Shahzad, M. F., Xu, S., & Baheer, R. (2024a). Assessing the factors influencing the intention to use information and communication technology implementation and acceptance in China's education sector. 1–15. https://doi.org/10.1057/s41599-024-02777-0

Shahzad, M. F., Xu, S., Lim, W. M., Hasnain, M. F., & Nusrat, S. (2024b). Cryptocurrency awareness, acceptance, and adoption: the role of trust as a cornerstone. *Humanities & Social Sciences Communications.*

Singh, M., Mittal, M., Mehta, P., & Singla, H. (2021). Personal values as drivers of socially responsible investments: A moderation analysis. *Review of Behavioral Finance, 13*(5), 543–565.

Suandi, E., Herri, Yulihasri, & Syafrizal. (2023). Evaluating the relationship between entrepreneurial marketing, competitive advantage, and Islamic banks performance. *Journal of Financial Services Marketing, 28*(3), 599–614.

Taçoğlu, C., Ceylan, C., & Kazançoğlu, Y. (2019). Analysis of variables affecting competitiveness of SMEs in the textile industry. *Journal of Business Economics and Management, 20*(4), 648–673.

Tjahjadi, B., Soewarno, N., Jermias, J., Hariyati, H., Fairuzi, A., & Anwar, D. N. (2022). Does engaging in global market orientation strategy affect HEIs' performance? The mediating roles of intellectual capital readiness and open innovation. *Journal of Open Innovation: Technology, Market, and Complexity, 8*(1), 29.

Udriyah, T. J., & Ferdous Azam, S. M. (2019). The effects of market orientation and innovation on competitive advantage and business performance of textile smes. *Management Science Letters, 9*(9), 1419–1428.

Wagner, M., & Schaltegger, S. (2004). The effect of corporate environmental strategy choice and environmental performance on competitiveness and economic performance: An empirical study of EU manufacturing. *European Management Journal, 22*(5), 557–572.

Wong, D. T. W., & Ngai, E. W. T. (2023). The impact of advanced manufacturing technology, sensing and analytics capabilities, and planning comprehensiveness on sustained competitive advantage: The moderating role of environmental uncertainty. *International Journal of Production Economics, 265*(August), 109007.

Yayla, S., Yeniyurt, S., Uslay, C., & Cavusgil, E. (2018). The role of market orientation, relational capital, and internationalization speed in foreign market exit and re-entry decisions under turbulent conditions. *International Business Review, 27*(6), 1105–1115.

Zhang, X., Abbas, J., Shahzad, M. F., Shankar, A., Ercisli, S., & Dobhal, D. C. (2024). Association between social media use and students' academic performance through family bonding and collective learning: The moderating role of mental well-being. *Education and Information Technologies*, 0123456789.

Muhammad Farrukh Shahzad is a Ph.D. student in the School of Economics and Management, Beijing University of Technology, majoring in Management Sciences and Engineering. His area of interest is Entrepreneurship, Technology Management, Supply Chain Management, and Technology Innovation. As a young emerging researcher, he has recently published several articles in reputable international journals.

Iqra Javed is Lecturer in the University of Engineering and Technology, Narowal Campus, Pakistan. She is pursing Ph.D. degree in Business Administration from the University of Engineering and Technology main campus. Her research interests include the integration of Advanced Econometrics Techniques with Financial Management Decisions, and Bibliometric Analysis.

Imran Zahid is working as an Assistant Professor in the department of Mechanical Engineering and Technology at Government College University Faisalabad. He received a Ph.D. degree from University of Engineering and Technology Lahore Pakistan. His research interests include Engineering Management, Scientific Fronts Detection, and Knowledge Management.

Knowledge and Innovation Management as a Driver of Competitiveness in Textile Firms

José Moleiro Martins

Abstract This study aims to comprehensively investigate the role of knowledge and innovation management as crucial drivers of competitiveness in textile firms. Employing a rigorous bibliometric analysis, the study seeks to explore the scholarly landscape, identify trends, and unravel patterns related to knowledge and innovation management in the textile industry. By systematically reviewing academic literature from Scopus and Web of Science data bases using R and Biblioshiny, the study identifies key themes, explores collaborative networks, and evaluates gaps in existing knowledge. The analysis includes a thematic map, meta-analysis of research trends, and an examination of sources, authors, affiliations, and countries contributing to the literature. The study reveals a multidimensional exploration of the textile industry, emphasizing innovation, knowledge management, competition, and sustainable development. The thematic map illustrates the interconnectedness of diverse topics, emphasizing the complexity of factors influencing competitiveness. Clusters such as the textile industry, market, sustainable development, knowledge-based systems, and product design provide a nuanced understanding of knowledge and innovation management in textiles. The analysis highlights a pivotal interplay between knowledge and innovation management in shaping competitiveness, with a growing scholarly focus on the subject. China emerges as a leading contributor, reflecting global engagement.

J. M. Martins (✉)
Department of Management, Lisbon Accounting and Business School, Av. Miguel Bombarda 20, 1069-035 Lisboa, Portugal
e-mail: zdmmartins@gmail.com

Instituto Universitário de Lisboa (ISCTE-IUL), Business Research Unit (BRU-IUL), Lisboa, Portugal

Lisbon Accounting and Business School, Polytechnic Institute of Lisbon (ed.), *Industry and Innovation: Textile Industry*, SDGs and Textiles,
https://doi.org/10.1007/978-3-031-57804-5_6

1 Introduction

Knowledge and innovation management play pivotal roles in shaping the competitiveness of firms, particularly within the dynamic landscape of the textile industry. In the contemporary business environment, characterized by rapid technological advancements and global competition, the ability of textile firms to effectively manage knowledge and foster innovation has emerged as a critical determinant of their success. This study embarks on a comprehensive exploration of the intricate interplay between knowledge and innovation management as key drivers of competitiveness in textile firms. Through a meticulous bibliometric analysis, we delve into the scholarly landscape to unravel trends, identify gaps and provide valuable insights that can inform strategic decisions for textile industry stakeholders.

The textile industry stands at the intersection of tradition and innovation, with a rich history rooted in craftsmanship and an evolving future shaped by technological breakthroughs. In this context, the efficient management of knowledge becomes a linchpin for firms seeking to navigate the complexities of modern business. By understanding the existing body of knowledge, firms can harness the collective wisdom of the industry, build on past successes, and strategically position themselves for future challenges. Innovation, on the other hand, serves as the engine driving progress and competitiveness. Textile firms that embrace a culture of innovation not only respond adeptly to market demands but also set trends and redefine industry standards. Innovation is not limited to technological advancements; it encompasses novel approaches to design, sustainable practices, and responsive strategies to changing consumer preferences. This study explores how the synergy between knowledge and innovation management empowers textile firms to not only survive but thrive in an era of constant disruption.

The bibliometric analysis conducted in this study offers a systematic examination of the scholarly contributions to the field of knowledge and innovation management in the context of textile firms. By analyzing the academic literature, we aim to identify key themes, prominent authors, and seminal works that have shaped the discourse. This approach provides a comprehensive overview of the current state of knowledge, enabling us to pinpoint areas that have received extensive attention and those that warrant further exploration.

Understanding the dynamics of competitiveness in the textile industry requires a nuanced examination of the factors that contribute to or hinder the effective management of knowledge and the fostering of innovation. Through an interdisciplinary lens, we draw on insights from management, technology, and textile studies to construct a holistic understanding of the challenges and opportunities faced by textile firms. This integrative approach allows us to bridge disciplinary boundaries and offer a nuanced perspective that goes beyond siloed analyses.

The implications of this study extend beyond academic inquiry, reaching into the practical realm of industry strategy and policymaking. As textile firms grapple with the complexities of a rapidly changing landscape, evidence-based insights become indispensable for informed decision-making. By synthesizing the existing body of knowledge, the study provides a roadmap for industry practitioners, policymakers, and researchers to navigate the intricate relationship between knowledge management, innovation, and competitiveness. The primary aim of this study is to comprehensively investigate the role of knowledge and innovation management as crucial drivers of competitiveness in textile firms. Through a rigorous bibliometric analysis, the study seeks to achieve the following objectives:

- Explore the scholarly landscape to identify trends and patterns related to knowledge and innovation management in the context of textile firms. This includes understanding the evolution of key concepts and the emergence of novel ideas.
- Systematically analyze academic literature to identify and categorize key themes associated with knowledge and innovation management in the textile industry. This will provide a nuanced understanding of the multifaceted nature of these concepts.
- Investigate prominent authors and collaborative networks within the field. This aspect aims to recognize thought leaders, influential contributors, and the dynamics of scholarly collaboration, contributing to the social network of knowledge creation.
- Identify gaps in the existing body of knowledge, pinpointing areas that have received extensive attention and those that require further exploration. This evaluation contributes to a more comprehensive understanding of the current state of research in the field.
- Translate insights derived from the bibliometric analysis into practical implications for industry stakeholders. By doing so, the study aims to inform strategic decision-making processes for textile firms, policymakers, and researchers.

This study sets out to unravel the intricate dynamics that underpin the competitiveness of textile firms through the lenses of knowledge and innovation management. The intersection of tradition and innovation in the textile industry offers a fascinating backdrop for the exploration, and the bibliometric analysis serves as a robust methodology to uncover patterns and trends within the scholarly discourse. As we embark on this intellectual journey, the overarching aim is to contribute valuable insights that empower textile firms to navigate the complexities of the contemporary business landscape and position themselves as leaders in a dynamic and competitive market.

2 Systematic Literature Review

The data in Table 1 reflects a comprehensive and dynamic dataset, encompassing a wide range of sources and document types. The collaborative nature of the research, international perspectives, and the mix of document types contribute to the richness and depth of the bibliometric analysis on knowledge and innovation management in textile firms.

Table 1 Main information about data used in the study

Description	Results
Timespan	1999:2023
Sources (Journals, Books, etc.)	98
Documents	121
Annual growth rate %	12.25
Document average age	7.54
Average citations per doc	7.719
References	1
Authors	312
Authors of single-authored docs	31
Single-authored docs	34
Co-authors per doc	2.79
International co-authorships %	20.66
Article	73
Book	2
Book chapter	7
Conference paper	30
Conference paper article	1
Conference review	3
Review	5

The study covers a considerable timespan from 1999 to 2023, providing a comprehensive view of the evolution of knowledge and innovation management in textile firms over the past two decades. The data is derived from a diverse set of sources, including journals, books, and other publications. The inclusion of multiple sources enhances the robustness and comprehensiveness of the analysis. A total of 121 documents were analyzed, indicating a substantial dataset. This sizeable sample allows for a more in-depth and reliable bibliometric analysis of knowledge and innovation management in textile firms. The impressive annual growth rate of 12.25% suggests a dynamic and evolving field. This growth rate indicates an increasing interest and focus on knowledge and innovation management in the textile industry over the years. The average age of the documents is 7.54 years. This metric provides insights into the recency of the literature, indicating that the majority of the documents are relatively current, contributing to the relevance of the analysis.

With an average of 7.719 citations per document, the literature under consideration appears to be influential and well-referenced. This suggests that the analyzed documents have made a meaningful impact on the field. The low number of references (1) might indicate a focus on recent and highly relevant sources, or it could be a limitation in terms of the depth of the literature review. Further examination is needed to understand the rationale behind this low reference count.

Involving 312 authors reflects a collaborative effort in researching and contributing to the subject matter. The diversity in authorship may bring different perspectives and expertise to the analysis. The presence of 31 authors who contributed to single-authored documents indicates a balance between collaborative and independent research efforts in the field. The 34 single-authored documents highlight the existence of standalone contributions, suggesting a range of individual perspectives on knowledge and innovation management in textile firms.

The average of 2.79 co-authors per document signifies a collaborative approach in the research, with multiple perspectives and expertise contributing to each analyzed document. The presence of international co-authorships at 20.66% indicates a degree of global collaboration in the field. This international dimension adds richness to the analysis, incorporating diverse viewpoints and experiences.

The distribution of document types provides a snapshot of the diverse scholarly output in the field, including a significant number of articles and conference papers, which are common in dynamic and evolving fields.

The network of co-occurrences, derived from examined publications using keywords such as "knowledge", "innovation", "management", "competitiveness", and "textile firms", provides valuable insights into the connections and importance of different concepts in this research area (Fig. 1). Through an examination of key metrics like betweenness, closeness, and PageRank, the analysis reveals the central significance of specific nodes in the network.

Extensive research has emphasized the significance of knowledge management in textile firms, as it enables organizations to capture, share, and apply valuable insights. By fostering a culture of continuous learning, organizations can adapt to changing market conditions and gain a competitive edge (Kayakutlu & Büyüközkan, 2010). The relationship between knowledge management and innovation has been a focal point of scholarly inquiry. Studies highlight that effective knowledge management processes can stimulate innovation in textile firms. This connection is vital for firms aiming to develop and implement novel solutions to address market demands

Fig. 1 Co-occurrence network based on studied publications

(Ahmed et al., 2020; Usman et al., 2020). The adoption of technology plays a crucial role in innovation management within textile firms. Research suggests that firms leveraging advanced technologies can enhance their innovation capabilities, leading to improved competitiveness. This underscores the importance of aligning technological investments with innovation goals (Kumarapeli et al., 2022).

Collaboration and knowledge sharing among different stakeholders in the textile industry are essential components of successful knowledge management strategies. Partnerships with suppliers, customers, and other industry players contribute to a more holistic approach to innovation (Zucchella & Siano, 2014). Such collaborative efforts foster the exchange of ideas and expertise, enriching the innovation ecosystem (Di Chiacchio et al., 2023a, 2023b). The role of leadership in driving knowledge and innovation management has been widely acknowledged in the literature. Effective leadership is essential for creating a culture that values and prioritizes knowledge creation and innovation (Iqbal & Piwowar-Sulej, 2023a, 2023b). Leaders who champion these initiatives can inspire their teams to proactively contribute to the organization's competitiveness (Nabi et al., 2023). Knowledge and innovation management also involve effective utilization of intellectual capital within textile firms. Studies suggest that organizations with a strong focus on managing their intellectual capital can leverage it for innovation and competitive advantage. This highlights the need for textile firms to recognize and optimize their intellectual assets (Sivalogathasan & Wu, 2015).

The integration of knowledge and innovation management into the overall strategic framework of textile firms is crucial for sustained competitiveness. Scholars argue that organizations should align their knowledge and innovation strategies with broader business objectives to achieve long-term success (Rese et al., 2022). This alignment ensures that knowledge and innovation efforts are purposeful and contribute directly to organizational goals (Azeem et al., 2021). The globalization of markets has added a new dimension to knowledge and innovation management in the textile industry. Firms operating in diverse markets must navigate cultural and contextual differences, requiring adaptive knowledge management approaches (Watchravesringkan et al., 2010). This underscores the importance of developing global competencies in knowledge and innovation management.

Challenges in implementing effective knowledge and innovation management processes in textile firms have also been explored. Research indicates that organizational culture, resistance to change, and inadequate infrastructure can hinder the successful implementation of knowledge and innovation initiatives (Gardas et al., 2018). Overcoming these challenges is imperative for realizing the full potential of knowledge and innovation in enhancing competitiveness (Wakeford et al., 2017). The role of education and training in building a knowledgeable and innovative workforce within textile firms is a recurring theme in the literature. Investments in employee development contribute to the creation of a skilled and adaptable workforce, essential for driving innovation (Prakash et al., 2020). Continuous learning initiatives ensure that employees stay abreast of industry trends and emerging technologies (Cooke & Schienstock, 2000).

In recent years, the textile industry has witnessed a paradigm shift in its competitive landscape, with a growing recognition of the pivotal role played by knowledge and innovation management. Scholars have increasingly focused their attention on understanding the intricate relationship between these two drivers and the overall competitiveness of textile firms. Studies underscore the significance of knowledge management in textile firms, emphasizing its positive impact on organizational performance (Danskin et al., 2005). The integration of knowledge management practices has been shown to enhance decision-making processes, streamline operations, and foster a culture of continuous learning within these organizations. Furthermore, scholars highlight the role of innovation as a catalyst for competitiveness in the textile industry (García-Álvarez, 2015; Hutahayan, 2021). Innovative practices, ranging from product development to process optimization, have been identified as crucial elements that distinguish successful textile firms in the market. A recurring theme in the literature is the interconnectedness of knowledge and innovation management, with scholars positing that effective knowledge sharing within organizations cultivates an environment conducive to innovation (Dangelico et al., 2013; Pal et al., 2018). The exchange of tacit and explicit knowledge among employees has been linked to increased innovation capabilities, allowing textile firms to adapt to dynamic market conditions (Durana et al., 2020). This symbiotic relationship is essential for sustaining a competitive edge in an industry characterized by rapid technological advancements and changing consumer preferences.

The literature underscores the role of technology in facilitating knowledge and innovation management in textile firms. Technological platforms and digital tools have emerged as enablers for efficient knowledge dissemination and collaborative innovation efforts. Textile firms leveraging these technologies demonstrate a propensity for quicker adaptation to market trends and a heightened ability to capitalize on emerging opportunities (Chen et al., 2023; Shen et al., 2022).

Despite the growing body of literature on knowledge and innovation management in the textile industry, gaps remain, necessitating further research. Scholars advocate for more nuanced examinations of contextual factors that influence the effectiveness of knowledge and innovation initiatives within specific organizational settings (Chaurasia et al., 2020). Additionally, there is a call for longitudinal studies that track the evolution of knowledge and innovation practices in textile firms over time, providing valuable insights into their sustained impact on competitiveness (Shahzad et al., 2020; Tu & Wu, 2021).

The literature highlights the integral role of knowledge and innovation management in shaping the competitive landscape of textile firms. As the industry continues to evolve, an ongoing commitment to research in this domain is essential for both scholars and practitioners seeking to navigate the complex dynamics of the global textile market.

Figure 2 provides valuable insights into the key themes and focus areas within the corpus of documents related to "Knowledge and innovation management as a driver of competitiveness in textile firms". The analysis of the most frequently occurring words sheds light on the predominant topics and concepts that have been emphasized in the literature.

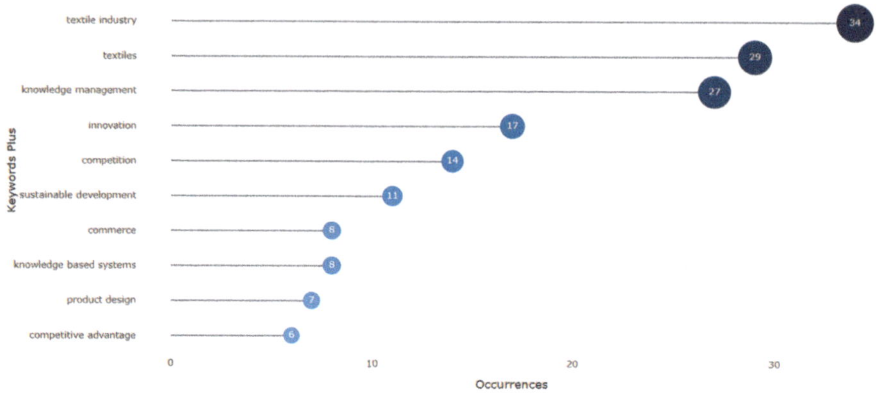

Fig. 2 Most frequently occurring words in the studied documents

The term "textile industry" stands out as the most frequently occurring phrase, with 34 mentions. This indicates a strong emphasis on the core industry under consideration, highlighting the centrality of textiles in the context of knowledge and innovation management. The terms "textiles" and "innovation" closely follow, with 29 and 17 occurrences, respectively. This suggests a keen interest in exploring the intersection of textiles and innovation, emphasizing the dynamic nature of the textile sector and the role of innovation in shaping its trajectory.

"Knowledge management" is a central theme, appearing 27 times. This underscores the significance of managing and leveraging knowledge within the textile industry, indicating a recognition of the role of intellectual capital in fostering competitiveness.

The terms "competition" and "competitive advantage" are mentioned 14 and 6 times, respectively. This indicates a focus on the competitive landscape of the textile industry and the strategies employed to gain a competitive edge through knowledge and innovation management.

"Sustainable development" is a noteworthy term with 11 occurrences. This suggests an awareness of the importance of sustainability in the textile industry and the integration of sustainable practices into knowledge and innovation management processes.

"Commerce" and "product design" are mentioned 8 and 7 times, respectively. This hints at a broader perspective that goes beyond technological aspects, encompassing aspects of business and design in the context of knowledge and innovation.

"Knowledge-based systems" appear 8 times, indicating a focus on the application of advanced technological systems in managing and utilizing knowledge within the textile sector.

While not explicitly mentioned in the word list, the presence of terms such as "competition" and "knowledge management" suggests a global perspective. The 20.66% international co-authorship rate supports this notion, emphasizing collaboration and knowledge exchange on a global scale.

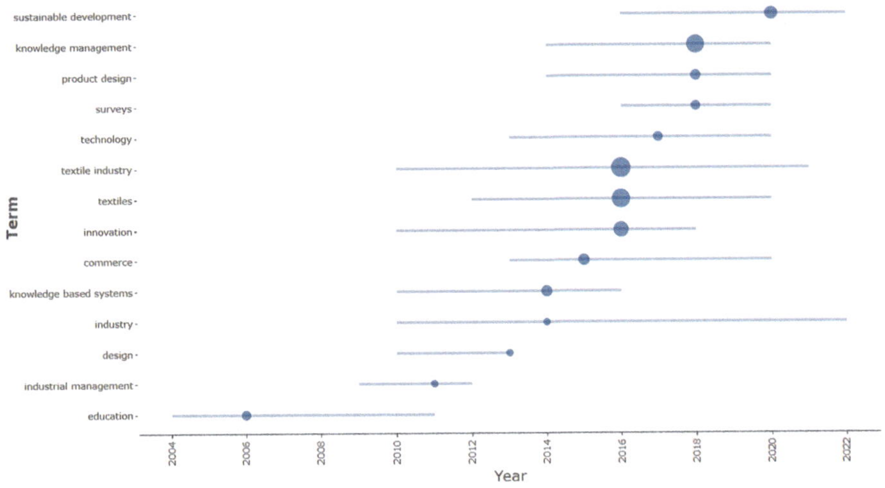

Fig. 3 Trend topics in the studied documents

The most frequently occurring words reflect a comprehensive exploration of the textile industry, with a focus on innovation, knowledge management, competition, and sustainable development. The prominence of these themes underscores the multidimensional nature of driving competitiveness in textile firms through effective knowledge and innovation management strategies.

The analysis of Fig. 3 provides valuable insights into the evolving trends and focus areas within the context of knowledge and innovation management in textile firms. The data is presented in terms of the frequency of specific topics, along with the years in which these topics gained prominence.

The topic of education shows a consistent presence throughout the years, with a peak in 2011. This suggests an ongoing interest in educational aspects related to knowledge and innovation management in the textile industry. Industrial management emerges as a significant topic, with a notable increase in frequency around 2009 and sustained interest until 2012. This indicates a period of emphasis on managerial practices within the textile sector. The focus on design experiences fluctuations, with a peak in 2010 and another in 2013. This suggests intermittent but significant attention to the role of design in the context of knowledge and innovation in textile firms. The frequency of knowledge-based systems displays a steady increase from 2010 to 2016, indicating a growing interest in the application of advanced systems to manage knowledge within the textile industry. The general topic of industry shows a consistent presence, with increased attention around 2010 and sustained interest until 2022. This may reflect an overarching interest in broader industrial trends influencing the textile sector. Commerce experiences a significant increase in frequency from 2013 to 2020, suggesting a heightened focus on commercial aspects related to knowledge and innovation in textile firms during this period.

Unsurprisingly, the term "textile industry" dominates the landscape, with a peak in 2016. This central theme underscores the article's focus on the textile industry and its various dimensions. Similar to the "textile industry", the term "textiles" is consistently prevalent, indicating sustained attention to the broader category encompassing various textile products and processes. The concept of innovation shows a consistent presence from 2010 to 2018, suggesting ongoing interest and exploration of innovative practices in the textile industry.

Knowledge management exhibits a steady increase from 2014 to 2020, reflecting a growing acknowledgment of the importance of effective knowledge management in maintaining competitiveness in textile firms. The frequency of product design peaks in 2018, indicating a specific focus on the role of design in product development within the textile sector during that period. The presence of surveys becomes notable from 2016 onwards, suggesting an increased use of survey methodologies to gather insights into knowledge and innovation management practices.

Sustainable development gains prominence from 2016 onwards, indicating a growing awareness and consideration of sustainability aspects in the context of knowledge and innovation within the textile industry.

The analysis of trend topics provides a nuanced understanding of the evolving research landscape in knowledge and innovation management within textile firms, showcasing a dynamic interplay of various themes over the years.

The thematic map presented in Fig. 4 provides insights into the clusters identified in the studied documents, along with relevant metrics such as Callon Centrality, Callon Density, Rank Centrality, Rank Density, and Cluster Frequency.

Textile industry cluster (frequency - 221) has the highest frequency, indicating a predominant focus on the textile industry within the analyzed documents (Table 2). The Callon Centrality and Callon Density values (16.44 and 99.14, respectively) suggest that the textile industry is a central and densely interconnected theme in the bibliometric analysis.

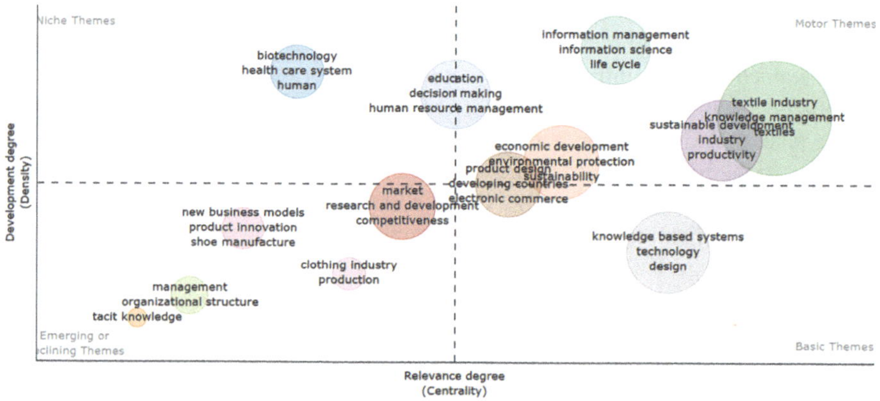

Fig. 4 Thematic map for the studied documents

Table 2 Cluster analysis according to the thematic map

Cluster	Callon centrality	Callon density	Rank centrality	Rank density	Cluster frequency
Market	1.7708	78.67063	6	6	20
Biotechnology	0.3889	115	4	12	11
Textile industry	16.4386	99.1430	13	10	221
Sustainable development	12.2854	90.1505	12	9	42
Tacit knowledge	0	50	1	1	2
Product design	3.9190	79.7619	8	7	19
New business models	0.25	75	3	5	6
Knowledge based systems	8.3735	65.6053	11	4	44
Information management	5.4389	120.1235	10	13	23
Economic development	4.9792	84.0278	9	8	31
Education	2.6042	100.9259	7	11	24
Clothing industry	0.5	62.5	5	3	4
Management	0.1111	58.3333	2	2	5

While not as dominant as the textile industry, the market cluster (frequency - 20) is still significant, with a relatively high Callon Centrality (1.77) and Callon Density (78.67). This suggests that the market theme is well-connected and central within the knowledge network.

The sustainable development cluster (frequency - 42) is notable, indicating a substantial focus on the environmental and social aspects of textile firms' competitiveness. The Callon Centrality (12.29) and Callon Density (90.15) values suggest a central and dense thematic connection within the documents.

The presence of Knowledge-based systems cluster (frequency - 44) highlights the importance of knowledge-based systems in the context of competitiveness in textile firms. The relatively high Callon Centrality (8.37) indicates that knowledge-based systems are a central theme in the knowledge network.

Product design cluster (frequency - 19) is another significant theme, with moderate Callon Centrality (3.92) and Callon Density (79.76) values. This suggests that product design plays a central and interconnected role in the discussions around knowledge and innovation management in textile firms.

Clusters such as biotechnology, tacit knowledge, new business models, information management, economic development, education, clothing industry, and management are also identified but with varying degrees of centrality and density.

The thematic map highlights the multidimensional nature of knowledge and innovation management in the textile industry. It suggests that researchers and practitioners are exploring diverse topics, ranging from market dynamics and sustainable development to specific areas like knowledge-based systems and product design. This diversity underscores the complexity of factors influencing competitiveness in textile firms, urging a holistic approach that considers various dimensions of knowledge and innovation management.

2.1 The Common Research Trends and Meta-Analysis

In the rapidly evolving landscape of the textile industry, the interplay between knowledge and innovation management has emerged as a pivotal factor influencing the competitiveness of firms. As textile companies strive to stay ahead in a globalized market, understanding the dynamics of knowledge creation and utilization becomes imperative. This article embarks on a comprehensive exploration, employing a bibliometric analysis to dissect the trends and patterns in scholarly contributions to the field. Through this lens, we seek to unravel the evolution of knowledge and innovation management within textile firms and its impact on their competitive positioning.

Figure 5 presents a compelling overview of the annual scientific production in the realm of knowledge and innovation management within textile firms. Notably, the data spans a considerable timeframe, offering insights into the ebb and flow of scholarly activity. The initial years, marked by a modest output in the late 1990s and early 2000s, give way to a more pronounced engagement from 2003 onwards. The intermittent spikes and troughs in the ensuing years signify the dynamism and evolving focus within the research community.

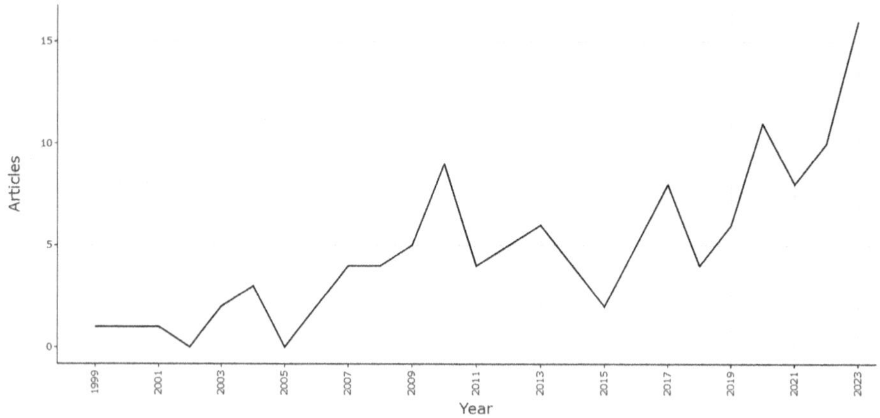

Fig. 5 Annual scientific production on the knowledge and innovation management in textile firms

A notable observation lies in the surge of articles from 2006 to 2010, reflecting a heightened interest and perhaps a critical juncture in the exploration of knowledge and innovation management in the textile sector. The subsequent years, particularly from 2011 to 2015, witness a fluctuating pattern, indicating potential shifts in research priorities or external influences impacting scholarly contributions. The most striking revelation emerges in the recent years post-2015, where a discernible upward trajectory is evident. The steep climb in annual scientific production, peaking at 16 articles in 2023, signals an intensified scholarly focus on the subject matter. This surge may be indicative of an increased recognition within the academic community regarding the pivotal role of knowledge and innovation management in bolstering the competitiveness of textile firms. The trends depicted prompt further exploration into the drivers behind these shifts, the emerging themes, and the potential implications for the textile industry. As we delve deeper into this bibliometric analysis, a richer understanding of the intricate relationship between knowledge, innovation, and competitiveness in the textile sector is poised to unfold.

In the initial years, particularly in 1999 and 2000, the mean citations per article were relatively low (Fig. 6), suggesting a limited influence of the research in those early stages. However, it's important to note that the number of citations per year and the mean citations per year varied significantly. The document's impact saw a notable increase in 2013, where the mean citations per article skyrocketed to 35.83, indicating a significant recognition and acknowledgment within the scholarly community. This spike in citations continued in 2014, reflecting sustained interest and impact.

The subsequent years demonstrate a fluctuating pattern, with some variations in mean citations per year. Notably, in 2016, there is a decline in the mean citations per year, suggesting a potential shift in the research landscape or focus during that period. As we move towards more recent years, particularly 2020 and 2021, there is a resurgence in the average citations per year, reaching 8.64 and 13.5, respectively. This uptrend indicates a sustained and growing interest in the research, possibly

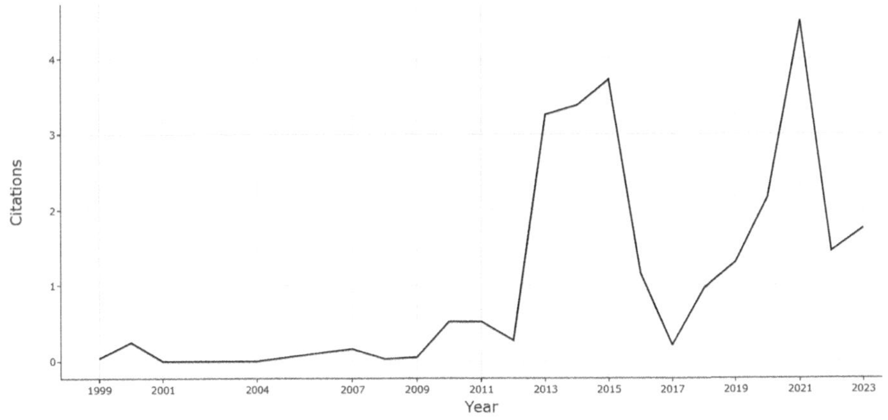

Fig. 6 Average citations per year for the studied documents

fueled by its relevance to current trends in knowledge and innovation management in textile firms. It's crucial to acknowledge the decline in mean citations per year in 2022 and 2023. This could be attributed to various factors such as changes in research direction, emerging topics in the field, or external influences impacting the visibility of the research. While the research has experienced periods of both acceleration and deceleration in terms of average citations per year, its overall trajectory suggests a positive impact and continued relevance in the scholarly community. Understanding these trends can be instrumental in shaping future research directions and maintaining the competitiveness of textile firms in the realm of knowledge and innovation management.

Figure 7 illustrates the publication trends of five different sources from 1999 to 2023. From 1999 to 2006, there is a noticeable absence of publications in all sources except for "Technical Textiles International", which starts contributing in 2004. The subsequent years witness a gradual increase in publications across all sources, with "Research Journal of Textile and Apparel" leading the trend from 2012 onwards.

Starting from 2017, there is a more diversified distribution of publications across the sources. The upward trajectory suggests a growing interest in the field, with an increasing number of sources contributing to the discourse on knowledge and innovation management in textile firms. It is noteworthy that by 2023, all sources, except "Industria Textila", exhibit a consistent and high level of production, indicating a robust and widespread engagement with the subject matter. "Research Journal of Textile and Apparel" stands out as the most prolific source in the given context. The analysis of Fig. 7 underscores the evolving landscape of sources contributing to the discussion on knowledge and innovation management in textile firms. The increasing diversity and volume of publications suggest a vibrant and expanding research

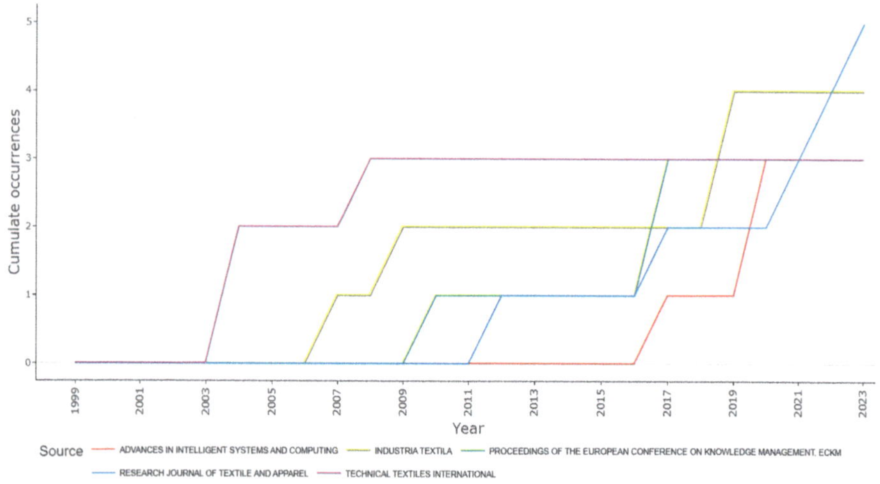

Fig. 7 Sources' production over time

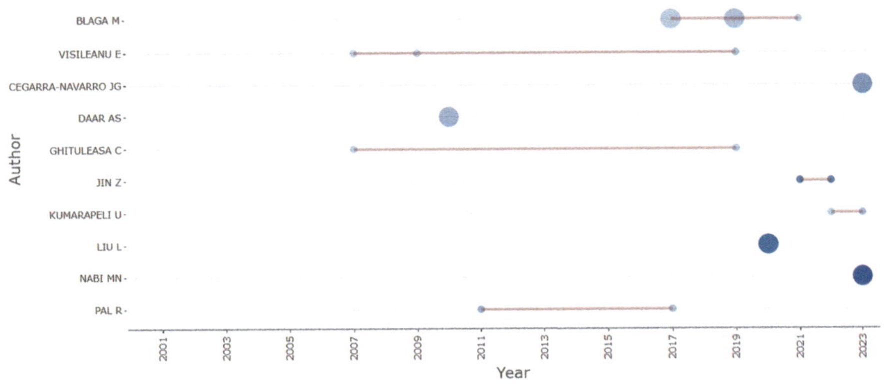

Fig. 8 Authors' production over time

domain, with multiple journals and conference proceedings actively participating in disseminating knowledge in this field.

The trends in authors' production over time suggest a progression from initial exploration to a more established and actively researched area within the field of knowledge and innovation management in textile firms (Fig. 8). The diversity in sources and the sustained interest in recent years point to the enduring significance of the topic.

Scholars like Blaga M. and Rădulescu I. have emphasized the importance of digital cataloging of medical devices for hernia repair, highlighting the potential for technological advancements in enhancing the industry's capabilities. Blaga's exploration of the knowledge matrix for innovation in textile enterprises further underscores the critical role of systematic knowledge application in fostering innovation within the industry (Rădulescu et al., 2019a, 2019b). In parallel, Visileanu E. has contributed to the discourse by developing logical and sequential e-learning content tailored to support specialists in textile enterprises. This innovative approach aligns with the growing recognition that continuous learning and skill development are vital components of a competitive textile sector. Additionally, Visileanu E. has proposed a strategic guideline for sustainable development in a knowledge-based society, emphasizing the concept of zero waste as a central tenet for achieving environmental and economic sustainability in the industry.

Moving beyond specific case studies, Nabi et al. (2023) investigated the effects of leaders' stewardship behavior on radical innovation, revealing a complex interplay involving knowledge management dynamic capability and environmental uncertainty. Similarly, Martínez-Martínez et al. (2023) explores the mediating effect of technology assimilation and embarrassment in small and medium-sized enterprises (SMEs), shedding light on the relationship between green skills and organizational reputation. In the context of the textile industry's environmental responsibilities, Cegarra-Navarro J. delves into eco-innovation, emphasizing the strategic importance of data privacy and environmental knowledge management. This underscores the multifaceted nature of innovation, encompassing both technological and sustainability dimensions.

In terms of risk management within the textile and apparel industries, Kumarapeli et al. (2023) proposed a risk model tailored to different innovator types. This contribution adds a nuanced perspective to the literature, acknowledging the diverse risk profiles associated with various innovation endeavors.

On the technological front, Jin Z. explores the application of graphene-based textiles for thermal management and flame retardancy, illustrating the industry's responsiveness to cutting-edge materials and their potential impact on product performance. Similarly, Jin Z. investigates the relationship between intellectual capital elements and firm performance in the Chinese textile and apparel industry, offering insights into the intangible assets that drive competitive advantage (Ge et al., 2022).

Liu and Ramakrishna introduces the concept of circular economy, providing a foundational understanding of sustainable economic models relevant to the textile industry. This work lays the groundwork for integrating circular economy principles into the fabric of textile enterprises, aligning with the broader global shift towards sustainable practices (Liu & Ramakrishna, 2021).

Moreover, Ghituleasa C. contributes to the discourse on e-learning content, specifically tailored to support specialists in textile enterprises. This reinforces the notion that continuous learning and skill development are pivotal for staying competitive in an ever-evolving industry (Rădulescu et al., 2019a, 2019b).

In the realm of health innovation, Daar A. examines science-based health innovation in Tanzania, showcasing the broader applicability of innovation principles beyond the conventional scope of the textile industry. Furthermore, Daar A. extends this exploration to sub-Saharan Africa, emphasizing the need for region-specific approaches to health innovation (Shah et al., 2010).

The literature reviewed here highlights the multifaceted nature of knowledge and innovation management in textile firms, emphasizing their role as crucial drivers of competitiveness (Fig. 9). The integration of digital technologies, sustainable practices, and strategic leadership emerges as common themes, underscoring the industry's dynamic and evolving nature.

The contemporary landscape of textile firms is intricately woven with the dynamics of knowledge and innovation management, playing a pivotal role in enhancing competitiveness. A bibliometric analysis reveals a rich tapestry of scholarly contributions in this domain. Dangelico et al. (2013) explores the intersection of knowledge and innovation management in the textile industry, emphasizing its profound impact on competitiveness. Zucchella and Siano (2014) complements this perspective, shedding light on international studies that underscore the strategic significance of effective management practices in textile organizations. Köhler and Som (2014) add a technological dimension to the discourse, emphasizing the role of innovation in shaping the competitive landscape. García-Álvarez (2015) contributes to the narrative by examining the behavioral aspects within textile firms, illustrating how effective knowledge and innovation management can influence human behavior and, consequently, competitiveness. Borowski (2021) brings a contemporary lens to

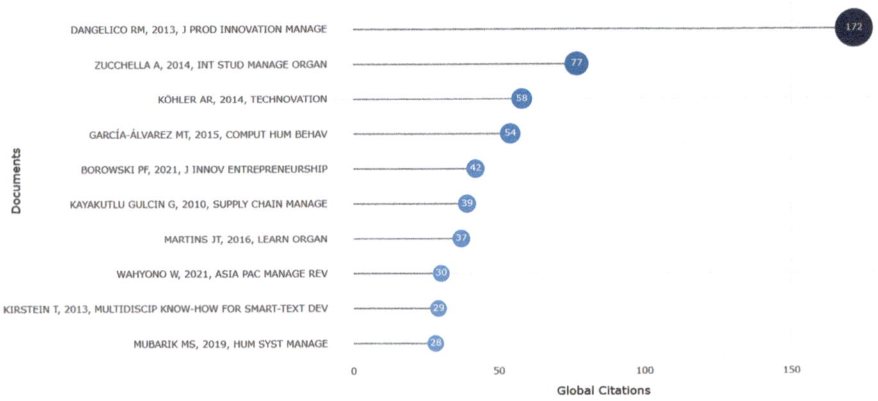

Fig. 9 Most global cited documents

the discussion, emphasizing the evolving nature of innovation and entrepreneurship in the textile sector. Kayakutlu and Büyüközkan (2010) broadens the scope by delving into supply chain management, highlighting its symbiotic relationship with competitiveness. Martins (2016) delves into organizational learning within textile firms, accentuating the role of knowledge management in fostering adaptive and competitive organizations. Hutahayan (2021) adds an Asia–Pacific perspective, enriching our understanding of the contextual nuances that influence knowledge and innovation management in the region (Hutahayan, 2021). Mubarik et al. (2019) contribute a systems thinking approach, underscoring the holistic management of human systems in the textile industry. These diverse scholarly contributions collectively paint a comprehensive picture of the intricate relationship between knowledge and innovation management and competitiveness in textile firms. The multidimensionality of the literature underscores the need for a nuanced and integrative approach to navigate the challenges and opportunities in this dynamic industry.

The data reveals that the majority of authors (939%) have contributed to a single document (Fig. 10). This indicates a highly skewed distribution, where a small proportion of authors are responsible for the majority of the published work in the field. The concentration of authors with only one document aligns with Lotka's law, emphasizing the prevalence of prolific authors.

A smaller proportion of authors (5.1%) have contributed to two documents, and an even smaller fraction (0.3%) of authors have been productive in three, four, or five documents each. This distribution conforms to Lotka's law, which predicts a decreasing number of authors as the number of documents they contribute to increases. The observed pattern highlights the presence of a core group of highly productive authors who significantly contribute to the scholarly output in the field of knowledge and innovation management in textile firms. Understanding and recognizing this distribution of author productivity can be valuable for identifying key contributors and trends in the research landscape.

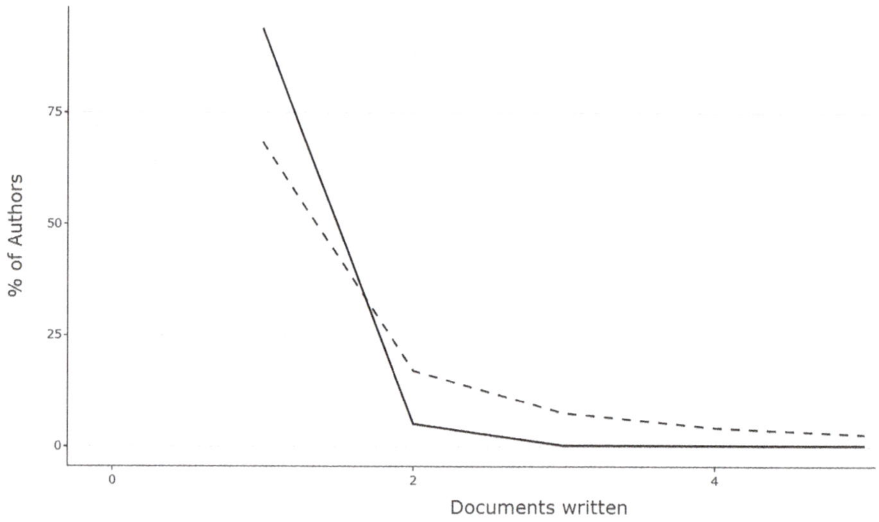

Fig. 10 Author productivity through Lotka's law

The figure illustrates the production of articles by different affiliations over time in the field of knowledge and innovation management within textile firms. Three prominent affiliations, Xi'an Polytechnic University, Zhejiang University, and University of Borås, have been consistently contributing to the research landscape (Fig. 11).

Xi'an Polytechnic University started with no publications until 2010 but has gradually increased its productivity, reaching three publications per year from 2016 onwards. This upward trend indicates a growing research focus or collaboration within the institution. Zhejiang University has maintained a steady output of four

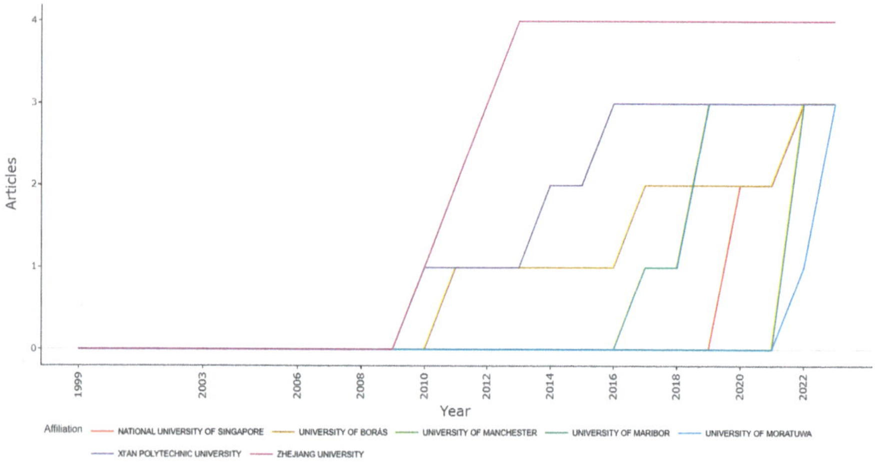

Fig. 11 Affiliations' production over time

publications annually since 2011. This suggests a consistent involvement in research activities related to knowledge and innovation management in textile firms. The University of Borås shows a different pattern, with a single publication in 2011 and a gradual increase in subsequent years. By 2023, the affiliation is contributing three publications per year. This may signify a deliberate effort to engage in research on this topic over time. Other institutions, such as the University of Maribor, National University of Singapore, University of Manchester, and University of Moratuwa, have also made contributions, with varying levels of productivity. The data suggests a diverse range of institutions participating in the exploration of knowledge and innovation management in textile firms, contributing to the overall body of literature in the field.

Figure 12 provides information on the number of articles, Single Country Publications (SCP), Multiple Country Publications (MCP), frequency, and MCP ratio for each country. The majority of articles (44) are attributed to an unspecified country (50). China, with 15 articles, is the leading contributor among identified countries, followed by the United Kingdom with 5 articles.

Several countries, including Romania, Indonesia, Spain, Sweden, USA, Greece, Italy, Malaysia, Sri Lanka, Switzerland, Albania, Argentina, Australia, Brazil, Colombia, Denmark, Germany, Iran, Netherlands, Pakistan, Poland, Portugal, Saudi Arabia, Serbia, Slovakia, Slovenia, Thailand, Turkey, and Ukraine, each contributed to one article. The SCP/MCP classification provides insights into the collaboration patterns. For instance, China and the United Kingdom have a higher proportion of SCP, indicating a greater tendency for these countries to publish articles independently. On the other hand, several countries, including Brazil, Colombia, Denmark, Malaysia, Netherlands, Saudi Arabia, Thailand, and Turkey, have a MCP ratio of 1.0, indicating that all their publications are collaborative.

The presence of articles from a diverse set of countries highlights the global nature of the research, with contributions from Europe (e.g., Romania, Spain, Sweden), Asia

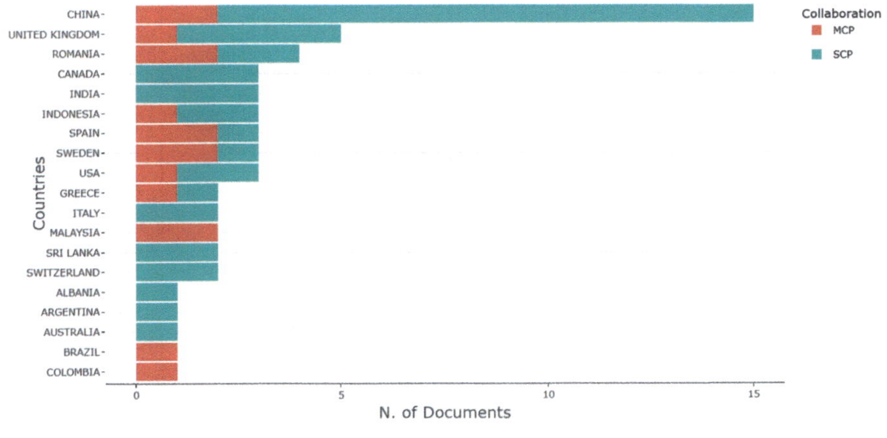

Fig. 12 Corresponding author's countries

(e.g., China, India, Indonesia, Malaysia), North America (Canada, USA), and other regions. The MCP ratio provides insights into the collaborative nature of the research from each country. For example, Malaysia has a MCP ratio of 1.0, suggesting that all its publications involve multiple countries, while China and the United Kingdom have relatively lower MCP ratios, indicating a higher proportion of single-country publications. This underscores the international collaboration and distribution of research efforts in the field of knowledge and innovation management in textile firms, with notable contributions from various countries across the globe.

Figure 14 presents a comprehensive overview of the scientific production in the field of knowledge and innovation management in textile firms, highlighting the countries that have contributed significantly to the body of literature. Here is an analysis of the data (Fig. 13).

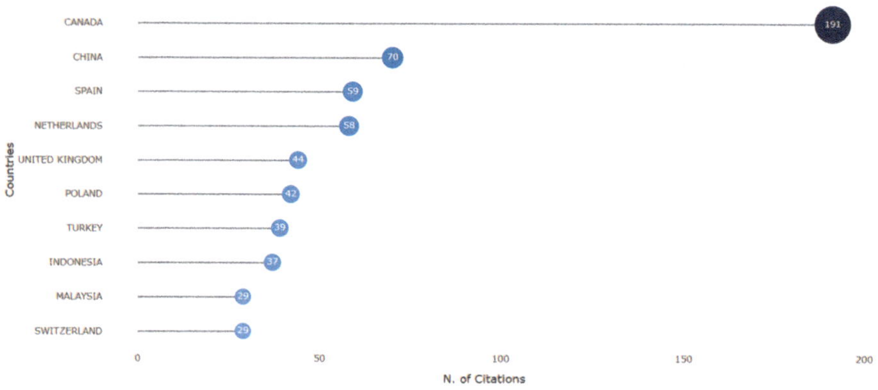

Fig. 13 Most cited countries

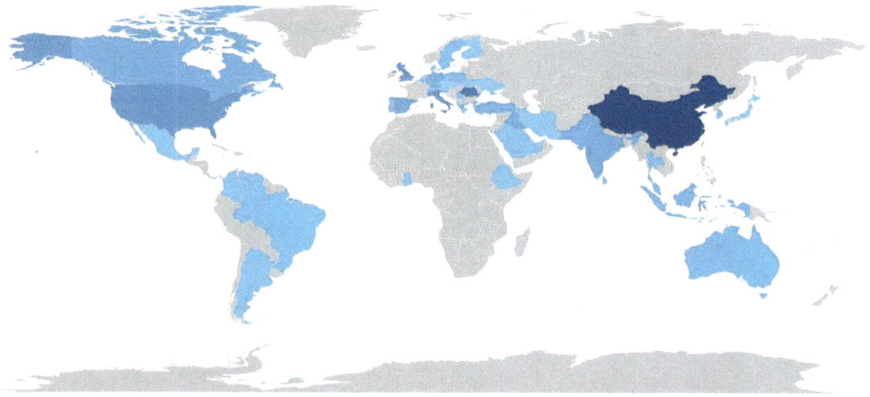

Fig. 14 Countries' scientific production

China emerges as the leading contributor to the scientific production in this domain, with a notable frequency of 36 articles. This dominance underscores China's growing influence and active participation in research related to knowledge and innovation management in textile firms. Romania and Italy follow closely behind China, with 19 and 12 articles, respectively. The European region, as a whole, demonstrates a substantial presence in this field. The United Kingdom (UK) also makes a noteworthy contribution with 12 articles, emphasizing the collaborative nature of research in this domain across European nations.

The dataset indicates a diverse collaboration among countries in scientific endeavors related to textile firms. The presence of countries like the USA, Canada, Germany, India, and Iraq signifies a global perspective and collaborative efforts in advancing knowledge and innovation management. The data showcases a mix of countries from different regions, reflecting a broad global interest in the subject. This diversity is essential for gaining a comprehensive understanding of knowledge and innovation management practices, considering potential variations in industrial contexts and approaches. This underscores the global nature of research in knowledge and innovation management in textile firms, with China leading the scientific production, closely followed by a diverse set of countries from Europe, North America, and Asia. This international collaboration contributes to a richer and more nuanced understanding of the challenges and opportunities in the textile industry's knowledge and innovation landscape.

In Fig. 15, a word map of the researched publications is presented, depicting the factorial analysis of key terms and their clusters.

Cluster 1, represented by words like "innovation", "knowledge management", "textile industry", "technological innovation", "innovation process", "sustainability", and "textiles", suggests a thematic focus on the intersection of innovation, knowledge management, and sustainability within the textile industry. Cluster 2, characterized by terms such as "absorptive capacity" and "knowledge", indicates

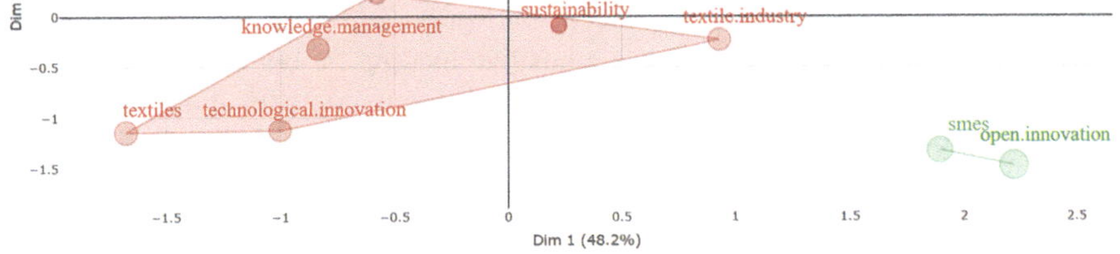

Fig. 15 Word map of the researched publications factorial analysis

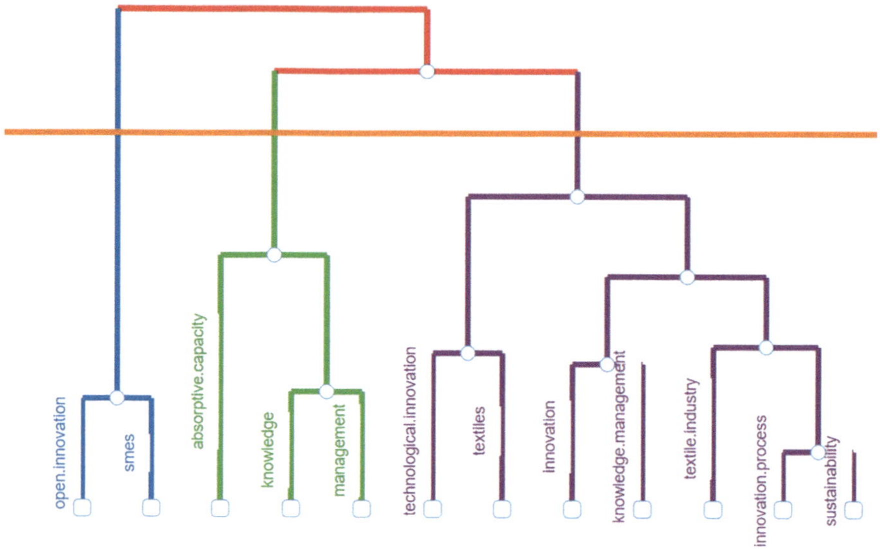

Fig. 16 Factorial analysis dendrogram of the researched publications

a concentration on the capacity of organizations to absorb and utilize knowledge for innovation. Cluster 3, associated with "open innovation" and "smes" (small and medium-sized enterprises), highlights a specific emphasis on open innovation practices and their relevance to small and medium-sized enterprises in the textile sector.

The factorial analysis showcases the multidimensional nature of knowledge and innovation management in the textile industry, emphasizing the interplay between innovation, knowledge absorption, management strategies, open innovation practices, and the unique challenges faced by small and medium-sized enterprises (Fig. 16).

The factorial analysis in Table 3 provides insights into the key themes and clusters within the studied documents on knowledge and innovation management in textile firms.

The word "innovation" has a negative loading on Dimension 1, indicating a focus on different aspects of innovation within the textile industry. Similarly, "knowledge.management" and "innovation.process" also contribute to this cluster, suggesting a strong association between innovation and knowledge management practices. The term "knowledge" has a positive loading on both Dimension 1 and Dimension 2, indicating its significance across different contexts in the documents.

Absorptive Capacity: "Absorptive.capacity" is positively loaded on Dimension 2, suggesting a focus on the ability of firms to assimilate and apply external knowledge.

"Open.innovation" and "SMEs" (small and medium-sized enterprises) are strongly associated and contribute to Cluster 3, suggesting a thematic emphasis on how open innovation practices relate to the context of SMEs in the textile industry.

Table 3 Factorial analysis indicators of the studied documents

Word	Dim.1	Dim.2	Cluster
Innovation	−0.58	0.23	1
Knowledge.management	−0.84	−0.32	1
Textile.industry	0.93	−0.23	1
Absorptive.capacity	−0.39	1.51	2
Knowledge	0.65	2.37	2
Technological.innovation	−1	−1.13	1
Innovation.process	0.22	−0.09	1
Management	0.87	2.03	2
Open.innovation	2.22	−1.48	3
Smes	1.89	−1.33	3
Sustainability	0.22	−0.09	1
Textiles	−1.68	−1.15	1

Sustainability: The term "sustainability" also contributes to this cluster, indicating a potential intersection between open innovation, SMEs, and sustainable practices. The term "textiles" has a negative loading on Dimension 1, indicating its association with the broader context of the textile industry. The factorial analysis reveals distinct clusters of terms, highlighting the interconnectedness of concepts related to innovation, knowledge management, absorptive capacity, open innovation, SMEs, sustainability, and the broader context of the textile industry.

3 Conclusion

The analysis uncovers core thematic areas, revealing a multidimensional exploration of the textile industry with a focus on innovation, knowledge management, competition, and sustainable development. The centrality of the textile industry in the thematic map underscores its pivotal role in knowledge and innovation management discussions. The prominence of terms such as "textiles", "innovation", and "knowledge management" reflects a comprehensive exploration of these concepts. The analysis also reveals a global perspective, supported by the significant international co-authorship rate, emphasizing collaboration on a global scale.

The evolving trends over the years highlight the dynamic nature of research focus, with topics such as education, industrial management, design, and knowledge-based systems gaining prominence in different periods. The thematic map further illustrates the interconnectedness of diverse topics, emphasizing the complexity of factors influencing competitiveness in textile firms.

Clusters identified in the thematic map, such as the textile industry, market, sustainable development, knowledge-based systems, and product design, provide a nuanced understanding of the multidimensional nature of knowledge and innovation management in the textile industry. These clusters, with varying degrees of centrality and density, underscore the importance of considering multiple dimensions in research and practice.

The analysis of common research trends and meta-analysis reveals a pivotal interplay between knowledge and innovation management in shaping the competitiveness of textile firms. The upward trajectory in annual scientific production, especially in recent years, suggests an intensified scholarly focus on the subject. The fluctuations in mean citations per year highlight periods of increased recognition and impact, emphasizing the relevance of the research to current trends.

The examination of sources, authors, affiliations, and countries contributing to the literature provides a comprehensive view of the global research landscape. The presence of prolific authors, diverse affiliations, and contributions from various countries underscores the collaborative and widespread engagement with the subject matter. China emerges as a leading contributor, reflecting its growing influence in research related to knowledge and innovation management in textile firms.

The analysis illuminates the complexity, diversity, and global nature of research on knowledge and innovation management in textile firms. The identified thematic areas, trends, and contributors offer valuable insights for future investigations. Recognizing the interconnectedness of concepts and the evolving landscape, researchers and practitioners are encouraged to adopt a holistic approach, considering the multidimensional factors influencing competitiveness in the dynamic textile industry. Future research could explore the specific mechanisms through which knowledge and innovation management practices impact sustainable development, market dynamics, and product design in the textile sector. Investigating the role of emerging technologies and digitalization in shaping knowledge and innovation processes could provide further avenues for exploration. Efforts to enhance international collaboration and knowledge exchange in this field could contribute to a more comprehensive and globally informed understanding of effective strategies for driving competitiveness in textile firms.

References

Ahmed, W., Najmi, A., & Ikram, M. (2020). Steering firm performance through innovative capabilities: A contingency approach to innovation management. *Technology in Society, 63*, 101385.

Azeem, M., Ahmed, M., Haider, S., & Sajjad, M. (2021). Expanding competitive advantage through organizational culture, knowledge sharing and organizational innovation. *Technology in Society, 66*, 101635.

Borowski, P. F. (2021). Innovation strategy on the example of companies using bamboo. *Journal of Innovation and Entrepreneurship, 10*(1), 1–17.

Martínez-Martínez, A., Cegarra-Navarro, J. G., Garcia-Perez, A., & De Valon, T. (2023). Active listening to customers: eco-innovation through value co-creation in the textile industry. *Journal of Knowledge Management, 27*(7), 1810–1829.

Chaurasia, S. S., Kaul, N., Yadav, B., & Shukla, D. (2020). Open innovation for sustainability through creating shared value-role of knowledge management system, openness and organizational structure. *Journal of Knowledge Management, 24*(10), 2491–2511.

Chen, A., Lin, Y., Mariani, M., Shou, Y., & Zhang, Y. (2023). Entrepreneurial growth in digital business ecosystems: An integrated framework blending the knowledge-based view of the firm and business ecosystems. *The Journal of Technology Transfer, 48*(5), 1628–1653.

Cooke, P., & Schienstock, G. (2000). Structural competitiveness and learning regions. *Enterprise and Innovation Management Studies, 1*(3), 265–280.

Dangelico, R. M., Pontrandolfo, P., & Pujari, D. (2013). Developing sustainable new products in the textile and upholstered furniture industries: Role of external integrative capabilities. *Journal of Product Innovation Management, 30*(4), 642–658.

Danskin, P., Englis, B. G., Solomon, M. R., Goldsmith, M., & Davey, J. (2005). Knowledge management as competitive advantage: Lessons from the textile and apparel value chain. *Journal of Knowledge Management, 9*(2), 91–102.

Di Chiacchio, L., Martínez-Caro, E., Cegarra-Navarro, J.G. and Garcia-Perez, A. (2023). Eco-innovation in the textile industry: the strategic importance of data privacy and environmental knowledge management. *Kybernetes*, ahead-of-print No. ahead-of-print.

Durana, P., Zauskova, A., Vagner, L., & Zadnanova, S. (2020). Earnings drivers of Slovak manufacturers: Efficiency assessment of innovation management. *Applied Sciences, 10*(12), 4251.

García-Álvarez, M. T. (2015). Analysis of the effects of ICTs in knowledge management and innovation: The case of Zara Group. *Computers in Human Behavior, 51*, 994–1002.

Gardas, B. B., Raut, R. D., & Narkhede, B. (2018). Modelling the challenges to sustainability in the textile and apparel (T&A) sector: A Delphi-DEMATEL approach. *Sustainable Production and Consumption, 15*, 96–108.

Ge, Q., Chu, J., Cao, W., Yi, F., Ran, Z., Jin, Z., ... Novoselov, K. S. (2022). Graphene-based textiles for thermal management and flame retardancy. *Advanced Functional Materials, 32*(42), 2205934.

Hutahayan, B. (2021). The relationships between market orientation, learning orientation, financial literacy, on the knowledge competence, innovation, and performance of small and medium textile industries in Java and Bali. *Asia Pacific Management Review, 26*(1), 39–46.

Iqbal, Q., & Piwowar-Sulej, K. (2023). Sustainable leadership and heterogeneous knowledge sharing: the model for frugal innovation. *European Journal of Innovation Management*, Vol. ahead-of-print No. ahead-of-print.

Kayakutlu, G., & Büyüközkan, G. (2010). Effective supply value chain based on competence success. *Supply Chain Management, 15*(2), 129–138.

Köhler, A. R., & Som, C. (2014). Risk preventative innovation strategies for emerging technologies the cases of nano-textiles and smart textiles. *Technovation, 34*(8), 420–430.

Kumarapeli, U., Ratnayake, V., & Jayawardana, T.S.S. (2022). Technological innovation management through root cause prioritization. *Research Journal of Textile and Apparel*, Vol. ahead-of-print No. ahead-of-print.

Kumarapeli, U., Ratnayake, V., & Jayawardana, S. S. (2023). Development of a risk model for different innovator types in textile and apparel industries. *Research Journal of Textile and Apparel.*

Liu, L., & Ramakrishna, S. (2021). Introduction and overview. In L. Liu & S. Ramakrishna (Eds.), *An introduction to circular economy*. Springer.

Martins, J. T. (2016). Relational capabilities to leverage new knowledge: Managing directors' perceptions in UK and Portugal old industrial regions. *The Learning Organization, 23*(6), 398–414.

Mubarik, M. S., Naghavi, N., & Mahmood, R. T. (2019). Intellectual capital, competitive advantage and the ambidexterity liaison. *Human Systems Management, 38*(3), 267–277.

Nabi, M. N., Liu, Z., & Hasan, N. (2023). Investigating the effects of leaders' stewardship behavior on radical innovation: A mediating role of knowledge management dynamic capability and moderating role of environmental uncertainty. *Management Research Review, 46*(2), 173–195.

Pal, R., Harper, S., & Vellesalu, A. (2018). Competitive manufacturing for reshoring textile and clothing supply chains to high-cost environment: A Delphi study. *The International Journal of Logistics Management, 29*(4), 1147–1170.

Prakash, Y., Charwak, B., & Kumar, P. V. (2020). Textile industry in new India: Challenges and opportunities. *International Journal of Indian Culture and Business Management, 21*(4), 435–458.

Rădulescu, I. R., Almeida, L., Vannucci, R., Blaga, M., Dufkova, P., & Stjepanovič, Z. (2019a). Texmatrix-The knowledge matrix for innovation applied to textile enterprises. *Industria Textila, 70*(2), 197–202.

Rădulescu, I. R., Ghiţuleasa, C., Visileanu, E., Dias, A., Stjepanovic, Z., Vannucci, R., & Blaga, M. (2019). Logical and sequential e-Learning content for supporting specialists in textile enterprises. *eLearning & Software for Education, 3*, 418–424.

Rese, A., Baier, D., & Rausch, T. M. (2022). Success factors in sustainable textile product innovation: An empirical investigation. *Journal of Cleaner Production, 331*, 129829.

Shah, R., Singer, P.A. & Daar, A.S. Science-based health innovation in Tanzania: bednets and a base for invention. *BMC Int Health Hum Rights* 10 (Suppl 1), S4 (2010).

Shahzad, M., Qu, Y., Zafar, A. U., Rehman, S. U., & Islam, T. (2020). Exploring the influence of knowledge management process on corporate sustainable performance through green innovation. *Journal of Knowledge Management, 24*(9), 2079–2106.

Shen, L., Zhang, X., & Liu, H. (2022). Digital technology adoption, digital dynamic capability, and digital transformation performance of textile industry: Moderating role of digital innovation orientation. *Managerial and Decision Economics, 43*(6), 2038–2054.

Sivalogathasan, V., & Wu, X. (2015). Impact of organization motivation on intellectual capital and innovation capability of the textile and apparel industry in Sri Lanka. *International Journal of Innovation Science, 7*(2), 153–168.

Tu, Y., & Wu, W. (2021). How does green innovation improve enterprises' competitive advantage? The role of organizational learning. *Sustainable Production and Consumption, 26*, 504–516.

Usman, I., Maupa, H., Idrus, M., Haerani, S., & Nurjanna, N. (2020). Moderation effect of competence of knowledge and innovation: Case of Bali. *Business Process Management Journal, 26*(6), 1307–1327.

Wakeford, J. J., Gebreeyesus, M., Ginbo, T., Yimer, K., Manzambi, O., Okereke, C., ... Mulugetta, Y. (2017). Innovation for green industrialisation: An empirical assessment of innovation in Ethiopia's cement, leather and textile sectors. *Journal of Cleaner Production, 166*, 503-511.

Watchravesringkan KT, Karpova E, Nelson Hodges N, Copeland R (2010) The competitive position of Thailand's apparel industry: Challenges and opportunities for globalization. *Journal of Fashion Marketing and Management, 14*(4), 576–597.

Zucchella, A., & Siano, A. (2014). Internationalization and innovation as resources for SME growth in foreign markets: A focus on textile and clothing firms in the Campania Region. *International Studies of Management & Organization, 44*(1), 21–41.

José Moleiro Martins is a Coordinator/Associate Professor at the Polytechnic Institute of Lisbon in Strategic Management, Entrepreneurship, and Knowledge and Innovation Management. He holds a Ph.D. in Management from ISCTE-IUL Business School, a Master's degree in Business Strategy and Development, a Postgraduate degree in Business Management, and a Bachelor's degree in Economics. He has business experience in the manufacturing industry and the aerospace sector, as well as in Strategic Management consulting across various industries. He has authored scientific books in both Portuguese and English and has publications as a sole author and co-author in reputable journals, including the Journal of Knowledge Management, Journal of the Knowledge Economy, Journal of Innovation and Entrepreneurship, Journal of Innovation & Knowledge, International Journal of Innovation and Technology Management, International Journal of Training and Development, Journal of Business Strategy, Journal of East-West Business, European Business Review, Industrial Management, International Journal of Business Performance Management, African Journal of Business Management, and Industrial Management & Data Systems. His main research interests encompass Knowledge Management, Innovation, Technological Innovation, Entrepreneurship, Strategic Management, Sustainability, Energy, Sub-Saharan Africa, and Asia.

How Knowledge Management Capabilities Enhance Textile Firms' Innovative Performance

Syed Arslan Haider and Sonia Tabassum Anees

Abstract The role of knowledge management capabilities and innovative performance within textile firms enhance the organization performance. The textile industry has become a critical component of the global economy and daily life. Consequently, an assessment of input utilization that considers innovation, technology, and efficacy is imperative. This study aims to adapt knowledge management capabilities and innovative performance within textile firms of Portugal. This study seeks to answer the two important questions: (1) Why Does Firm's Need Knowledge Management Capabilities? (2) How It Influences Textile Firm's Innovative Performance? According to findings the results reveals that there is a scarcity of scholarly investigations concerning the knowledge management capabilities and innovation performance of businesses operating in this industry. Through internal cooperation, organizations gain access to a vast array of tacit and explicit knowledge that exists within the organization. By strengthening internal engagement with the workforce, organizations gain knowledge of what, where, and to what extent information is stored within the organization. The potential for innovation to be stimulated within an organization exists through internal collaboration. There is a positive correlation between increased information exchange among employees and their greater capacity for innovation.

1 Introduction

In order to adapt to the new knowledge-based economy, the modern corporation must transform itself into an organization that generates, integrates, and protects knowledge (Cordeiro et al., 2023). To sustain their competitive edge, organizations

---

S. A. Haider (✉)
Department of Management, Sunway Business School, Sunway University, Bandar Sunway, Malaysia
e-mail: haidershah24@gmail.com

S. T. Anees
Department of Management, Government Sadiq College Women University Bahawalpur, Bahawalpur, Pakistan

Lisbon Accounting and Business School, Polytechnic Institute of Lisbon (ed.),
Industry and Innovation: Textile Industry, SDGs and Textiles,
https://doi.org/10.1007/978-3-031-57804-5_7

must perpetually prioritize the cultivation of their unique competencies, including dynamic capabilities. The importance of information is apparent in numerous spheres of society, encompassing the business and social spheres. Human knowledge shapes attitudes, exerts an impact on behavior, and illuminates the human intellect. To foster a more civilized society, it is generally recognized that the acquisition of knowledge is crucial and ought to be regarded as a fundamental responsibility. This acknowledgement is also present within the business sector. Duryan et al. (2020) posit that knowledge is acquired through the integration of structured experiences, values, contextual information, and expert insights. It functions as a structural framework for evaluating and incorporating novel experiences and information.

1.1 Textile Industry of Portugal

The textile and garment sectors are integral components of Portugal's industrial framework and contribute significantly to the country's global trade. In 2006, the combined contributions of these two sectors comprised 2.0% of the gross value added, 4.3% of the employment force, and 11.8% of the overall industrial exports of Portugal (Lima et al., 2021). The relative significance of these sectors, nevertheless, has diminished over the past two decades due to substantial disruptions. Even though textiles and apparel are distinct industries with their own characteristics, their technological and trade policy operations are intricately intertwined. The vertical connection between the two sectors is intrinsic, given that textiles are utilized as the principal inputs in the production of apparel products. However, these interconnections also encompass the sales and distribution processes, as retailers in the apparel industry increasingly supervise the supply chains of the textile and clothing industries (Amoako, 2019).

The textile and garment industry employs 35 million individuals worldwide and contributes to 7% of global exports (Hossian et al., 2019). The industry's relevance to European economies cannot be overstated. Portugal holds the notable status of being the preeminent exporter in Europe, whereas this sector provides substantial employment opportunities in the United Kingdom. The COVID-19 pandemic nonetheless exerted a significant influence on the supply chain. It was anticipated that fashion sales in the United States would decrease by nearly fifty percent by 2020 (Prakash et al., 2020). Organizations that conducted business in China or South Korea encountered significant increases in transportation and logistics costs, in addition to labor and raw material shortages (Johnson & Haug, 2021). Impediments to travel and blockades have severely impeded the operations of the corporate entity, leading to diminished consumer demand and impeding customers' ability to make purchases in fashion capitals (Ndzana et al., 2021). Furthermore, the lack of efficient crisis management presented difficulties for businesses in guaranteeing adequate financial resources during the pandemic (Mariev et al., 2023). Organizations encounter difficulties in sustaining their financial sustainability as a result of unanticipated expenses such as payment setbacks, order voids, fixed costs, and additional charges.

Moreover, the textile sector has encountered substantial fluctuations in its financial performance as a consequence of a multitude of factors. The aforementioned factors comprise fierce competition from economies characterized by inexpensive labor, volatility in worldwide supply and demand, and obstacles stemming from technological progress and changes in consumer inclinations, including the implementation of circular economy methodologies. The convergence of these unfavorable variables requires a novel approach to competition and flexibility concerning products, procedures, and management structures.

The textile industry has become a critical component of the global economy and daily life (Abdelmeguid et al., 2022). Consequently, an assessment of input utilization that considers innovation, technology, and efficacy is imperative. In the modern era of technology, economic advancement is propelled by innovation, which increases the efficiency of production elements and, consequently, the productivity of an organization. Enterprises require innovation to sustain their global competitiveness. Critical are the contributions that innovation makes to the growth of competitive businesses on the international market and to the enhancement of labor productivity. Additionally, it contributes to the growth of prosperity and improves social welfare. Furthermore, an increase in exports contributes to the enhancement of innovation output efficiency (Freixanet et al., 2021).

2 Why Does Firm's Need Knowledge Management Capabilities: How It Influences Textile Firm's Innovative Performance?

2.1 Knowledge Management

Knowledge management (KM) is an organizational procedure that aims to establish a centralized repository of organizational knowledge. The knowledge contained in this repository is explicit and tacit, internal and external, and has undergone the following processes: acquisition, assimilation, distribution, integration, sharing, retrieval, and reuse. Knowledge management's ultimate objective is to foster innovation within the organization, which will be demonstrated through enhanced products, improved personnel, and optimized organizational procedures. According to the research of Haider and Kayani (2021), knowledge is a critical strategic objective for businesses, and the success of an organization is contingent on implementing a knowledge-based strategy. Organizations can acquire fundamental capabilities simply by acquiring new information. The fundamental principle underlying the concept is utilizing knowledge as the primary input and source. As a result, organizations that rely on information can manage their resources more efficiently. Knowledge management is defined by Migdadi (2022), as a sequence of procedures designed to transform data into actionable information or knowledge that propels the organization's advancement. All of these activities are encompassed within the knowledge generation, acquisition, storage, sharing, and utilization processes (Obeso et al., 2020). Agile management facilitates the generation of high-caliber knowledge. It is possible to conclude that knowledge management has an influence on the quality of choices and behaviours. Bratianu et al. (2023) posit that the implementation of knowledge management practices within an organization can potentially bolster its ability to generate significant long-term profits. Gujral (2020) initially identified the duality of knowledge and divided knowledge into two distinct categories.

- Tacit knowledge
- Explicit knowledge

2.1.1 Tacit Knowledge

According to Gujral (2020), tacit knowledge comprises abilities, aptitudes, and conceptual reasoning. Furthermore, he argued that tacit knowledge is attributed to the individual possessing that information as well as to that which is known. At times, individuals possessing substantial knowledge may encounter difficulties in expressing their comprehension in an effective manner, or they might be unable to disseminate their expertise to those who require it. Acquiring tacit knowledge is exceedingly difficult on account of its intrinsic character as concepts, abilities, and capabilities that are profoundly engrained in the minds of individuals. The elusive nature of tacit information stems from the fact that its visibility is limited to its practical implementation. As a result, the task of capturing, utilizing, and distributing tacit knowledge among the personnel of an organization becomes extremely difficult.

2.1.2 Explicit Knowledge

According to Gujral (2020), the transmission and exchange of explicit knowledge are possible through the use of tangible information, well-defined procedures, and standardized criteria. Explicit knowledge was defined by Baporikar (2019) as "knowledge of rationality". Knowledge that is explicit is easily captured, administered, distributed, and shared with individuals.

2.1.3 Importance of Knowledge Management

The significance of the concept of KM has been extensively acknowledged in recent decades. Knowledge management has been widely recognized as a critical element of the industrial economy, in the same way that intellectual assets and natural resources are fundamental components. The level of focus has been significantly heightened since 1995 (Jordão et al., 2020). The implementation of knowledge management has become a prominent area of interest for numerous corporate entities. The recognition by the management that the value of their organization is dependent on their ability to produce and supervise knowledge has occurred. To enhance firm performance and attain a competitive edge through the implementation of knowledge management methodologies, the presence of capable individuals is crucial. Knowledge management places considerable importance on individuals, their behavior, and the professional culture concerning both technical and social dimensions. The central concept is that information is generated and distributed through interpersonal exchanges at work. While concepts may be developed, teams and operational entities are the primary contributors to knowledge production within an organization (Hayaeian et al., 2022).

2.2 Innovation

Innovation could be described in a variety of ways. The primary and fundamental purpose of innovation, according to Azeem et al. (2021), is to produce novel knowledge that may advance society and reveal feasible options. Innovation encompasses the methodology and procedure of obtaining, accumulating, organizing, and distributing information to generate novel insights that empower the manufacturing and provision of exceptional and distinct products and services (Xie et al., 2023). Ndemezo and Kayitana (2020) stated that innovation is the generation of unique knowledge that increases the financial viability of a nascent company. Innovation aims to enhance the structure and internal business processes of an organization so that it can produce products and services that are acknowledged by the market. Innovation encompasses the undertakings and procedures entailed in generating and executing novel insights in order to produce distinctive goods, services, and procedures that satisfy the varied demands and inclinations of consumers (Jie et al., 2022). Additionally, it strives to improve the level of complexity in systems, frameworks, and innovations, thus promoting well-being among individuals, collectives, and the broader community.

Organizations can no longer sustain competitiveness and innovation solely through the utilization of their own resources and concepts. Companies gain advantages by venturing beyond their internal organization and industry boundaries, in the midst of a vast environment where information and concepts are widely distributed (Tian et al., 2021). Considering the predominantly external nature of innovation sources in well-established sectors such as the textile industry, this shift in emphasis from variations within sectors to variations between sectors is especially relevant in these particular domains. (Giannini et al., 2019), who used the innovation model of enterprises to classify industries, previously identified and classified the textile industry as one of the supplier-dominated sectors.

Supplier-dominated enterprises are typically characterized by their small-scale and constrained internal capabilities in research and development (R&D) and engineering. These companies contribute minimally to the development of their processes or product technology. Giannini et al. (2019) *states that apparatus and material suppliers are the primary originators of innovations.*

To remain competitive in a volatile market and diversify their product offerings, businesses are compelled to implement and allocate resources towards new technologies as a result of evolving market conditions, technological advancements, and the expanding impact of globalization (Haider & Tehseen, 2022). The precise economic ramifications of innovation in developing countries remain ambiguous and require a comprehensive understanding of their economic complexities (Valdez-Juárez & Castillo-Vergara, 2021). The relationship between innovation and productivity is investigated using the CDM model (Ramírez et al., 2020). Several scholarly investigations make a distinction between process innovation and product innovation. Additionally, certain researchers recognize the existence of other types of innovation, including incremental and radical innovation (Gui et al., 2022). Hence, for an organization to be deemed innovative, its products or services must be, at minimum, novel about the organization's operations. The minimal requirement is implemented to motivate the organization to adopt innovative practices. Companies may be incentivized to innovate, which may start the economy's innovation adoption process. The occurrence of innovation diffusion within the economy serves to stimulate economic activity, resulting in heightened levels of productivity and export performance (Ray & Ray, 2021).

2.2.1 Scope of Innovation

It is possible to categorize innovation as radical or gradual. A radical innovation is defined as a novel or substantial improvement to an established characteristic or service that leads to enhanced performance and cost-effectiveness (Silva, 2020). Businesses face substantial risks when attempting to implement radical innovation as a result of the heightened intricacy involved in commercializing such products. Nevertheless, the organization cannot survive without radical innovation in process, product, or service, as it involves the development and deployment of novel technologies. One crucial characteristic of radical innovation is the extent to which the novel technology surpasses the complexity and sophistication of the existing technology (Ukobitz & Faullant, 2022).

Silva (2020) introduced the notion of fostering radical innovation through the utilization of multiple nodes. A notable center among these is the epicenter of idea generators. Idea generators are individuals who generate novel concepts, while idea hunters are those who actively pursue and execute these concepts. The primary function of idea-gatherers is to receive ideas. They exhibit the requisite capabilities, expertise, discernment, and enthusiasm to tackle these unique concepts effectively. The incorporation of hunters, gatherers, and generators into large organizations is vital to the development of revolutionary innovations. Radical innovation is pursued by organizations for two fundamental reasons. Research conducted by Bhardwaj (2020) demonstrates that competitors also have a higher capacity to design and manufacture exceptionally innovative products. Product modification is what (Pacheco & Paul, 2023) refers to as incremental innovation; it is also referred to as line extension or market draw innovation. Innovation on an incremental basis does not require a significant departure from the current organization. This type of innovation enhances the competencies and knowledge of the personnel within the organization. The implementation of incremental innovation is of paramount importance for organizations due to its capacity to bolster market share and sustain a sustained industry presence (Chaubey & Sahoo, 2022).

2.2.2 Innovative Performance

Instead of being regarded as an independent concept, innovative performance is aggregated into a set of indicators that assess the capacity of an organization to innovate. A shared standard is generally utilized by multiple approaches to measuring innovation. This criterion prioritize the development of original products, procedures, or business models that are a direct consequence of effective organizational processes and activities that foster innovation (Teng et al., 2023). Therefore, the capacity for innovation is pivotal in generating original or improved outcomes, which ultimately determine the degree of innovation attained (Ferreira et al., 2020). Innovation performance indicators serve to demonstrate innovation capability. Innovation is vital to the success and survival of businesses. Innovation contributes to the resolution of social discord and economic downturns, propels economic growth and development, and transforms scientific and technological domains. The economy, the environment, and society all benefit from it overall. Analysis of the variables that influence innovation performance has been a managerial and academic priority.

To remain competitive and endure in rapidly evolving global markets, organizations must now prioritize innovation (Akter et al., 2022). It is widely recognized among experts and professionals in the field that exceptional innovation performance is primarily determined by two factors: the acquisition of specialized knowledge within an organization and its capacity to efficiently manage its current knowledge resources (Masoka, 2019). In light of recent developments in economic growth theory and the expanding accessibility of data, the critical importance of innovation in sustaining long-term productivity and production growth is underscored. Innovation plays a pivotal role in stimulating economic growth through various mechanisms. These include augmenting productivity, reducing costs, creating superior employment prospects, expanding industry diversity, increasing earnings, refining marketing strategies, and implementing efficient management reorganizations within businesses.

2.2.3 Importance of Innovation and Knowledge Management

Innovation is principally intended to create value for the organization. Innovation is critical to business in the current era of intense competition because it empowers organizations to develop unique products and services. Due to the rapid evolution of consumer preferences and inclinations in both mature and emerging markets, innovation is vital. According to one study, 75% of chief executive officers (CEOs) of organizations experiencing accelerated growth cite their innovative products and processes as their most effective competitive advantage. In any given context, organizations that fail to generate new products and services will be eliminated by competitors, as innovation acts as a catalyst for organizational expansion (Idrees et al., 2023). Given that it has the capacity to grant businesses and organizations a sustainable competitive advantage, innovation is critical. A considerable number of organizations are facing considerable obstacles in maintaining competitiveness within their respective industries due to the rapid pace of change, particularly in technological advancements. In order to effectively respond to swift transformations in manufacturing processes, production lines, and management methodologies, staff and managers must employ ingenuity and innovation (Ogbari et al., 2022).

Innovation within an organization is frequently defined as the ability to produce value, products, services, and concepts. Furthermore, innovation serves as a highly efficacious strategy to foster organizational transformations and progress in terms of outcomes. Organizational innovation is attained through the application of knowledge and abilities to the formulation of concepts and strategies that empower the establishment to execute efficacious measures under specific conditions. Additionally, this process entails inspiring individuals and augmenting their capabilities and aptitudes (Mutonyi et al., 2020). Creative innovation, as well as other forms of innovation, are critical success factors and competitive advantages for robust economies and businesses. At this time, virtually every organization is operating in a highly dynamic business environment due to the rapid pace of technological progress, limited product life cycles, and the process of globalization. A company's ability to survive, compete, and maintain an industry-leading position depends more and more on its innovation and originality, particularly in the technology sector (Rua & Santos, 2022). In a capitalist system, innovation is crucial for sustained economic development, which includes the creation of novel technologies and the enhancement of living standards. The coordination of innovation, as an intricate and diverse undertaking, presents an intrinsic difficulty.

3 Relationship Between Knowledge Management and Innovation

The integration of knowledge components within organizations is the source of innovation (Hock-Doepgen et al., 2021). According to (Guo et al., 2020) concur that knowledge is the essential resource for innovation. In the field of innovation research, a multiplicity of studies (Bai et al., 2020; Rui et al., 2019) have repeatedly demonstrated a strong correlation between an organization's knowledge resources and its innovation performance. Knowledge assets are regarded as invaluable assets in the pursuit of sustaining a competitive edge. However, in order to generate value, knowledge is a static asset that must be transferred and disseminated throughout the organization. A collection of interrelated processes—including generation, transmission, and integration, and application— can effectively direct and control knowledge (Garousi Mokhtarzadedeh et al., 2022; Naowakhoaksorn & Suwannarat, 2021). Acquiring knowledge from internal and external sources, including employees, customers, suppliers, and competitors, constitutes knowledge creation. The converted knowledge is subsequently archived in the knowledge repository of the organization in a format that is functional. Conversely, knowledge transfer pertains to the distribution and interchange of information among members of the organization, covering various levels from departments to personnel.

Knowledge integration encompasses the systematic arrangement, reconfiguration, and interpretation of pre-existing knowledge through the implementation of directives, organizational protocols, and self-sufficient task forces. Creating new, significant knowledge and maximizing knowledge synergy are the objectives. Knowledge application pertains to the pragmatic implementation of strategic and valuable knowledge in order to bolster the capabilities of an organization and attain intended results, including but not limited to modifications in the behavior of recipients, the creation of new products, enhanced innovation and operational excellence, strengthened external relationships, strategic direction adjustments, and adaptations to evolving competitive circumstances (Idrees et al., 2023; Sahoo et al., 2023). This application of knowledge may assist the organization in maintaining a competitive edge. Several researchers (Garousi Mokhtarzadedeh et al., 2022; Naowakhoaksorn & Suwannarat, 2021) contend that there exists a robust correlation between knowledge management capabilities (KMC) and innovation and creativity-related performance outcomes. Habib and Bao (2019) presented empirical findings that lend support to the proposition that organizations that possess the capacity to utilize KMC efficiently experience greater levels of innovation.

An investigation carried out by Liao et al. (2023) across 222 organizations operating in the telecommunications and biotechnology sectors of Spain revealed that the implementation of knowledge management capability positively impacted the entrepreneurial performance and innovation of the companies. The process of generating and implementing new information or modifying existing knowledge, according to (Chawla et al., 2022), ultimately results in the creation of innovative products and services. Research conducted by (Anser et al., 2021; Imran et al., 2022) has demonstrated that organizations can significantly increase their competitive advantage and innovativeness through the discovery and exploitation of knowledge. The function of KM in the innovation process has been the subject of numerous studies. The findings of Migdadi (2022) provide further support for the critical importance of knowledge management in improving the capabilities of knowledge processing.

Taghizadeh et al. (2020) and Tung (2021) have both recently completed studies examining the relationship between KMC and innovation performance. According to research by Taghizadeh et al. (2020) on German companies, information management capability (KMC), which enables the discovery, acquisition, and assimilation of pertinent data from internal or external sources, positively impacts innovation performance. Furthermore, Tung (2021) utilized a longitudinal dataset comprising all U.S. enterprises listed in Standard & Poor's COMPUSTAT database from 1976 to 2004 to analyze the effect of KMC on innovation performance. The result revealed that KMC positively influence innovation performance. This is because KMC enables organizations to modify their knowledge base combinations and integrate new and existing knowledge domains efficiently. Consequently, this results in the development of more valuable inventions.

Moreover, the proliferation of open innovation models has substantially expedited the pace of innovation in recent times (Tirmizi et al., 2020). These concepts are predicated on the understanding that organizations can no longer afford to maintain a limited emphasis on their internal processes and depend solely on their resources and ideas to foster innovation and sustain competitiveness (Shehzad et al., 2022). Hence, this represents a substantial departure from the traditional method of innovation, wherein TI firms predominantly relied on relationships with suppliers and customers to implement incremental enhancements to products and procedures (Riaz et al., 2023). In essence, according to the innovation model, enterprises operating in the textile industry (TI) must establish collaborations with non-TI entities to obtain or adapt novel technological knowledge to their specific needs.

Knowledge management capabilities pertain to the aptitudes that enable an organization to effectively oversee knowledge resources across its business divisions through a variety of procedures, including but not limited to development, transfer, integration, and application (Azeem et al., 2021). Previous research, Gürlek and Cemberci (2020), has established that KMC and innovation performance are highly correlated. Moreover, the meta-analysis performed by Imran et al. (2022) validated the facilitative effect of KMC on the innovativeness of businesses. Due to the prospective benefits it provides, including increased innovation and creativity in the provision of products and services, organizations are motivated to pursue knowledge management (Mohamed, 2019). Knowledge has a significant role in fostering creative thinking and facilitating invention (Füller et al., 2022). Therefore, innovation is seen as the domain with the most potential benefit from KM (Ngoc Thang & Anh Tuan, 2020).

3.1 *Effects of Knowledge Management on Innovation*

Exploration, testing, and development of novel technologies, products and services, manufacturing processes, and organizational structures are examples of innovative endeavors. The pragmatic execution of concepts constitutes innovation (Füller et al., 2022). According to the research (Arranz et al., 2019), innovation can be described as a novel administrative system or structure, an internally acquired component, a policy, an innovative manufacturing method, or a new product or service introduced to an organization. The innovation process is incredibly dependent on information, particularly tacit knowledge (Acevedo & Diaz-Molina, 2023; Marco-Lajara et al., 2023). By transforming general knowledge into specific knowledge, new and practical information is generated that can be applied to the development of products, services, and procedures (Weck et al., 2022). The research conducted by Ng et al. (2020), on knowledge production, underscores the criticality of knowledge for competitiveness and innovation. By facilitating quicker access to and awareness of new information, an innovative KM system is thought to facilitate the innovation process (Papa et al., 2022). In addition, one of the most important factors in the success of launching new products is the execution of effective information management. Knowledge management is frequently regarded as a substantial determinant of innovation.

3.2 Innovation Effects on Textile Firms Performance

The significance of innovation in enabling organizations to create value and sustain a competitive advantage in the dynamic and complex business environment is widely recognized (Azeem et al., 2021). In general, innovation not only increases the organization's potential value and efficiency while optimizing the use of existing resources but also provides the establishment with new intangible assets. Organizations with a greater propensity for innovation are more likely to successfully satisfy customer needs and improve financial performance and success through the development of novel competencies (Vargo et al., 2020). In the era of time-based competition, contemporary enterprises have come to recognize that the swift response of their competitors to the launch of new products presents a substantial obstacle to overall success. As a result, their objective is to accelerate the implementation of innovative products, services, or procedures (Shin & Pérez-Nordtvedt, 2020).

In general, three distinct approaches to innovative firm can be identified:

- Initially, innovation is regarded as an essential element for the growth and enhanced financial performance of the organization. The principal focus of this approach is the formulation of a business strategy that revolves around innovation and financial investments to bolster the company's ability to generate new products.
- The second approach posits that innovation is an outcome of dynamic organizational development and promotes decision-making exclusively based on innovation management, which includes establishing work conditions and priorities.
- The third approach acknowledges innovation as a critical component.

4 Conclusion

Despite the considerable importance of the TI in numerous countries, there is a scarcity of scholarly investigations concerning the knowledge management capabilities and innovation performance of businesses operating in this industry. The suppliers-dominated' innovation paradigm (Zhang et al., 2023) defines the TI as one in which businesses significantly rely on the acquisition of external knowledge for innovation. The essential factors that contribute to effective information dissemination within an organization are critical to fortifying the organization's knowledge culture as a whole. Firms must incorporate the innovation determinants, which are the fundamental elements that propel innovation, to accelerate the innovation process. The exchange of knowledge is vital to promoting innovation. Collaboration can take place either within or outside of an organization. Through internal cooperation, organizations gain access to a vast array of tacit and explicit knowledge that exists within the organization. By strengthening internal engagement with the workforce, organizations gain knowledge of what, where, and to what extent information is stored within the organization. The potential for innovation to be stimulated within an organization exists through internal collaboration.

This is because nurturing integration and cooperation among internal employees may provide a reservoir of creativity and knowledge, both of which are essential for implementing innovation in the workplace. External entities, including competitors (Medase & Abdul-Basit, 2020) and customers (Aliasghar et al., 2022), are important contributors to external cooperation's innovation-promoting effect. At this time, establishing relationships with external organizations is crucial for acquiring the knowledge and abilities necessary for innovation (Wang et al., 2020). Collaboration with external entities and strategic collaborators is an essential approach to cultivating innovation. For any organization to innovate effectively in a highly complex environment, (Srisathan et al., 2023) asserts that an open and adaptable structure is required to facilitate the innovation process. Organizations must integrate external expertise and capabilities. By fostering integration and cooperation, the organization can amass an increased quantity of novel ideas, which are fundamental to the process of innovation.

Further, for an organization to prosper in a dynamic and rapidly evolving business environment, it is critical that it place a high value on ongoing education. This is imperative due to the rapid evolution of client preferences and demands, which necessitates that organizations proactively pursue and obtain new competencies and knowledge to meet these changing standards. The integration of these innovative methodologies and strategies into the organization is contingent upon the knowledge acquisition process. Firms learning enhances the capacity of the organization to obtain and apply knowledge. Firms learning facilitates knowledge transformation by requiring employees to impart their expertise and experiences to those who require current information in order to close the disparity in knowledge (Haider & Kayani, 2021; Khaksar et al., 2023). Alrowwad et al. (2020), inspires individuals to contribute to the expansion of the organization's knowledge base. Firms learning promotes interaction among employees, which in turn facilitates the dissemination, integration, and exchange of knowledge. Yeboah (2023) posits that organizational learning facilitates the accumulation of knowledge and improves the quantity and quality of information within a dynamic environment. Learning within an organization will increase its capacity to generate and apply new information.

According to this study, innovation is an indispensable factor in bolstering the operational effectiveness of textile companies. It enhances the organization's capacity to operate effectively. Therefore, innovation possesses the capacity to enhance the efficiency of businesses. It is of the utmost importance to assist businesses with employee training and research and development (R&D) expenditures, as doing so has a direct positive impact on output. In the textiles sector, optimal performance is more probable among personnel who have undergone comprehensive training and possess a profound capacity for information recall. Employees who participated actively in information sharing at work had enhanced creative thinking abilities, which in turn motivated them to develop novel products and services (Jorfi & Heidari, 2017). For this reason, we suggest that organizations establish an appropriate learning environment and provide staff with continuous learning and hands-on training opportunities to facilitate the adoption and enhancement of knowledge management processes.

Furthermore, it is imperative to place greater emphasis on and allocate greater priority to the cultivation of creativity, innovation, and the encouragement of creative endeavors among personnel. This objective can be realized using managers' proactive engagement and the allocation of essential resources, including financial support, organizational frameworks, and other facilities that streamline knowledge management processes. There is a positive correlation between increased information exchange among employees and their greater capacity for innovation. Migdadi (2022) have conducted research that is consistent with these findings. Based on this premise, we propose organizing conferences and symposiums to foster the improvement and enrichment of information sharing as well as facilitate the exchange of experiences, including both success and failure tales, among colleagues. Furthermore, it is critical to implement protocols that provide incentives for employees to participate in the dissemination of knowledge gained from their experiences and expertise to other members of the organization.

References

Abdelmeguid, A., Afy-Shararah, M., & Salonitis, K. (2022). Investigating the challenges of applying the principles of the circular economy in the fashion industry: A systematic review. *Sustainable Production and Consumption, 32*, 505–518.

Acevedo, J., & Diaz-Molina, I. (2023). Learning organizations in emerging economies: The effect of knowledge management on innovative culture in Chilean companies. *The Learning Organization, 30*(1), 37–54.

Akter, S., Babu, M. M., Hossain, M. A., & Hani, U. (2022). Value co-creation on a shared healthcare platform: Impact on service innovation, perceived value and patient welfare. *Journal of Business Research, 140*, 95–106.

Aliasghar, O., Rose, E. L., & Asakawa, K. (2022). Sources of knowledge and process innovation: The moderating role of perceived competitive intensity. *International Business Review, 31*(2), 101920.

Alrowwad, A. A., Abualoush, S. H., & Masa'deh, R. E. (2020). Innovation and intellectual capital as intermediary variables among transformational leadership, transactional leadership, and organizational performance. *Journal of Management Development, 39*(2), 196–222.

Amoako, A. G. (2019). *An assessment of local factors influencing firm performance and facilitating integration into global value chains: A case study of textile and apparel firms in Ghana* [Doctoral dissertation, University of Johannesburg, South Africa].

Anser, M. K., Yousaf, Z., Khan, A., & Usman, M. (2021). Towards innovative work behavior through knowledge management infrastructure capabilities: Mediating role of functional flexibility and knowledge sharing. *European Journal of Innovation Management, 24*(2), 461–480.

Arranz, N., Arroyabe, M. F., Li, J., & de Arroyabe, J. F. (2019). An integrated model of organisational innovation and firm performance: Generation, persistence and complementarity. *Journal of Business Research, 105*, 270–282.

Azeem, M., Ahmed, M., Haider, S., & Sajjad, M. (2021). Expanding competitive advantage through organizational culture, knowledge sharing and organizational innovation. *Technology in Society, 66*, 101635.

Bai, O., Wei, J., Yang, X., & Chen, R. R. (2020). Third-party relational governance and collaborative innovation performance: The role of IPR protection. *International Journal of Innovation Studies, 4*(1), 1–15.

Baporikar, N. (2019). Knowledge management dynamics and public sector organization development. In *Handbook of research on implementing knowledge management strategy in the public sector* (pp. 55–71). IGI Global.

Bhardwaj, B. R. (2020). Influence of knowledge management on product innovation by intrapreneurial firms. *Global Knowledge, Memory and Communication, 69*(1/2), 38–57.

Bratianu, C., Stănescu, D. F., & Mocanu, R. (2023). The mediating role of customer knowledge management on the innovative work behavior and product innovation relationship. *Kybernetes, 52*(11), 5353–5384.

Campos Lima, M. D. P., Martins, D., Costa, A. C., & Velez, A. (2021). Internal devaluation and economic inequality in Portugal: Challenges to industrial relations in times of crisis and recovery. *Transfer: European Review of Labour and Research, 27*(1), 47–73.

Chaubey, A., & Sahoo, C. K. (2022). The drivers of employee creativity and organizational innovation: A dynamic capability view. *Benchmarking: An International Journal, 29*(8), 2417–2449.

Chawla, A. S., Kundu, S. C., Kumar, S., Gahlawat, N., & Kundu, H. (2022). The effect of knowledge management capacity on firm performance through sequential mediations of strategic HRM, administrative and technical innovations. *Journal of Asia Business Studies, 16*(6), 923–942.

Cordeiro, M., Puig, F., & Ruiz-Fernández, L. (2023). Realizing dynamic capabilities and organizational knowledge in effective innovations: The capabilities typological map. *Journal of Knowledge Management, 27*(10), 2581–2603.

Duryan, M., Smyth, H., Roberts, A., Rowlinson, S., & Sherratt, F. (2020). Knowledge transfer for occupational health and safety: Cultivating health and safety learning culture in construction firms. *Accident Analysis & Prevention, 139*, 105496.

Ferreira, J., Coelho, A., & Moutinho, L. (2020). Dynamic capabilities, creativity and innovation capability and their impact on competitive advantage and firm performance: The moderating role of entrepreneurial orientation. *Technovation, 92*, 102061.

Freixanet, J., Monreal, J., & Sánchez-Marín, G. (2021). Family firms' selective learning-by-exporting: Product vs process innovation and the role of technological capabilities. *Multinational Business Review, 29*(2), 210–236.

Füller, J., Hutter, K., Wahl, J., Bilgram, V., & Tekic, Z. (2022). How AI revolutionizes innovation management–Perceptions and implementation preferences of AI-based innovators. *Technological Forecasting and Social Change, 178*, 121598.

Garousi Mokhtarzadedeh, N., Jafarpanah, I., & Zamani Babgohari, A. (2022). Knowledge management capability, entrepreneurial creativity, entrepreneurial intensity and firm performance: The mediating role of ambidexterity. *British Food Journal, 124*(7), 2179–2208.

Giannini, V., Iacobucci, D., & Perugini, F. (2019). Local variety and innovation performance in the EU textile and clothing industry. *Economics of Innovation and New Technology, 28*(8), 841–857.
Gui, L., Lei, H., & Le, P. B. (2022). Determinants of radical and incremental innovation: The influence of transformational leadership, knowledge sharing and knowledge-centered culture. *European Journal of Innovation Management, 25*(5), 1221–1241.
Gujral, G. (2020). Knowledge management: A breakthrough for preserving implicit knowledge in libraries and information centres. *KIIT Journal of Library and Information Management, 7*(2), 111–116.
Guo, J., Guo, B., Zhou, J., & Wu, X. (2020). How does the ambidexterity of technological learning routine affect firm innovation performance within industrial clusters? The moderating effects of knowledge attributes. *Technological Forecasting and Social Change, 155*, 119990.
Gürlek, M., & Cemberci, M. (2020). Understanding the relationships among knowledge-oriented leadership, knowledge management capacity, innovation performance and organizational performance: A serial mediation analysis. *Kybernetes, 49*(11), 2819–2846.
Habib, A., & Bao, Y. (2019). Impact of knowledge management capability and green supply chain management practices on firm performance. *International Journal of Research in Business and Social Science (2147–4478), 8*(6), 240–255.
Haider, S. A., & Kayani, U. N. (2021). The impact of customer knowledge management capability on project performance-mediating role of strategic agility. *Journal of Knowledge Management, 25*(2), 298–312.
Haider, S. A., & Tehseen, S. (2022). Role of decision intelligence in strategic business planning. In *Decision intelligence analytics and the implementation of strategic business management* (pp. 125–133).
Hayaeian, S., Hesarzadeh, R., & Abbaszadeh, M. R. (2022). The impact of knowledge management strategies on the relationship between intellectual capital and innovation: Evidence from SMEs. *Journal of Intellectual Capital, 23*(4), 765–798.
Hock-Doepgen, M., Clauss, T., Kraus, S., & Cheng, C. F. (2021). Knowledge management capabilities and organizational risk-taking for business model innovation in SMEs. *Journal of Business Research, 130*, 683–697.
Hossian, M. S., Kabir, R., & Latifee, E. H. (2019). Export competitiveness of Bangladesh readymade garments sector: Challenges and prospects. *International Journal of Research in Business and Social Science (2147–4478), 8*(3), 45–63.
Idrees, H., Xu, J., Haider, S. A., & Tehseen, S. (2023). A systematic review of knowledge management and new product development projects: Trends, issues, and challenges. *Journal of Innovation & Knowledge, 8*(2), 100350.
Imran, M. K., Fatima, T., Sarwar, A., & Amin, S. (2022). Knowledge management capabilities and organizational outcomes: Contemporary literature and future directions. *Kybernetes, 51*(9), 2814–2832.
Jie, W., Poulova, P., Haider, S. A., & Sham, R. B. (2022). Impact of internet usage on consumer impulsive buying behavior of agriculture products: Moderating role of personality traits and emotional intelligence. *Frontiers in Psychology, 13*, 951103.
Johnson, J. E., & Haug, P. (2021). Modifications to global supply chain management strategies resulting from recent trade disruptions: An exploratory study. *Journal of Global Operations and Strategic Sourcing, 14*(4), 701–722.
Jordão, R. V. D., Novas, J., & Gupta, V. (2020). The role of knowledge-based networks in the intellectual capital and organizational performance of small and medium-sized enterprises. *Kybernetes, 49*(1), 116–140.
Jorfi, S., & Heidari, S. (2017). The relationship between knowledge management and organizational innovation (Case study: Ahwaz Industrial Estates Company). *International Journal of Business Management, 2*(2), 96–108.
Khaksar, S. M. S., Chu, M. T., Rozario, S., & Slade, B. (2023). Knowledge-based dynamic capabilities and knowledge worker productivity in professional service firms The moderating role of organisational culture. *Knowledge Management Research & Practice, 21*(2), 241–258.

Liao, S., Wei, J., & Hu, Q. (2023). Politics or markets: The dual role of the motivation to achieve organizational legitimacy in the development of knowledge management capabilities and business model innovation. *Frontiers in Psychology, 14*, 1112240.
Marco-Lajara, B., Úbeda-García, M., Zaragoza-Sáez, P., & Manresa-Marhuenda, E. (2023). The impact of international experience on firm economic performance. The double mediating effect of green knowledge acquisition & eco-innovation. *Journal of Business Research, 157*, 113602.
Mariev, O., Davidson, N., Nagieva, K., & Pushkarev, A. (2023). Innovations and their complementarities: Which types of innovations drive export performance? *Post-Communist Economies*, 1–36.
Masoka, T. R. (2019). *Knowledge in modern organisations: Establishing the link between knowledge management, organisational learning, environmental dynamism and innovation* [Doctoral dissertation, University of the Witwatersrand, Faculty of Commerce, Law and Management].
Medase, S. K., & Abdul-Basit, S. (2020). External knowledge modes and firm-level innovation performance: Empirical evidence from sub-Saharan Africa. *Journal of Innovation & Knowledge, 5*(2), 81–95.
Migdadi, M. M. (2022). Knowledge management processes, innovation capability and organizational performance. *International Journal of Productivity and Performance Management, 71*(1), 182–210.
Mohamed, E. E. (2019). Knowledge management: An integrative approach literature review. *International Journal of Knowledge Management Studies, 10*(2), 175–204.
Mutonyi, B. R., Slåtten, T., & Lien, G. (2020). Empowering leadership, work group cohesiveness, individual learning orientation and individual innovative behaviour in the public sector: Empirical evidence from Norway. *International Journal of Public Leadership, 16*(2), 175–197.
Naowakhoaksorn, P., & Suwannarat, P. (2021). *The influences of knowledge management capability and knowledge-oriented leadership on public organizational innovativeness: An empirical study from the tax administrative organizations in Thailand* [Doctoral dissertation, Mahasarakham University].
Ndemezo, E., & Kayitana, C. (2020). *Innovation and firms' performance in the Rwandese manufacturing industry: A firm-level empirical analysis* (pp. 99–114). Springer Singapore.
Ndzana, M., Cyrille, O., Mvogo, G., & Bedzeme, T. (2021). Innovation and small and medium enterprises' performance in Cameroon. *Journal of Small Business and Enterprise Development, 28*(5), 724–743.
Ng, H. S., Kee, D. M. H., & Ramayah, T. (2020). Examining the mediating role of innovativeness in the link between core competencies and SME performance. *Journal of Small Business and Enterprise Development, 27*(1), 103–129.
Ngoc Thang, N., & Anh Tuan, P. (2020). Knowledge acquisition, knowledge management strategy and innovation: An empirical study of Vietnamese firms. *Cogent Business & Management, 7*(1), 1786314.
Obeso, M., Hernández-Linares, R., López-Fernández, M. C., & Serrano-Bedia, A. M. (2020). Knowledge management processes and organizational performance: The mediating role of organizational learning. *Journal of Knowledge Management, 24*(8), 1859–1880.
Ogbari, M. E., Chima, G. U. K., Olarewaju, F. O., Arasomwan, E. J., & Nwunde, C. (2022). Innovative leadership styles: Innovative entrepreneurial leadership and sustainable economy-a case study of Asia, Europe and America. *World Review of Entrepreneurship, Management and Sustainable Development, 18*(5–6), 656–674.
Pacheco, C., & Paul, B. (2023). Applying complexity theory perspective to knowledge management in the innovation context. *VINE Journal of Information and Knowledge Management Systems*.
Papa, A., Mazzucchelli, A., Ballestra, L. V., & Usai, A. (2022). The open innovation journey along heterogeneous modes of knowledge-intensive marketing collaborations: A cross-sectional study of innovative firms in Europe. *International Marketing Review, 39*(3), 602–625.
Prakash, Y., Charwak, B., & Kumar, P. V. (2020). Textile industry in new India: Challenges and opportunities. *International Journal of Indian Culture and Business Management, 21*(4), 435–458.

Ramírez, S., Gallego, J., & Tamayo, M. (2020). Human capital, innovation and productivity in Colombian enterprises: A structural approach using instrumental variables. *Economics of Innovation and New Technology, 29*(6), 625–642.

Ray, S., & Ray, P. K. (2021). Innovation strategy of latecomer firms under tight appropriability regimes: The Indian pharmaceuticals industry. *Journal of International Management, 27*(1), 100820.

Riaz, M., Jie, W., Ali, Z., Sherani, M., & Yutong, L. (2023). Do knowledge-oriented leadership and knowledge management capabilities help firms to stimulate ambidextrous innovation: Moderating role of technological turbulence. *European Journal of Innovation Management.*

Rua, O. L., & Santos, C. (2022). Linking brand and competitive advantage: The mediating effect of positioning and market orientation. *European Research on Management and Business Economics, 28*(2), 100194.

Rui, Z., & Lyytinen, K. (2019). How do ventures become more innovative? The effect of external search and ambidextrous knowledge integration. *European Journal of Innovation Management, 22*(5), 845–865.

Sahoo, S., Kumar, A., & Upadhyay, A. (2023). How do green knowledge management and green technology innovation impact corporate environmental performance? Understanding the role of green knowledge acquisition. *Business Strategy and the Environment, 32*(1), 551–569.

Shehzad, M. U., Zhang, J., Alam, S., & Cao, Z. (2022). Determining the role of sources of knowledge and IT resources for stimulating firm innovation capability: A PLS-SEM approach. *Business Process Management Journal, 28*(4), 905–935.

Shin, K., & Pérez-Nordtvedt, L. (2020). Knowledge acquisition efficiency, strategic renewal frequency and firm performance in high velocity environments. *Journal of Knowledge Management, 24*(9), 2035–2055.

Silva, T. F. P. D. (2020). *Innovation, creativity and new product development: A human central design case study* [Doctoral dissertation].

Srisathan, W. A., Ketkaew, C., & Naruetharadhol, P. (2023). Assessing the effectiveness of open innovation implementation strategies in the promotion of ambidextrous innovation in Thai small and medium-sized enterprises. *Journal of Innovation & Knowledge, 8*(4), 100418.

Taghizadeh, S. K., Karini, A., Nadarajah, G., & Nikbin, D. (2020). Knowledge management capability, environmental dynamism and innovation strategy in Malaysian firms. *Management Decision, 59*(6), 1386–1405.

Teng, Z. L., Guo, C., Zhao, Q., & Mubarik, M. S. (2023). Antecedents of green process innovation adoption: An AHP analysis of China's gas sector. *Resources Policy, 85*, 103959.

Tian, H., Dogbe, C. S. K., Pomegbe, W. W. K., Sarsah, S. A., & Otoo, C. O. A. (2021). Organizational learning ambidexterity and openness, as determinants of SMEs' innovation performance. *European Journal of Innovation Management, 24*(2), 414–438.

Tirmizi, S. M. A., Malik, Q. A., & Hussain, S. S. (2020). Invention and open innovation processes, and linkages: A conceptual framework. *Journal of Open Innovation: Technology, Market, and Complexity, 6*(4), 159.

Tung, F. (2021). Do lenders still monitor?: Leveraged lending and the search for covenants. *Journal of Corporation Law, 47*, 153.

Ukobitz, D. V., & Faullant, R. (2022). The relative impact of isomorphic pressures on the adoption of radical technology: Evidence from 3D printing. *Technovation, 113*, 102418.

Valdez-Juárez, L. E., & Castillo-Vergara, M. (2021). Technological capabilities, open innovation, and eco-innovation: Dynamic capabilities to increase corporate performance of SMEs. *Journal of Open Innovation: Technology, Market, and Complexity, 7*(1), 8.

Vargo, S. L., Akaka, M. A., & Wieland, H. (2020). Rethinking the process of diffusion in innovation: A service-ecosystems and institutional perspective. *Journal of Business Research, 116*, 526–534.

Wang, C., Chin, T., & Lin, J. H. (2020). Openness and firm innovation performance: The moderating effect of ambidextrous knowledge search strategy. *Journal of Knowledge Management, 24*(2), 301–323.

Weck, M., Humala, I., Tamminen, P., & Ferreira, F. A. (2022). Knowledge management visualisation in regional innovation system collaborative decision-making. *Management Decision, 60*(4), 1017–1038.

Xie, X., Khan, S., Rehman, S., Naz, S., Haider, S. A., & Kayani, U. N. (2023). Ameliorating sustainable business performance through green constructs: A case of manufacturing industry. In *Environment, development and sustainability* (pp. 1–33).

Yeboah, A. (2023). Knowledge sharing in organization: A systematic review. *Cogent Business & Management, 10*(1), 2195027.

Zhang, X., Chu, Z., Ren, L., & Xing, J. (2023). Open innovation and sustainable competitive advantage: The role of organizational learning. *Technological Forecasting and Social Change, 186*, 122114.

Syed Arslan Haider is a Ph.D. scholar in the Department of Management at Sunway University Business School, Malaysia. He received his B.S. in Computer Science and Master in Project Management from the Capital University of Science and Technology, Pakistan. He is active in research in the areas of Knowledge management, Innovation, Leadership, Organizational Culture and Project Complexity. His research appeared in good journals such as Journal of Knowledge Management, Journal of Innovation & Knowledge, European Journal of Innovation Management, International Journal of Business Communication and many others Scopus and ABDC indexed journals. He is a Project Manager of Soulmate Construction Company in Pakistan.

Sonia Tabassum Anees received her B.S. (BBA) in the Department of Management at Government Sadiq college women university Bahawalpur. Her research interests are Leadership, Knowledge Management, Human Resources, and Innovation. Her research appeared in good journal such as Imperial Journal of Interdisciplinary Research.

Innovation, the Clothing Industry, and Textile Manufacturing in Portugal: An Exploratory Analysis of the Sector Between 2003 and 2022

Ricardo de Moraes e Soares

Abstract The study examines the average technical efficiency of the textile and clothing industry in Portugal in terms of the relationship between innovation and development (I&D) spending and return on assets. Data envelopment analysis (DEA) methodology was used for this purpose. The technical efficiency levels of the different sub-sectors were calculated for the period 2003–2022. The data suggests that there are significant variations in efficiency between the various sub-sectors, with the textile sector showing better levels of technical efficiency compared to the clothing sector in terms of the efficiency between the ratio of I&D spending and the return on assets. The results also highlight the need to invest in I&D in order to boost the industry's overall efficiency. On the other hand, they highlight the importance of developing a closer relationship between production and innovation in order to optimise the allocation of resources to innovation and achieve better levels of efficiency. The research corroborates the conclusions of existing studies, emphasising the need to invest in I&D in order to boost industry efficiency. The study points to the need to define strategies aimed at increasing the efficiency of economic activities and promoting a strategy based on competitiveness, adaptability, and greater efficiency in the textile and clothing industry in Portugal.

1 Introduction

In Portugal, the problem of the manufacturing sector, specifically the textile and clothing sub-sectors, has been discussed at length (Gonçalves & Caraça, 1986). However, according to the doctrine, Portugal is a part of a group of countries with late industrialization and peripheral production activities (Fortuna, 1993). In fact, this characteristic is a direct result of the development model adopted, which led the country to adopt a set of policies and experiences from other countries, such as Japan

R. de Moraes e Soares (✉)
Department of Finance, Lisbon Accounting and Business School, Av. Miguel Bombarda 20, 1069-035 Lisboa, Portugal
e-mail: rmsoares@iscal.ipl.pt

Lisbon Accounting and Business School, Polytechnic Institute of Lisbon (ed.), *Industry and Innovation: Textile Industry*, SDGs and Textiles,
https://doi.org/10.1007/978-3-031-57804-5_8

and Germany, in order to design its own national development policies (Marques, 2023).

Without going into detail, the countries that adopted this type of strategy gave rise to the so-called "*economic miracles*", which, individually, proved to be short-lived since the foundations that supported the models changed over time (Gonçalves & Caraça, 1986). However, it must be recognised, as happened in other countries (Nassif, 2023), that Portuguese industrial development, based on the adoption of international models, led to accelerated rates of industrial development, especially between the 1970s and the 1990s.

In the case of Portugal, it is possible to name two factors that supported the development strategy of the manufacturing industry, dating back to the 1950s. The first is the Industrial Conditioning Law, which, together with the colonial empire, made a real industrial revolution in the textile market possible through the possibility of overseas industries being able to compete directly with those in the metropolis (Neto, 1991). The second and even more important factor was the development of the technological component. The Portuguese textile industry's development process was based, above all, on the intensive use of energy. This, in turn, proved to be an unsustainable source as soon as the sector suffered high-cost increases (CIP, 2017), forcing companies to modernise and innovate technologically.

Currently, development dynamics are particularly related to production capacities and existing technologies, as they are fundamental to the competitiveness of manufacturing companies since they interfere with the modernization of organisations, the type of equipment, and the ability to process raw materials and respond to markets (ATP, 2019). The competitive performance of the textile and clothing industries depends to a large extent on the technologies adopted, the innovation achieved, and also their suitability to the needs of demand. However, adapting and/or modernising new production technologies almost always involves investing in new equipment and/or support infrastructures to meet needs (ATP, 2019), obtaining intellectual property (WIPO, 2005), and identifying competitive advantages in the market (Proença, 2018).

In addition to the national framework, the industry's relationship with European organisations, especially at the level of the common market, is also an important factor today. At the European level, the textiles and clothing industry currently accounts for around 4.0% of total production and 7.0% of available employment in the manufacturing industry and faces a wide range of challenges. These challenges, which are the same as those faced by the Portuguese sector, relate to changes in the global environment, particularly as a result of the elimination of import quotas in 2005 and the evolution of competitive factors, which are increasingly linked to innovation (Montenegro, 2011), research (Sousa, 2020), employment and production (Ferreira, 2021), quality (Cardoso, 2013), the creation of added value (Pereira, 2021), and gains in organisational skills (Merle, 2023). But in fact, for the European Commission (2003), the most relevant factors are restructuring and modernization.

On the other hand, for the first time in five decades and following the abolition of import quotas, the textile and clothing sector was subject to the same trade and import rules as other economic sectors (Silva, 2013). The changes come after an increase in the available labour force in the sector, which meant that processes had to be carried out to adapt to the new circumstances, which coincided with other structural changes in the international market, the economic situation of states, and technological developments (European Commission, 2003).

While Portuguese companies in the textile and clothing sector currently have the main responsibility for responding to the current challenges, it is also true that it is up to public players to define a legal framework that is favourable to the development of the industry so that it can develop and benefit from opportunities for internal, European, and international competition on a level playing field (European Commission, 2003).

Regardless of the size of the sector, worth 200 billion euros, most of which are small and medium-sized enterprises, the sector plays an essential role in some regions of the European Union and especially in Portugal. In the case of Portugal, the industry has shown a great capacity for innovation, the product of a centuries-old tradition (Nogueira, 2020), empirical knowledge (Santos, 2011), quality, and creativity (Proença et al., 2008).

The aim of this article is to analyse the contribution of innovation to the efficiency of invested capital, regardless of origin, which finances the net assets of the textile and clothing sectors in Portugal between 2003 and 2022.

2 State-of-the-Art

The textiles and clothing sector in Portugal is one of the most important manufacturing industries and plays a strategic role in the economy in terms of employment and the creation of added value.

Traditionally, Portugal has a strong tradition in the sector, and it is one of the main export areas. In fact, Portugal has a long history of textile production, with the application of techniques and processes passed down through several generations. This art has been fundamental to the development of the manufacturing industry, especially in the textile and clothing sectors.

Due to the importance of the sector, the textile industry in Portugal is able to offer a wide range of job opportunities, which makes a significant contribution to the country's economy, producing income through exports and trade on the domestic market. On the other hand, Portuguese companies are known for the quality of their products. It is due to this fact that Portuguese textiles are exported to various parts of the world, mainly to European countries. The factors of quality and innovation have been important elements in increasing the competitiveness of the textile industry.

In terms of innovation, the Portuguese textile industry has endeavoured to obtain new, more advanced technologies and sustainable production practices. However, the challenges of global competition and the need to adapt to changes in consumption patterns have pushed the Portuguese textile industry to reform and adopt new forms of innovation in order to remain competitive.

The efficiency of activities in the non-financial sector, such as the textile and clothing sector, is traditionally measured by the operating return on assets (ROA), the economic return on assets (REA), and the net return on assets (RLA). According to financial theory, the operating return on assets measures the efficiency of the use of assets (Marques, 2013). For this purpose, the indicator is obtained by applying the following expression:

$$ROA = \frac{Operating\ profit}{Total\ assets} * 100$$

The economic return on assets (REA) reflects the comparison between earnings before interest and taxes and total assets and aims to analyse the level of return on invested capital, regardless of the origin of the capital that finances the investment of funds (Breia et al., 2014). The economic return on assets ratio is measured as follows:

$$REA = \frac{Earnings\ before\ interest\ and\ taxes}{Total\ assets} * 100$$

The net return on assets (RLA), in turn, measures the efficiency of the application of assets and the level of net return they produce (Leite, 2020). It is calculated as follows:

$$RLA = \frac{Net\ profit\ for\ the\ period}{Total\ assets} * 100$$

Several authors point to ROA, REA, and RLP as fundamental indicators for analysing the efficiency of invested capital. Other authors refer to the importance of investment and development spending in investments (Carcangu, 2016), in results and in the creation of added value (Martins, 2012), and, above all, in terms of asset efficiency (Ferreira, 2021). Thus, the efficiency of research and development spending (EGI&D) can be noted using the following expression:

$$EGI\&D = \frac{Research\ and\ development\ spending}{Gross\ value\ added} * 100$$

Jajri and Ismail (2006) desired to examine the trend of technical efficiency, technological change, and growth in the Malaysian manufacturing sector. The analysis is based on data collected by the Malaysian Department of Statistics, and data envelopment analysis (DEA) was applied. The results suggest that the sectors with the highest technical efficiency are food products, wood, chemicals, and iron products. The authors conclude that technical efficiency in the food, wood, and textile sectors is low.

Khalil (2011) sought to measure the relative efficiency of the most water-polluting sector in Pakistan (textile processing). Data envelopment analysis was applied to estimate the relative efficiency of decision-making units that use various factors of production to produce desirable and undesirable results. The results suggest that the environmental efficiency of all production units is low, which makes it essential to develop new, more effective measures and instruments to control the increasing levels of pollution in water resources discharged by the textile processing industry.

In the study by Rakhmawan et al. (2012), they aim to analyse efficiency and productivity as performance indicators and how efficiency affects the productivity of the textile industry. For the analysis of efficiency and productivity, the methodology of data envelopment analysis was adopted, and then linear regression was used to analyse the effects of efficiency on imports and productivity in the textile industry. The results suggest that the textile industry's average level of efficiency is 40%, with an average productivity growth rate of 2.40%, and indicate that increasing efficiency has a clear positive effect on the sector's productivity.

Another study by Fang et al. (2013) looked at the technical efficiency of the Chinese manufacturing industry and analysed how environmental variables, measured through energy consumption, affect the performance of different industrial sectors under the constraint of the low-carbon economy. In order to measure efficiency, the authors applied the DEA model in three stages. The results suggest that the textile sector has a higher technical efficiency and, when environmental variables are not taken into account, the efficiencies in the machinery and electronics manufacturing industries have a significant increase in their efficiency.

Pérez et al. (2017) present an analysis of energy efficiency and greenhouse gas emissions in the Chilean manufacturing industry, by region and sector. To this end, data envelopment analysis models were applied, and three ways of analysing unwanted outputs were compared. The source of inefficiency for each decision-making unit (DMU) was calculated using scale efficiency, and its evolution was analysed using the Malmquist index. The authors concluded that the industries in the Chilean regions of Coquimbo, La Araucania, and Aysen were the most efficient, while the industries in the regions of Tarapaca, Antofagasta, and Biobio were the least efficient.

Li and Tsai (2018) published a study on the efficiency of the distribution relationship between the clothing industries, from the manufacture of fibres and fabrics to clothing and ultimately to the end consumers (known as the textile/clothing pipeline). The authors conclude that performance depends above all on the efficiency of costing and retailing.

Fahmy-Abdullah et al. (2018) analysed the level of technical efficiency of the textile industry in Malaysia. To this end, the authors defined training expenses, educational level, cost of capital, company size, wage level, innovation and technology expenses, and research and development expenses as the variables to be analysed. A total of 1,010 companies were studied, and stochastic frontier analysis was used. The results obtained suggest that the overall level of technical efficiency of the companies is high and conclude that the textile industry needs to develop a better approach to high-tech production through a closer relationship between production, research, level of education, motivation, and level of rewards.

Sueyoshi and Goto (2019), meanwhile, seek to develop the best way to improve the sustainability of companies by comparing operational and environmental results.

The authors therefore propose applying data-envelopment analysis to measure environmental results. The results suggest that Japanese industry has a high level of efficiency in resource allocation and energy utilisation. Among the high-performing industries, only the pharmaceutical industry has a high performance in terms of environmental protection. With regard to the pulp and paper industry and the textile products industry, the authors conclude that the sectors have the potential to improve efficiency by investing in new technologies.

Sudri et al. (2019) argue that the efficiency of companies can be maximised using data envelopment analysis. With the aim of achieving maximum efficiency, with the relative efficiency constraints of all units equal to one, the results suggest that there are 5 DMUs that are less efficient, while 7 DMUs are efficient (efficiency equal to one). For the authors, the best way to improve the efficiency of the production process is by adding or reducing operators, reducing working hours (by increasing the number of machines), evaluating unproductive machines, and adjusting raw material requirements.

Another relevant study was carried out by Popker and Raju (2020). The authors sought to assess the financial performance and efficiency of 15 textile companies. The Z-Score and data envelopment analysis (DEA) methodologies were applied in the analysis. The results suggest that the financial performance of the companies, according to the Z-Score, is indicated as very good, while the DEA analysis suggests that four companies are efficient and that the others are not efficient.

Du et al. (2022) analysed the weight of innovation in the manufacturing industry of Wuhan city using DEA-BCC models, the DEA-Malmquist index, and the location quotient method. In order to analyse the level of innovation, they defined a performance evaluation index system comprising two input indicators: innovation and development expenditure and personnel related to innovation and development, and two output indicators: revenue from sales of new products and the number of patent applications. The results obtained suggest that the level of innovation and the average value of technological evolution are low. Technical efficiency is low, but it is not the main factor affecting the level of innovation in Wuhan's textile industry. On the other hand, the level of innovation and technology in the cities of Xianning, Xiantao, and Tianmen is higher. The authors conclude that increasing the level of investment in innovation in the manufacturing industry can improve innovative efficiency.

Darji and Dahiya (2023) wanted to investigate the operational efficiency of textile production units in Haryana, India. To this end, the input-oriented data envelopment approach (DEA) was applied. The findings suggest that the textile units had remarkably low operational performance and that, in order to increase efficiency, companies should reduce the size of their production factors.

This study aims to analyse the contribution of innovation to the efficiency of the textile and clothing sectors in Portugal, taking into account the doctrine produced and the statistical data available.

3 Methodology

This research is based on data from the financial statements of entities in the textile and clothing sector with commercial headquarters in Portugal.

The data collected was sorted into two sectors of economic activity (I-textiles and II-clothing) and then into eighteen sub-sectors (I.1-preparation and spinning of textile fibres; I.2-textile weaving; I.3-finishing of textiles; I.4-manufacture of knitted and crocheted fabrics; I.5-manufacture of made-up textile articles, except apparel; I.6-manufacture of carpets and rugs; I.7-manufacture of cordage and netting; I.8-manufacture of nonwovens and articles thereof, except apparel; I.9-manufacture of textiles for technical and industrial use; I.10-manufacture of other textiles; II.1-manufacture of leather clothing; II.2-manufacture of workwear; II.3-manufacture of other outerwear; II.4-manufacture of underwear; II.5-manufacture of other articles and clothing accessories; II.6-manufacture of fur articles; II.7-manufacture of knitted and crocheted hosiery; and II.8-manufacture of other knitted and crocheted clothing). DEA analysis was applied to each of the sub-sectors and the output and input variables were calculated and used.

The information on the financial statements was obtained from the INE (National Statistics Institute) and Banco de Portugal databases. The data collected covers the period from 2003 to 2022.

The efficiency of the textile and clothing sectors was analysed using data envelopment analysis (DEA) in order to identify the most efficient sub-sectors. Each of the economic sub-sectors is presented as a process with its inputs and outputs, and a pivotal entity was defined, called a DMU (Decision-Making Unit). A non-parametric approach was used, and no explicit relationship was established between inputs and outputs. The ability to transform inputs into outputs was measured as an efficiency index, with the economic subsector being more efficient in a given period of time, the more outputs it obtains (results) with the fewest possible inputs.

Efficiency, in the DEA model, varies between 0 (not efficient) and 1 (efficient) and is achieved through the distance between the inefficient DMU and the efficient DMU, which translates into input surpluses or output deficits. This article uses input-oriented models, and DMU efficiency can be described as follows:

$$DMU\ efficient_{input} = \frac{\sum_{r=1}^{s} \mathrm{U_r Y_r}}{\sum_{i=1}^{m} \mathrm{V}_i \mathrm{X}_i}$$

where U_r = weight of output r; Y_r = output r; V_i = weight of input i; and X_i = input i.

According to the model, a DMU is efficient when the set of n DMU's ($\mathrm{DMU_j}$, $\mathrm{j} = 1, 2, \ldots, \mathrm{n}$) produces s outputs using m inputs. Let $x_j = \{x_{ij}, i = 1, 2, \ldots, m\}$ be the input vector, and $y_j = \{y_{rj}, r = 1, 2, \ldots, s\}$ be the output vector of DMU_j. In fact, the data set is described using the input matrix $X = (x_{ij}, i = 1, 2, \ldots, m;\ j = 1, 2, \ldots, n)$, and the output matrix $Y = (y_{rj}, r = 1, 2, \ldots, s;\ j = 1, 2, \ldots, n)$.

The methodology adopted is based on assessing the efficiency of $DMU_0, o \in \{1, 2, \ldots, n\}$, and aims to identify the pivot DMU in which the inputs and outputs are defined as a linear combination of the inputs and outputs of the other DMUs, namely $X\lambda$ and $Y\lambda$. For the method adopted, $\lambda = (\lambda_1, \lambda_2, \ldots, \lambda_n), \lambda > 0$ reveals the shares of the contributions of the efficient DMUs to the projections of DMU_0 on the efficiency frontier, with e translating the vector of lines with all elements equal to 1. On the other hand, the pivot DMU must be the most efficient compared to DMU_0. Thus, input-oriented models can generally be deduced using the following linear expression:

$$min\theta; \theta, \lambda$$

Subject to the following restrictions:

$$\theta x_0 - X\lambda \geq 0$$

$$Y\lambda \geq y_0$$

$$\lambda \geq 0$$

$$e\lambda = 1$$

The above restrictions reflect the DEA-CCR and DEA-BCC systems, with their respective limitations. The formulae, in turn, also correspond to the excess inputs and outputs, i.e., the respective vectors:

$$s^- = \theta x_0 - X\lambda; e$$

$$s^+ = Y\lambda - y_0$$

where the efficiency of the DMU (subsectors) is manifested in θ. Therefore, DMU_0 is considered efficient when it is the optimal solution $(\theta^*, \lambda^*, s^{-*}, s^{+*})$, dipping minimization of θ and minimising the sum of s^{-*} and s^{+*} achieve $\theta^* = 1$ and $s^{-*} = 0; s^{+*} = 0$, otherwise the DMU are not efficient. In terms of traditional models, efficiency in the DEA-CCR system is referred to as technical efficiency, while in the DEA-BCC model it is indicated as pure technical efficiency.

The calculations were carried out using Excel software, and the DEA methodology was selected for analysing the efficiency of the textile and clothing industry due to its flexibility in selecting inputs and outputs, as well as the model's orientation, i.e., the research preference is to maintain inputs (level of expenditure on innovation) and increase outputs (results through the efficiency of net fixed applications).

4 Analysis and Discussion

Taking official statistical data and the methodology adopted into account, the technical efficiency levels of each of the sub-sectors being analysed (textiles and clothing) were calculated. Table 1 shows the results obtained.

Table 1 shows that in the textiles sector, DMU I.1, which translates the economic activity of preparing and spinning textile fibres, the level of participation of research and development spending in the efficiency of results is 14.36%. In other words, in 2003, 2006–2010, 2013, and 2019–2021, research and development costs did not contribute to the operating, economic, or net returns on assets. In the 2004, 2005, 2011, 2012, 2016, 2018, and 2022 financial years, economic activity, according to the doctrine, is inefficient, since the results suggest technical efficiency levels of less than 1. According to the methodology, research and development costs contributed little to the efficiency of the application and the respective return on assets. On the other hand, for 2015 and 2017, the results show that research and development spending contributed significantly to the efficiency of economic activity, since the efficiency levels were equal to 1.

With regard to DMU I.2, the textile weaving activity, the results indicate an average efficiency level of 31.22% (inefficient activity). The results suggest that in 2008, 2009, 2011, and 2012, the share of I&D spending in the efficiency of results and the respective return on assets was zero, or rather, the contribution of this spending to the return on assets was zero. Alternatively, in the years 2003 to 2007, 2010, 2013 to 2016, and 2018 to 2022, the results show that DMUs are inefficient. In other words, the relationship between I&D spending and return on assets is not efficient. The results also show that in 2017, I&D spending contributed positively to the activity being technically efficient.

With regard to DMU I.3, which refers to the economic activity of textile finishing, the results point to an average efficiency level of 41.43% (inefficient activity). Table 1 shows that in the financial years 2006, 2008, 2009, 2011, and 2012, the relationship between research and development spending and return on assets (operating, economic, and net) was zero. From 2003 to 2005, 2007, 2010, 2013 to 2019, and 2021 and 2022, the relationship between I&D spending and return on assets is inefficient, as the results suggest an average technical efficiency of less than 1. According to the results, the 2020 period is the one in which the relationship between I&D spending and return on assets cooperated positively to make the activity technically efficient.

With regard to DMU I.4, which concerns the economic sector of knitted fabric manufacturing, the results suggest an average technical efficiency level of 43.45%, which corresponds to an inefficient activity. Between 2003 and 2019, according to the DEA methodology, the sector exhibits an inefficient relationship between I&D spending and return on assets, due to the results being lower than 1. However, the results suggest an efficient relationship between I&D spending and return on assets for the financial years 2021 and 2022. In other words, I&D spending contributed efficiently to the efficiency of operating, economic, and net results.

Table 1 Results of the DEA model in the textile and clothing sector for the period 2003–2022

DMU	Years										
	2003	2004	2005	2006	2007	2008	2009	2010	2011	2012	2013
Textile industry											
I.1	0.00%	1.86%	3.34%	0.00%	0.00%	0.00%	0.00%	0.00%	2.56%	5.36%	0.00%
I.2	14.72%	20.36%	40.64%	29.24%	10.81%	0.00%	0.00%	4.73%	0.00%	0.00%	25.20%
I.3	10.92%	41.03%	77.46%	0.00%	22.84%	0.00%	0.00%	13.68%	0.00%	0.00%	63.19%
I.4	18.67%	15.98%	31.92%	7.29%	35.73%	18.25%	5.70%	18.19%	20.70%	17.44%	55.86%
I.5	51.63%	34.66%	70.03%	84.42%	81.22%	29.77%	40.92%	35.26%	0.00%	0.00%	0.84%
I.6	17.98%	27.81%	62.01%	16.53%	25.85%	0.00%	0.00%	20.12%	0.00%	0.00%	17.71%
I.7	40.58%	52.58%	100.00%	17.17%	71.39%	24.03%	100.00%	49.16%	22.61%	15.40%	35.17%
I.8	34.01%	55.98%	100.00%	9.41%	32.59%	18.40%	10.99%	89.71%	41.25%	65.29%	15.53%
I.9	55.59%	53.66%	100.00%	74.47%	69.24%	44.03%	46.48%	67.90%	56.63%	46.20%	62.57%
I.10	61.46%	54.10%	96.30%	69.99%	62.92%	69.73%	42.46%	40.25%	10.96%	20.19%	58.60%
Average	30.56%	35.80%	68.17%	30.85%	41.26%	20.42%	24.66%	33.90%	15.47%	16.99%	33.47%
	2014	2015	2016	2017	2018	2019	2020	2021	2022	Average DMU	
I.1	23.30%	100.00%	7.49%	100.00%	13.06%	0.00%	0.00%	0.00%	30.24%	14.36%	
I.2	33.72%	34.73%	44.79%	100.00%	60.36%	42.66%	35.40%	75.41%	51.67%	31.22%	
I.3	82.63%	48.79%	71.94%	90.28%	53.50%	55.73%	100.00%	61.61%	34.91%	41.43%	
I.4	54.10%	62.43%	80.01%	79.12%	56.62%	44.20%	44.69%	100.00%	100.00%	43.35%	
I.5	22.10%	40.35%	44.43%	61.55%	45.43%	35.66%	38.12%	100.00%	92.85%	45.46%	
I.6	100.00%	44.83%	30.48%	67.19%	45.72%	76.76%	29.86%	72.04%	100.00%	37.74%	

(continued)

Table 1 (continued)

I.7	42.04%	64.17%	54.51%	43.16%	21.00%	46.63%	100.00%	100.00%	100.00%	54.98%	
I.8	69.81%	100.00%	88.38%	100.00%	81.45%	45.86%	100.00%	100.00%	67.45%	61.31%	
I.9	88.08%	100.00%	97.60%	86.32%	100.00%	97.05%	72.30%	93.88%	84.22%	74.81%	
I.10	58.91%	59.57%	80.44%	100.00%	93.99%	95.99%	64.65%	78.76%	100.00%	65.96%	
Average	57.47%	65.49%	60.01%	82.76%	57.11%	54.05%	58.50%	78.17%	76.13%	–	
DMU	Year										
	2003	2004	2005	2006	2007	2008	2009	2010	2011	2012	2013
Clothing industry											
II.1	0.00%	0.00%	0.00%	0.00%	0.00%	0.00%	0.00%	0.00%	0.00%	0.00%	0.00%
II.2	37.63%	33.02%	33.22%	40.28%	36.08%	79.98%	67.48%	26.34%	45.84%	26.69%	68.06%
II.3	58.93%	53.19%	61.20%	91.77%	55.62%	48.43%	18.99%	8.97%	26.24%	32.11%	71.93%
II.4	27.64%	22.59%	26.67%	25.38%	31.99%	36.88%	0.00%	15.73%	0.59%	0.00%	25.23%
II.5	16.08%	23.89%	21.82%	33.44%	20.34%	0.00%	10.27%	5.97%	6.35%	0.00%	0.00%
II.6	0.00%	19.85%	31.17%	3.13%	0.00%	0.00%	0.00%	4.37%	0.00%	12.17%	49.95%
II.7	0.00%	3.53%	7.24%	0.00%	0.00%	0.00%	100.00%	0.00%	24.79%	8.91%	100.00%
II.8	6.50%	24.61%	31.93%	5.46%	0.00%	0.00%	0.00%	0.00%	32.70%	3.56%	41.30%
Average	18.35%	22.59%	26.66%	24.93%	18.00%	20.66%	24.59%	7.67%	17.06%	10.43%	44.56%
	2014	2015	2016	2017	2018	2019	2020	2021	2022	Average DMU	
II.1	0.00%	15.12%	100.00%	100.00%	76.42%	0.00%	0.00%	100.00%	0.00%	19.58%	
II.2	100.00%	79.59%	76.88%	89.47%	100.00%	73.04%	100.00%	55.17%	66.54%	61.77%	

(continued)

Table 1 (continued)

II.3	88.07%	100.00%	65.84%	86.83%	100.00%	68.81%	52.12%	74.80%	100.00%	63.19%
II.4	27.47%	49.05%	37.26%	53.65%	68.78%	46.25%	42.86%	100.00%	100.00%	36.90%
II.5	23.92%	32.19%	48.06%	51.38%	38.80%	40.35%	45.83%	57.79%	100.00%	28.82%
II.6	100.00%	73.72%	20.01%	11.99%	75.91%	50.81%	44.35%	100.00%	0.00%	29.87%
II.7	23.21%	0.00%	51.29%	9.31%	0.00%	0.00%	0.00%	59.81%	87.34%	23.77%
II.8	60.48%	74.20%	100.00%	78.87%	90.51%	55.56%	2.02%	37.27%	100.00%	37.25%
Average	52.89%	52.98%	62.42%	60.19%	68.80%	41.85%	35.90%	73.11%	69.24%	–

With regard to the economic activity of manufacturing made-up textile articles, except clothing (DMU I.5), we see an average efficiency of 45.46%, which means that the sector is, in the light of the methodology, inefficient in allocating I&D spending in relation to the return provided by the assets. Taking Table 1 into account, it can be seen that the years 2011 and 2012 are periods in which the relationship between the return on assets and I&D spending is zero (spending on research and development by companies in the made-up textile manufacturing sector does not lead to better results). The years 2003 to 2010, 2013 to 2020, and 2022 represent moments in the activity's time series of technical inefficiency, i.e., periods in which the activity exhibits inefficiencies in the allocation of I&D spending in relation to the return on assets obtained ($e\lambda < 1$). According to the estimated results, 2021 is the year in which the ratio of I&D spending to return on assets is most efficient ($e\lambda = 1$).

With regard to the activity of manufacturing carpets and rugs (DMU I.6), the results suggest an average technical efficiency in the allocation of I&D spending of 37.74% (the result thus points to an inefficient sector). For the 2008, 2009, 2011, and 2012 economic periods, we observed zero efficiency, i.e., I&D spending did not contribute to the efficiency of the sector's results. For the financial years 2003–2007, 2010, 2013, and 2015–2021, the carpet and rug manufacturing sector are technically inefficient. Conversely, the 2014 and 2022 economic periods are technically efficient, i.e., I&D spending has contributed favourably to the relationship between return on assets and innovation spending.

With regard to the rope and net manufacturing sector (DMU I.7), the data indicates an average technical efficiency of 54.98% (inefficient activity). For the years 2003, 2004, 2006–2008, 2010–2019, according to the rules of the model, we found an inefficient sector ($e\lambda < 1$). However, the financial years 2005, 2009, 2020, and 2022 are technically efficient, i.e., spending on innovation and development contributed positively to the efficiency and profitability of the assets of the sector ($e\lambda = 1$).

With regard to DMU I.8 (manufacture of nonwovens and articles thereof, except apparel), the average level of technical efficiency for the financial years in question was 61.31%. In other words, the relationship between the sector's I&D spending and the level of return on assets is inefficient. More specifically, the years 2003, 2004, 2006, 2014, 2016, 2018, 2019, and 2022, according to the results, were technically inefficient. On the other hand, the financial years 2005, 2015, 2017, 2020, and 2021 were technically efficient. In other words, the periods indicate an efficient relationship between innovation and development spending and the level of return on assets of companies in the sector.

In DMU I.9 (manufacturing of textiles for technical and industrial use), we examined an average technical efficiency level of 74.81%. Among the various sectors under review, the manufacturing sector of textiles for technical and industrial use is the one that has a higher level of efficiency between the ratio of I&D expenditure and the profitability of assets. In 2003, 2004, 2006, 2014, 2016, 2017, 2019, and 2022, the results suggest technical efficiency less than 1, that is, correspond to inefficient economic years. On the contrary, the years 2005, 2015, and 2018 represent years of technical efficiency.

As regards the manufacturing sector of other textiles (DMU I.10), the results indicate an average efficiency level of 65.96%. In the economic periods 2003–2016 and 2018–2021, I&D expenditure in relation to asset profitability was technically inefficient ($e\lambda < 1$). In turn, the results suggest that the periods of 2017 and 2022 were technically efficient ($e\lambda = 1$).

With regard to the garment industry, it can be noted from Table 1 that the leather garment manufacturing sector (DMU II.1) is the second sector that has a lower level of technical efficiency in the ratio between I&D expenditure and asset profitability (19,58%). The economic periods 2003–2010, 2013, 2019, 2021, and 2021 show zero technical efficiency ($e\lambda = 0$). That is, there is no efficiency relationship between asset returns and expenditure on innovation and development. In the years 2015 and 2018, the data indicate a level of efficiency between I&D expenditure and asset profitability of less than 1 (inefficient activity). In relation to the fiscal years 2016, 2017, and 2021, the results obtained suggest that the ratio between I&D expenditure and asset profitability was technically efficient.

For DMU II.2 (working clothing manufacturing activity), the results point to an average technical efficiency of 61.77%. The periods from 2003 to 2013, 2015 to 2017, 2019, 2021, and 2022 have a relationship between I&D expenditure and inefficient asset profitability. Meanwhile, the results suggest that the years 2014, 2018, and 2020 are periods of efficiency (benchmarking years), that is, spending on I&D competed efficiently for the positive results of asset profitability. The results obtained, according to the methodology adopted, do not suggest the existence of periods with zero efficiency.

As for the construction activity of other outer clothing (DMU II.3), the data point to an average technical efficiency of 63.19%. The results suggest that for the years 2003–2014, 2016, 2017, 2019–2021, the technical efficiency was less than 1 ($e\lambda < 1$), that is, the ratio between the expenditure on innovation and development and the results obtained from the application of the assets in the activity is inefficient. The periods for the years 2015, 2018, and 2022 correspond to the years of average technical efficiency equal to 1 ($e\lambda = 1$), i.e., the ratio between I&D expenditure and the results of the use of assets in the activity is efficient.

With regard to the underwear manufacturing sector (DMU II.4), the results suggest an average technical efficiency of 36.90%. This indicates that the sector, at the level of I&D expenditure in relation to the profitability of assets, exhibits a reduced level of technical efficiency. With regard to the economic periods of 2009 and 2012, we note zero efficiency, while in the years 2003–2008, 2010, 2011, 2013–2020, we observe a technical efficiency between I&D expenditure and asset profitability of less than 1. The data suggest that spending on innovation does not contribute significantly to asset profitability. In the years 2021 and 2022, the results indicate that they represent benchmarking years, as they represent economic periods in which I&D expenditure has converged efficiently to obtain asset profitability.

With regard to the manufacture of other clothing articles and accessories (DMU II.5), it is possible to note that the results point to an average technical efficiency level of 28.82%. The results thus indicate that it is the fourth most inefficient sector in the allocation of I&D expenditure for the purposes of asset profitability in the textile and clothing industry in Portugal. At the level of economic years, the results suggest that the fiscal years 2008, 2012, and 2013 are zero-efficiency years ($e\lambda = 0$). That is, there is no relationship between I&D expenditure and asset returns. The years 2003 to 2007, 2009 to 2011, 2014 to 2021 correspond to periods of technical inefficiency ($e\lambda < 1$). On the other hand, the 2022 financial year translates into a reference period, i.e., the ratio between I&D expenditure and asset profitability, which is a reference period for the rest of the sector.

In the manufacturing sector of fur products, relating to DMU II.6, we analysed a technical efficiency average of 29.87%. With regard to the series under review, the periods of 2003, 2007–2009, 2011, and 2022 indicate that they present years of zero inefficiency. The years 2004 to 2006, 2010, 2012, 2013, and 2015 to 2020 represent technically inefficient economic years (the relationship between innovation and development expenditure and the return on assets is reduced). On the contrary, the data suggest that they translate years of technical efficiency, which reflect benchmarks.

In the activity of manufacturing socks and similar mesh (DMU II.7), we observed that the technical efficiency, on average, of the allocation of I&D expenditure by the sector was 23.77%. In the years 2003, 2006 to 2008, 2010, 2015, 2018 to 2020, we find that they translate periods of zero technical efficiency. For the periods of 2004, 2005, 2011, 2012, 2014, 2016, 2017, 2021, and 2022, the data suggest that they correspond to moments of technical inefficiency ($e\lambda < 1$). In reverse, the economic periods of 2009 and 2013 translate moments of technical efficiency in the allocation of I&D expenditure in relation to the profitability of sector assets ($e\lambda = 1$).

With regard to the manufacturing sector of other mesh clothing (DMU II.8), the results suggest an average level of technical efficiency in the allocation of I&D expenditure for the purposes of asset profitability of 37.25%. Between 2007 and 2010, the data suggests zero efficiency in the sector. For the years 2003–2006, 2011–2015, and 2017–2021, the results point to an inefficiency in the allocation of I&D expenditure in relation to the profitability of the assets applied in the sector. Finally, the 2016 and 2022 periods translate into periods of technical efficiency ($e\lambda = 1$), in which I&D expenditure participates efficiently in asset profitability.

At the level of the textile industry, in the series under review, the results suggest that the period with the highest level of technical efficiency was 2017 (82.76%) and the lowest in 2011 (15.47%).

As for the clothing industry, the results in Table 1 indicate that the period showing the highest level of efficiency in the allocation of I&D spending was 2021 (73.11%), and the lowest efficiency was 2010 (7.67%).

The years of higher technical efficiency reflect periods after the financial crisis (2011–2015) in Portugal, while the years of lower efficiency reflect the period of the start of the Portuguese financial crisis. That is, the results suggest that the textile and clothing industry in Portugal has been able to adapt and understand the importance of I&D spending.

One of the objectives of the DEA methodology, if not the most important, is the elaboration of an order or classification of the average efficiency level of the DMUs in order to present the reference DMU. Table 2 presents the classification of the DMUs under review and their level of technical efficiency.

After the calculation of the relative efficiency of non-financial economic activities, according to the methodology adopted, the results were ordered according to the average efficiency of each of the economic sectors for the period under review (Table 2). The activities embedded in the textile industry prove to be the most efficient, since they show the best results in the relationship between I&D expenditure and asset profitability and have higher levels of ROA, REA, and RLA. On the other hand, the activities that exhibit a higher average technical efficiency are the manufacture of textiles for technical and industrial use (DMU I.9), as well as the manufacturing of other textiles (DMU I.10), followed by the manufacture of other outer clothing (DMU II.3) and the manufacture of working clothing (DMU II.2). Conversely, the activities showing lower levels of average technical efficiency were the manufacture of other articles and accessories for clothing (DMU II.5), the manufacture of socks and the like of mesh (DMU II.7), the manufacture of leather clothes (DMU II.1), and the textile fibre preparation and wiring sector (DMU I.1), as they show low levels of

Table 2 Classification of average efficiency for the textile and clothing industries for the period 2003–2022

Rank	Economic activities	Average (%)
1	I.9- Manufacture of textiles for technical and industrial use	74.81
2	I.10- Manufacture of other textiles	65.96
3	II.3- Confection of other outer clothing	63.19
4	II.2- Working clothing manufacture	61.77
5	I.8- Manufacture of non-woven fabrics and articles thereto, other than clothing	61.31
6	I.7- Manufacture of cords and nets	54.98
7	I.5- Manufacture of confectioned textiles, other than clothing	45.46
8	I.4- Manufacture of mesh fabrics	43.35
9	I.3- Finishing of textiles	41.43
10	I.6- Carpets and carpets manufacturing	37.74
11	II.8- Manufacture of other mesh garments	37.25
12	II.4- Interior clothing decoration	36.90
13	I.2- Textile weaving	31.22
14	II.6- Manufacture of fur products	29.87
15	II.5- Manufacture of other clothing articles and accessories	28.82
16	II.7- Manufacture of socks and similar mesh	23.77
17	II.1- Leather clothing manufacture	19.58
18	I.1- Preparation and wiring of textile fibres	14.36
	Total average efficiency	**42.88**

I&D cost share in obtaining the returns of the assets applied in each of the production sectors. The sectors that presented, according to the methodology, consistent results of technical efficiency throughout the period from 2003 to 2022 were the manufacture of mesh fabrics (DMU I.4), textile finishes (DMU I.3), the manufacture of carpets (DMU I.6), and, furthermore, the manufacturing of other mesh clothing. (DMU II.8). The results of these sectors reflect, despite their constant average technical efficiency, that they are inefficient sectors and are heavily dependent on factors other than innovation, such as the evolution of the market and the respective choices of customers.

The results of the study are in line with the findings of the authors mentioned, mainly because the results obtained provide an insight into the efficiency and performance of the textile and clothing industry in Portugal. It is in this sense that we intend to analyse the results in light of the context and the conclusions of the authors mentioned.

The results obtained show that it is possible to analyse the technical efficiency of the textile and clothing sectors, and, as Jajri and Ismail concluded, at the level of food products, wood, and chemicals, we also find that the technical effectiveness of the Portuguese industry is reduced. In the same way, the results match Khalil's research, which sought to analyse the efficiency of Pakistan's textile processing sector and concluded that the level of efficiency in Pakistan's production units is inefficient and that they lack the adoption of innovative, more efficient instruments.

At the level of the analysis of the efficiency of performance indicators, the results obtained coincide with those of the authors Rakhmawan, Hartono, and Awirya, who sought to analyse the effectiveness of performance indicators and how efficiency affects the productivity of the textile industry. According to the authors, the efficiency of the textile industry is 40.0%, while the results of the Portuguese industry vary between 14.36% and 74.81%, depending on the sub-sector of activity. In another study by Darji and Dahiya on the efficiency of the textile sector in India, the authors conclude that textile units exhibit significantly reduced operational efficiency. The authors' conclusions coincide with the results of the present study, as they suggest that the technical efficiency of the sector is low.

At the level of comparison between sectors, the results obtained suggest that the textile industry is technically more efficient than the clothing sector. The results are conducive to the conclusions of Fang, Hong, Li, Song, and Zhang, who conducted an analysis between the textile and machinery manufacturing and electronics industries. Like the authors, the results of the study also suggest that the textile industry is more efficient than the clothing industry. Another study, by Li and Tsai, on the efficiency of distribution between the garment industries concludes that efficiency depends on the type of inputs in the system. In the present investigation, the results indicate the existence of DMUs in which the level of efficiency is zero ($e\lambda = 0$), that is, it is not possible to demonstrate any relationship between inputs (I&D expenditure) and outputs (asset profitability).

Sueyoshi and Goto also conducted a comparative study between the textile, paper, and paste industries and the pharmaceutical industry in order to evaluate the efficiency of new technologies and concluded that the drug industry has a high level of efficiency in resource allocation compared to the other sectors. In the Portuguese case, we have found that most of the textile sectors have a higher level of efficiency than that of the clothing sectors.

In terms of innovation spending, Fahmy-Abdullah, Sieng, and Isa studied, in the context of Malaysia, how much I&D spending contributes to the efficiency of the textile industry. The authors conclude that the share of I&D spending for technical efficiency is reduced and that companies should promote a closer relationship between production and research. The results of this research also suggest that the technical efficiency of the textile industry is reduced, and, as in Malaysia, we argue that Portuguese companies should also invest in innovation in the sector. Another essay on the subject was carried out by Du, Wan, Long, and Xue, who observed the weight of innovation in the textile industry. The authors concluded that the level of innovation participation in the efficiency of the sector is reduced and that companies need to encourage investment in innovation in order to improve efficiency. Once again, the conclusions of the authors indicated are consistent with the results of the present study, since the companies in the Portuguese textile sector also present levels of I&D expenditure share, for the profitability of assets, quite low.

5 Conclusion

The aim of the research is to analyse in an isolated manner the subsectors of the textile and clothing industry in Portugal, with support in I&D expenditure, and their relationship with the efficiency of asset profitability over time. We draw the following conclusions from the research:

We can conclude that the technical efficiency of the textile and clothing sectors is variable. The results point to different levels of technical efficiency and show that some of the textile and clothing industry segments exhibit zero, moderate, and high efficiency, showing reference periods (DMU benchmarking). However, the clothing sector has higher levels of inefficiency between I&D spending and asset profitability.

We also concluded that there is a remarkable need for investment in I&D. The results reflect the need for Portuguese companies to invest in innovation in the textile and clothing industries. Existing studies reflect identical conclusions and corroborate this need, stressing that increased investment is essential to improving the overall efficiency of the sector.

It can be concluded, supported by the results, that there is a small relationship between innovation and the return on investment. The results suggest that there is a low ratio between I&D spending and asset profitability in all subsectors. Indeed, the research highlights the need to establish a closer relationship between I&D spending and the profitability of assets in order to ensure that the resources invested in innovation translate into more efficient results in terms of profitability.

We also conclude that the textile industry tends to be more technically efficient than the clothing industry. This underlines the need to adopt a more efficient vision of the clothing subsectors with a view to improving the ratio between I&D expenditure and asset profitability.

References

ATP (Associação Têxtil e Vestuário de Portugal). (2019). A fileira têxtil e vestuário no horizonte 2025. ATP, Lisboa.

Breia, A., Mata, M., & Pereira, V. (2014). Análise económica e financeira: aspetos teóricos e casos práticos. Rei dos Livros, Lisboa.

Carcangu, M. (2016). As atividades de investigação e desenvolvimento: um estudo sobre o setor farmacêutico [Dissertação de Mestrado, Universidade de Coimbra, Faculdade de Economia, Coimbra]. Repositório científico da UC. https://hdl.handle.net/10316/32765

Cardoso, R. (2013). A relação entre a gestão da qualidade a inovação e a performance no setor do têxtil em Portugal [Dissertação de Mestrado, Instituto Politécnico de Leiria, Escola Superior de Tecnologia e Gestão, Leiria]. Repositório IC-Online. http://hdl.handle.net/10400.8/1104

CIP (Confederação Empresarial de Portugal). (2017). O conceito de reindustrialização, indústria 4.0 e política industrial para o século XXI: o caso português. CIP, Lisboa.

Comissão Europeia. (2003). O futuro do sector dos têxteis e do vestuário na União Europeia alargada. Retrieved December 17, 2023, from https://eur-lex.europa.eu/LexUriServ/LexUriServ.do?uri=COM:2003:0649:FIN:PT:PDF

Darji, I., & Dahiya, S. (2023). Measuring operational efficiency of the textile industry using DEA technique: A study of the northern state of India. *Measuring Business Excellence, 27*, 421–432.

Du, X., Wan, B., Long, W., & Xue, H. (2022). Evaluation of manufacturing innovation performance in Wuhan city circle based on DEA-BCC model and DEA-Malmquist index method. In *Discrete dynamics in nature and society* (pp. 1–9)

Fahmy-Abdullah, M., Sieng, L., & Isa, I. (2018). Technical efficiency in Malaysian textile manufacturing industry: A stochastic frontier analysis (SFA) approach. *International Journal of Economics and Management, 12*, 407–419.

Fang, K., Hong, X., Li, S., Song, M., & Zhang, J. (2013). Choosing competitive industries in manufacturing of China under low-carbon economy: A three-stage DEA analysis. *International Journal of Climate Change Strategies and Management, 5*, 431–444.

Ferreira, L. (2021). Valor acrescentado bruto e ativos intangíveis: a sua relação no setor têxtil português [Dissertação de Mestrado, Instituto Politécnico de Santarém, Escoa Superior de Gestão e Tecnologia, Santarém]. Repositório Científico do Instituto Politécnico de Santarém. http://hdl.handle.net/10400.15/3842

Fortuna, C. (1993). O fio da meada: o algodão de Moçambique, Portugal e a economia mundo, 1860–1960. Edições Afrontamento, Porto.

Gonçalves, F., & Caraça, J. (1986). A indústria transformadora nacional na encruzilhada: Potencial inovador e competitividade. *Análise Social, 22*, 93–108.

Jajri, I., & Ismail, R. (2006). Technical efficiency, technological change and total factor productivity growth in the Malaysian manufacturing sector. *Munich Personal RePEc Archive, 2*, 2–19.

Khalil, S. (2011). Relative efficiency of decision-making units producing both desirable and undesirable outputs: A case of textile processing units in Pakistan. *Pakistan Development Review, 50*, 685–698.

Leite, C. (2020). Determinantes da rendibilidade e estrutura de capitais: evidência nas empresas da Euronext Lisboa [Dissertação de Mestrado, Instituto Superior de Administração e Gestão, Porto]. Repositório Comum. http://hdl.handle.net/10400.26/39535

Li, S., & Tsai, S. (2018). Efficiency of apparel retail at the firm level: An evaluation using data envelopment analysis (DEA). *Journal of Textile Engineering & Fashion Technology, 4*, 131–138.
Marques, A. (2023). Comércio internacional e crescimento a longo prazo: o contexto mundial, notas sobre Portugal. Almedina, Coimbra.
Marques, W. (2013). Análise de custos. Clube de Autores, Lisboa.
Martins, A. (2012). Avaliação da eficiência e identificação dos fatores determinantes da eficiência do setor bancário em Portugal [Doutoramento em Ciências Económicas e Empresariais, Universidade do Algarve, Faculdade de Economia]. Repositório da Universidade do Algarve. http://hdl.handle.net/10400.1/6077
Merle, K. (2023). Competências para inovar na indústria 4,0 do setor têxtil e de vestuário brasileiro: Uma revisão sistemática da literatura. *Revista s&g, 18*, 105–111.
Montenegro, R. (2011). Padrões concorrenciais têxtil-confecção sob o enfoque da aglomeração local de Nova Friburgo. Acta Scientiarum. *Human and Social Sciences, 33*, 33–43.
Nassif, A. (2023). Desenvolvimento e estagnação: o debate entre desenvolvimentistas e liberais neoclássicos. Editora Contracorrente, São Paulo.
Neto, A. (1991). Industrialização de Angola: reflexão sobre a experiência da administração portuguesa, 1961–1975. Escher, Taschen.
Nogueira, P. (2020). A ciência e a tecnologia na indústria têxtil e do vestuário: Uma abordagem historiográfica ao setor em Guimarães [Dissertação de Mestrado, Universidade de Coimbra, Faculdade de Ciências e Tecnologia]. Repositório científico da UC. https://hdl.handle.net/10316/95071
Pereira, B. (2021). Análise dos determinantes da produtividade na indústria têxtil e de vestuário portuguesa [Dissertação de Mestrado, Universidade do Minho, Escola de Economia e Gestão, Braga]. Repositório UM. https://hdl.handle.net/1822/76100
Pérez, K., González-Araya, M. C., & Iriarte, A. (2017). Energy and GHG emission efficiency in the chilean manufacturing industry: Sectoral and regional analysis by DEA and Malmquist indexes. *Energy Economics, 66*, 290–302.
Popker, S., & Raju, G. (2020). Financial performance and efficiency of cotton textile companies-an dea approach. *Journal of Advanced Research in Dynamical and Control Systems, 12*, 1753–1761.
Proença, A., Amaro, L., & Fragoso, R. (2008). Sector têxtil e vestuário. In J. Leitão, J. Ferreira, & S. Azevedo (Eds.), Dimensões competitivas de Portugal: contributos dos territórios, sectores, empresas e logística. Centro Atlântico, Lisboa.
Proença, C. (2018). Estudo de caso aplicado a empresa do sector têxtil e do vestuário [Dissertação de Mestrado, Universidade da Beira Interior, Faculdade de Engenharia, Covilhã]. Repositório uBibliorum. http://hdl.handle.net/10400.6/9457
Rakhmawan, C., Hartono, D., & Awirya, A. (2012). Efficiency and import penetration on the productivity of textile industry and textile products. *International Research Journal of Business Studies, 5*, 51–63.
Santos, D. (2011). Dinâmicas territoriais de inovação: o caso do sector têxtil. Edições Húmus, Ribeirão.
Silva, P. (2013). As redes de relacionamento e as capacidades dinâmicas: a indústria têxtil e de vestuário portuguesa [Dissertação do Mestrado, Universidade do Minho, Escola de Economia e Gestão]. Repositório UM. https://hdl.handle.net/1822/24951.
Sousa, A. (2020). A inovação na indústria têxtil e vestuário: o papel da colaboração [Dissertação de Mestrado, Universidade de Lisboa, Instituto Superior de Economia e Gestão, Lisboa]. Repositório da Universidade de Lisboa. http://hdl.handle.net/10400.5/20912
Sudri, N., Theresia, L., Widianty, Y., Puspiptek, J., Selatan, K., & Banten, P. (2019). Increasing textile industry production process efficiency with data development analysis. *Iptek, 3*, 239–244.
Sueyoshi, T., & Goto, M. (2019). DEA non-radial approach for resource allocation and energy usage to enhance corporate sustainability in Japanese manufacturing industries. *Energies, 12*, 1–22.
WIPO—World Intellectual Property Organization. (2005). *Um ponto no tempo: O uso inteligente da propriedade intelectual por empresas do setor têxtil*. WIPO.

Innovation and Green Finance Management in the Textile Industry: Emerging Trends and Challenges

Mário Nuno Mata

Abstract This article aims to conduct a comprehensive bibliometric analysis to examine the evolving landscape of innovation and green finance in the textile industry. The primary objectives are to map the intellectual structure, identify influential authors and institutions, and uncover emerging themes in sustainable textile management. The study employs bibliometric analysis, a quantitative approach, to review an array of scholarly articles, using R and Biblioshiny. The analysis reveals a clear upward trajectory in academic interest in innovation and green finance within the textile industry, with a noticeable increase in publications. Despite a slight dip in 2020 due to the COVID-19 pandemic, academic interest remained resilient, culminating in a peak in 2023. Research topics include biotechnology, nanotechnology, consumer behavior, and clean production methodologies, highlighting the multidisciplinary nature of challenges and solutions. China leads in total publications, India, the UK, and the USA. This bibliometric analysis provides valuable insights into the current state of research in innovation and green finance management in the textile industry. It informs researchers, industry stakeholders, and policymakers about emerging trends and opportunities to advance sustainable practices within the sector.

1 Introduction

The textile industry, historically renowned for its significant contributions to global economies, is simultaneously a major contributor to environmental degradation. The dual challenge of maintaining economic growth while ensuring environmental sustainability has never been more pressing. In this milieu, innovation and green finance have emerged as pivotal elements in steering the industry towards a sustainable future. This article conducts a comprehensive bibliometric analysis to unravel the

M. N. Mata (✉)
ISCAL—Instituto Superior de Contabilidade e Administração de Lisboa, Instituto Politécnico de Lisboa, Avenida Miguel Bombarda, 20, 1069-035 Lisboa, Portugal
e-mail: mnmata@iscal.ipl.pt

Insight: Piaget Research Center for Ecological Human Development, Lisboa, Portugal

Lisbon Accounting and Business School, Polytechnic Institute of Lisbon (ed.), *Industry and Innovation: Textile Industry*, SDGs and Textiles,
https://doi.org/10.1007/978-3-031-57804-5_9

evolving landscape of innovation and green finance in the textile sector, highlighting the emerging trends and pinpointing the challenges that lay ahead.

The textile industry, characterized by its intensive use of water, chemicals, and energy, has long been under scrutiny for its environmental footprint. The unsustainable practices prevalent in the industry not only deplete natural resources but also pose significant threats to ecosystems and public health. In response, there has been a growing emphasis on adopting innovative technologies and sustainable finance mechanisms. These endeavors are geared towards reducing the environmental impact, enhancing resource efficiency, and promoting circular economy principles within the textile sector.

Innovation in the textile industry encompasses a broad spectrum, ranging from the development of eco-friendly materials and sustainable production processes to the implementation of circular economy models. The adoption of these innovative practices is not just a matter of environmental imperative but also a strategic move to stay competitive in a market increasingly driven by sustainability-conscious consumers and stringent regulatory frameworks.

Green finance plays a crucial role in this transformative journey by mobilizing the necessary financial resources. Investments in sustainable textile practices are pivotal for enabling technological advancements and facilitating the shift towards environmentally friendly operations. Green bonds, sustainability-linked loans, and impact investing are among the financial instruments gaining traction in promoting eco-friendly initiatives within the industry.

Despite the criticality of these developments, there remains a gap in comprehensively understanding how innovation and green finance are shaping the textile industry. This bibliometric analysis aims to bridge this gap by systematically reviewing the existing literature. The objectives of this study are multifold:

- to map the intellectual structure and research trends in the domain of innovation and green finance in the textile industry;
- to identify the most influential authors, institutions, and countries contributing to this research area;
- to uncover the emerging themes and potential research directions in the context of sustainable textile management.

This study employs bibliometric analysis, a quantitative approach to the review of literature, enabling the identification of patterns, trends, and gaps in research. Through the analysis of a vast array of scholarly articles, this method provides insights into the evolution and current state of research on innovation and green finance in the textile industry.

The article is organized as follows: the next section reviews the literature, providing a comprehensive background on the integration of innovation and green finance in the textile industry. It is followed by a detailed description of the methodology employed for the bibliometric analysis. The subsequent section presents the results and discussion, delving into the key findings, trends, and thematic clusters identified through the analysis. Finally, the article concludes with a summary of the findings, implications for future research, and policy recommendations.

2 Systematic Literature Review

The intersection of innovation, green finance management, and the textile industry presents a dynamic and evolving landscape. The urgency of addressing environmental challenges, coupled with rapid technological advancements, has intensified the focus on sustainable practices within the textile sector. This bibliometric analysis aims to illuminate the emerging trends, challenges, and scholarly discourse surrounding these topics from 2014 to 2023. By examining a diverse array of sources, including journals, books, and conference papers, this study provides a comprehensive overview of the academic landscape as it relates to innovation and green finance in the textile industry.

The data collected span nearly a decade, from 2014 to 2023 in the Scopus and Web of Science databases, reflecting a period of significant change and growth in both green finance and the textile industry. A total of 100 sources, encompassing a variety of mediums like journals, books, and conference materials, have contributed to a corpus of 128 documents (Table 1). This breadth of sources indicates a multidisciplinary interest in the subject, highlighting its relevance across different academic and professional fields.

A remarkable annual growth rate of 49.81% in publications signals a rapidly increasing interest in these topics. This surge can be attributed to heightened environmental awareness, regulatory changes, and technological innovations that are reshaping the industry. The document average age of 2.16 years further underscores the timeliness and contemporary nature of the research, suggesting that the discourse is evolving rapidly in response to current global challenges and innovations. The

Table 1 Input data for the study

Description	Results
Timespan	2014:2023
Sources (journals, books, etc.)	100
Documents	128
Annual growth rate %	49,81
Document average age	2,16
Average citations per doc	12,95
Authors	495
Single-authored docs	13
Co-authors per doc	4,18
International co-authorships %	25,78
Article	77
Book	5
Book chapter	14
Conference paper	13
Conference review	4
Review	15

average citations per document stand at 12.95, indicating a robust engagement with these works within the academic community. This metric points to the relevance and impact of the research in this field, as well as its potential to influence future studies and industry practices.

In terms of document types, articles lead with 77 entries, followed by 15 reviews, 14 book chapters, 13 conference papers, 5 books, and 4 conference reviews. This distribution suggests a preference for disseminating findings and analyses through articles and reviews, which are typically more accessible and have a broader reach within the academic community. This bibliometric analysis reveals an active and growing field of study. The data reflect an interdisciplinary and international effort to address the challenges at the intersection of innovation, green finance, and the textile industry, emphasizing the need for continued research and collaboration in these areas.

To analyze the bibliometric co-occurrence network in the context of innovation and green finance management in the textile industry, we examine the relationships and importance of various nodes (terms or concepts) within the network. This analysis focuses on the emerging trends and challenges, which includes parameters like betweenness, closeness, PageRank, and the clustering of these nodes (Fig. 1).

The bibliometric analysis reveals a complex network where technical, environmental, and industry-specific nodes interact. The high betweenness and PageRank scores of certain nodes underscore their importance in the literature. The emerging trends suggest a strong focus on sustainability and innovation, while the challenges lie in integrating these concepts practically into the textile industry's operations. The network analysis, thus, provides valuable insights into the current state and future directions of research and practice in green finance management within the textile industry.

The textile industry, a significant contributor to global economic growth, has recently witnessed a paradigm shift towards sustainable practices, underpinned by

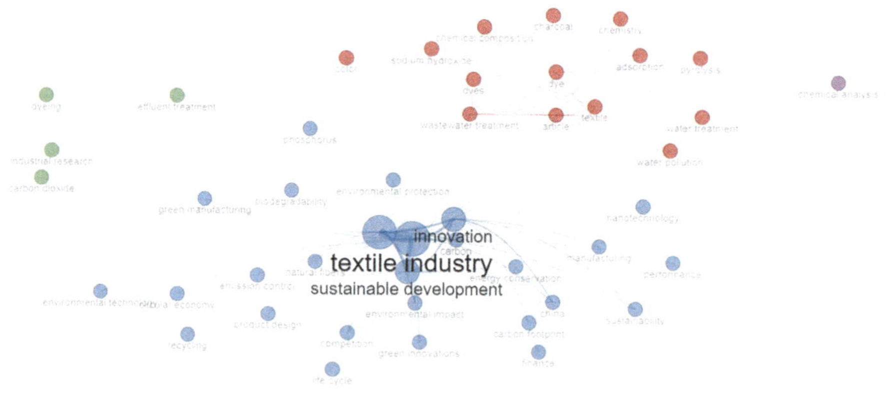

Fig. 1 Cluster analysis of publications on innovation and green finance management in the textile industry

innovation and green finance management. This evolution is critical in light of the industry's environmental footprint, characterized by substantial water consumption, chemical use, and carbon emissions (Guo et al., 2022; Huang, 2022). The integration of green finance and innovative strategies has emerged as a pivotal aspect of this transformation (Lu et al., 2022), offering a pathway to sustainable development while ensuring economic viability. This literature review synthesizes insights from various studies, shedding light on the emerging trends and challenges in this domain (Wang et al., 2021; Zhang et al., 2021).

Innovation in the textile industry has evolved from mere product and process improvements to encompassing broader aspects of sustainability and digital transformation. Recent studies highlight how technological advancements, such as the adoption of Industry 4.0, are revolutionizing production processes, leading to increased efficiency and reduced environmental impact (Bertola & Teunissen, 2018; Nouinou et al., 2023). Moreover, the development of sustainable materials and eco-friendly dyes has been instrumental in reducing the ecological footprint of textile products (Costa et al., 2019; Lara et al., 2022).

The role of green finance in the textile industry has gained significant traction, offering a means to support sustainable practices financially. Green bonds and sustainable loans are increasingly being leveraged to fund environmentally friendly projects within the industry. These financial instruments not only promote sustainable development but also encourage companies to adopt more responsible business practices (Wang, 2023; Xing et al., 2021). Furthermore, the integration of Environmental, Social, and Governance (ESG) criteria in investment decisions has been pivotal in directing capital towards sustainable textile initiatives (Liu et al., 2023; Sierdovski et al., 2022).

The intersection of innovation and green finance has led to several emerging trends in the textile industry. One such trend is the growing emphasis on circular economy models, which aim to minimize waste and promote the reuse and recycling of materials (Jia et al., 2020). Another trend is the increasing consumer demand for sustainable products, which is driving companies to innovate and invest in green practices (Desore & Narula, 2018; Wang & Yang, 2021). The rise of digital platforms for green finance is also facilitating greater access to funds for small and medium-sized enterprises in the textile sector (Bakos et al., 2020; Kumar et al., 2022).

Despite these advancements, the textile industry faces several challenges in fully realizing the potential of innovation and green finance. One of the primary challenges is the lack of standardized criteria and benchmarks for green investments, which can lead to inconsistencies in how sustainability is assessed and valued (Gonçalves & Silva, 2021). Additionally, the high cost of implementing sustainable technologies remains a significant barrier for many companies, especially in developing countries (Rahman et al., 2019).

The integration of innovation and green finance management in the textile industry is essential for fostering sustainable development. While there are challenges to be addressed, the emerging trends indicate a positive trajectory towards a more sustainable and responsible textile industry. Future research should focus on developing standardized frameworks for green finance and exploring innovative financing models to support small and medium enterprises in their sustainability journey.

A bibliometric analysis reveals key trends and challenges through the lens of the most frequently occurring words in the researched publications (Fig. 2). At the forefront, "textile industry" appears 40 times, underscoring the central subject of these studies. Closely following is "textiles" with 37 occurrences, highlighting the specific focus within the industry. "Sustainable development," mentioned 27 times, signals a significant emphasis on eco-friendly and long-term approaches in the textile sector. This aligns with current global trends towards sustainability and environmental stewardship. "Innovation," noted 19 times, suggests an active pursuit of new ideas and technologies in this field, reflecting the industry's response to evolving demands and challenges.

China emerges as a specific geographic focus, appearing 13 times, indicating the country's substantial role in the textile industry, possibly due to its large manufacturing base and market influence. "Carbon" is mentioned 12 times, pointing towards concerns related to carbon footprint and greenhouse gas emissions, crucial factors in the context of green finance and sustainable industry practices. "Wastewater treatment" is cited in 10 instances, illustrating the significance of water management and pollution control in textile manufacturing. This aligns with the growing awareness of water as a critical resource needing careful management, especially in industries with significant water usage like textiles. "Adsorption," "emission control," and "environmental impact," each mentioned 7 times, further emphasize the focus on environmental aspects. Adsorption could refer to the processes used in treating pollutants, while emission control indicates a direct concern with reducing harmful discharges. "Environmental impact" broadly encompasses the overall effects of the textile industry on the natural environment.

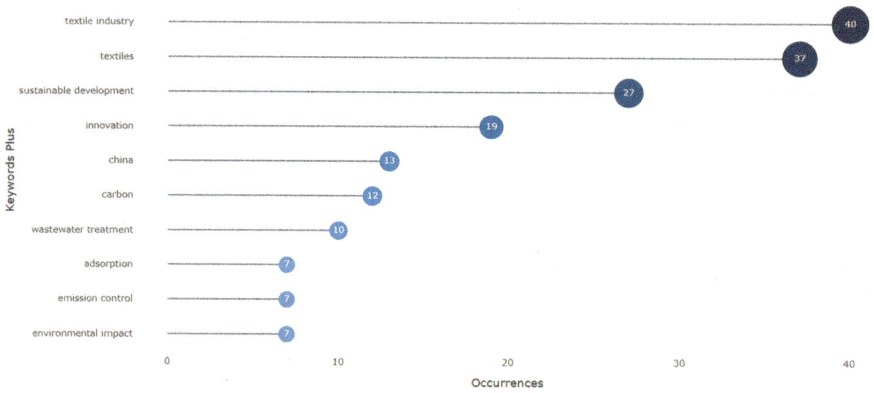

Fig. 2 The most frequently occurring words in the studied publications

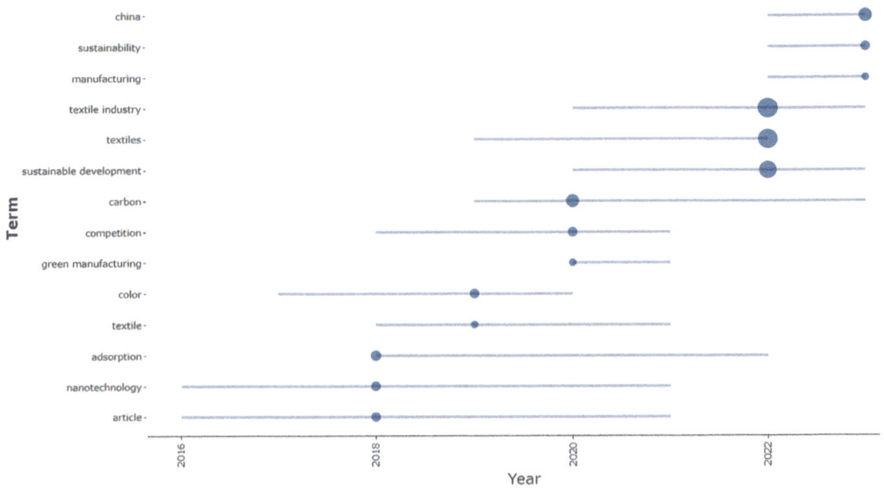

Fig. 3 Trend topics in the studied publications

This bibliometric analysis paints a picture of a textile industry deeply engaged with sustainability, innovation, and environmental issues, reflecting a shift towards greener practices influenced by global trends and specific regional focuses like China. The emphasis on wastewater treatment and carbon emissions highlights specific areas where innovation and green finance management are being actively pursued to mitigate environmental impacts.

The bibliometric analysis, as illustrated by Fig. 3, reveals several emerging trends and challenges through the frequency and temporal distribution of key topics in the researched publications.

Starting with "adsorption," this topic has been moderately frequent with seven mentions. Its emergence seems to have peaked around 2018, with a renewed interest in 2022, suggesting an ongoing relevance in the industry. "Nanotechnology" is also featured with the same frequency, emerging around 2016 and maintaining its presence until 2021. This shows its sustained importance in the textile industry, particularly in the context of innovation. The topic "color" with six mentions indicates a concentrated interest between 2017 and 2020. This could be related to the exploration of sustainable dyeing processes or the development of eco-friendly colorants in textiles. The keyword "textile" appeared 5 times between 2018 and 2021, signifying its core relevance in the broader context of the industry's evolution. "Carbon," with the highest frequency of 12 mentions, shows a rising trend from 2019 to 2023. This likely reflects the growing emphasis on carbon footprint reduction and carbon–neutral practices in the textile industry. The term "competition" appears 6 times, with its prominence between 2018 and 2021, highlighting the competitive dynamics within the industry, possibly influenced by green manufacturing and sustainability initiatives. Speaking of "green manufacturing," it has been a focal topic with five mentions, all clustered around 2020 and 2021, indicating a recent surge in interest likely tied to the industry's push towards sustainability. The "textile industry" as a keyword dominates with 40 mentions, peaking from 2020 to 2023. This overwhelming presence underscores the sector's critical examination in the context of innovation and green finance. Lastly, "textiles," with 37 mentions, has been consistently prominent from 2019 to 2022, suggesting its central role in the academic discourse around the sustainable transformation of the industry.

This analysis highlights a clear shift towards sustainability and innovation within the textile industry. The focus on carbon management, green manufacturing, and the integration of nanotechnology signals a transformative phase in the industry, grappling with environmental challenges while pursuing competitive advancement.

The thematic map for the researched publications reveals several emerging trends and challenges (Fig. 4). The map is structured into the clusters, each representing a different thematic focus within the broader context of the textile industry, sustainability, and innovation.

Cluster 1, labeled "Innovation," includes terms like "innovation" (6 occurrences), "textile" (4), and "green manufacturing" (3). This cluster is characterized by a high betweenness centrality, indicating that innovation serves as a pivotal concept connecting various aspects of the textile industry. The focus here is on integrating new technologies and approaches to enhance the manufacturing process.

"Sustainability," as Cluster 2, underscores the industry's shift towards sustainable practices. It encompasses "sustainability" (11 occurrences), "textiles" (4), "fashion" (3), and related terms. The high pagerank centrality of "sustainability" underscores its significance in current research, reflecting a growing emphasis on eco-friendly practices in the textile and fashion sectors.

Cluster 3, "Nanoparticles," with terms like "nanoparticles" and "patents," suggests a focus on advanced materials and intellectual property in textile manufacturing. This cluster may reflect research into new materials that enhance the environmental friendliness of textiles.

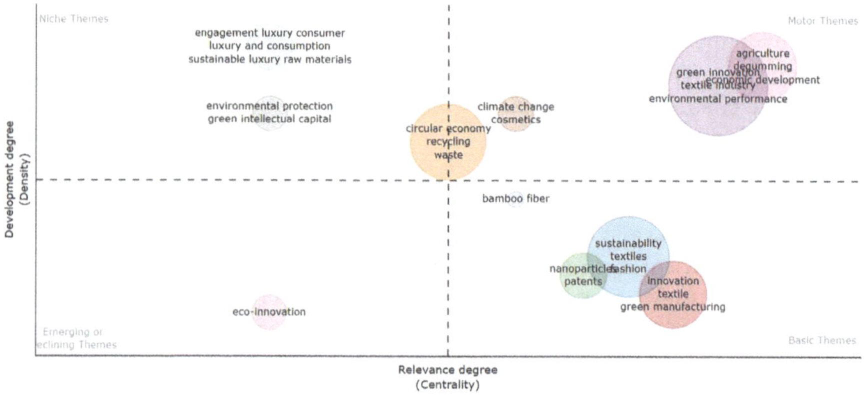

Fig. 4 Thematic map for the studied publications

The largest cluster, Cluster 4, "Green Innovation," includes terms such as "green innovation," "textile industry," "environmental performance," and "technological innovation." This cluster, with the highest occurrences and centrality measures, indicates an intense focus on integrating green practices within the industry's technological advancements. It highlights the industry's efforts to balance environmental concerns with technological growth.

Cluster 5, "Circular Economy," with terms like "circular economy," "recycling," and "waste," represents a paradigm shift in the industry towards sustainability. The focus here is on creating a closed-loop system where resources are reused and recycled, minimizing waste and environmental impact.

Interestingly, the map also includes clusters focused on broader societal and environmental issues. For instance, Cluster 6, "Climate Change," and Cluster 7, "Agriculture," indicate the textile industry's interconnection with global environmental and economic sectors. These clusters suggest an acknowledgment of the broader impact of textile manufacturing on climate and agriculture.

Clusters like "Environmental Protection" (8), "Corporate Social Responsibility" (9), and "Dyes" (10) further illustrate the multifaceted nature of the challenges faced by the textile industry. They reflect concerns over environmental impact, ethical practices, and the use of chemicals in production.

The presence of clusters like "Bamboo Fiber" (11), "Eco-Innovation" (12), and "Engagement Luxury Consumer" (13) indicate niche but significant areas of research focusing on alternative materials, consumer behavior, and luxury markets within the sustainable textile paradigm.

The thematic map reveals a complex and interconnected landscape of research in the textile industry. Innovation and sustainability are central themes, reflecting the industry's focus on integrating new technologies and green practices. The map highlights the multifaceted challenges facing the industry, ranging from material innovation and environmental protection to societal impact and corporate responsibility.

The bibliometric analysis of innovation and green finance management in the textile industry, as depicted in Table 2, reveals a complex landscape of emerging trends and challenges. This analysis is structured around a cluster analysis according to a thematic map, which encompasses various key themes such as green innovation, sustainability, and the circular economy.

The cluster of "green innovation" emerges with moderate Callon centrality (0.39) but high Callon density (87.39), suggesting that while it may not be the central theme in the literature, it is a densely packed and well-developed area. This is further supported by its relatively high rank in both centrality and density, and a notable frequency of 30, indicating its significant presence in the field. In contrast, "sustainability" shows lower centrality (0.20) but a considerable density (39.90), reflecting its importance as a well-established, though not central, theme in the discourse. It ranks moderately in centrality and higher in density, with a frequency of 16, underscoring its relevance but perhaps not its dominance in the literature. The "textile industry" cluster stands out with the highest centrality (1.14), indicating its core position in the discourse. Its high density (84.66) and frequency (25) further emphasize its critical

Table 2 Cluster analysis according to the thematic map

Cluster	Callon Centrality	Callon Density	Rank Centrality	Rank Density	Cluster Frequency
Green innovation	0,39	87,39	13	15	30
Sustainability	0,20	39,90	11	5	16
Textile industry	1,14	84,66	16	14	25
Circular economy	0,17	52,22	8,5	11	18
Innovation	0,18	16,67	10	1	6
Eco-innovation	0,00	25,00	4	2	4
Nanoparticles	0,44	43,75	14	6	6
Textile	0,38	33,33	12	3,5	7
Textiles	0,54	59,38	15	12	10
Technological innovation	0,17	33,33	8,5	3,5	3
Antibacterial	0,00	50,00	4	8,5	2
Bamboo fiber	0,00	66,67	4	13	6
Corporate social responsibility	0,00	50,00	4	8,5	2
Dyes	0,00	50,00	4	8,5	2
Engagement luxury consumer	0,00	100,00	4	16	6
Microbial enzymes	0,00	50,00	4	8,5	2

role and extensive coverage in the literature. This suggests that discussions around innovation and green finance in textiles largely revolve around the industry's core aspects.

"Circular economy" and "innovation" present as significant yet not dominant themes, with moderate centrality and density scores. Their respective frequencies (18 and 6) indicate a steady presence in the literature, though not as prominently as the core themes like the textile industry or green innovation. The clusters of "eco-innovation," "nanoparticles," "textile," and "textiles" show varied centrality and density, highlighting their specific but varied roles in the discourse. Particularly, "nanoparticles" and "textiles" exhibit higher centrality, suggesting their emerging importance in the field.

Clusters like "technological innovation," "antibacterial," "bamboo fiber," and others demonstrate lower scores in both centrality and density. These topics, while relevant, appear to be more niche or emerging, with "engagement luxury consumer" showing the highest density (100.00) but lowest centrality, indicating a very focused yet peripheral area in the literature.

This analysis paints a multifaceted picture of the textile industry's journey towards innovation and green finance. Central themes like the textile industry and green innovation are extensively covered and developed, while other emerging areas like nanoparticles and bamboo fiber present new avenues for exploration and growth. The diverse range of topics reflects a dynamic field, evolving to meet the challenges of sustainability and innovation in the textile sector.

2.1 *The Common Research Trends and Meta-analysis*

The textile industry, as an integral part of the global economy, has been undergoing significant transformations, particularly in the realms of innovation and green finance management. These shifts are essential in addressing the environmental challenges and sustainability goals of the twenty-first century. In this context, scholarly attention towards these areas has markedly increased, offering valuable insights into emerging trends and challenges. A bibliometric analysis provides a systematic approach to understand these trends by examining the scientific output over time. Figure 5 presents a chronological quantification of articles published from 2014 to 2023, offering a clear perspective on the evolving academic interest in this field.

The data in Fig. 5 reveal a compelling narrative of growing academic interest in the intersection of risk and innovation management within the textile industry. Starting in 2014, the scientific production was minimal, with only one article. This modest beginning indicates that the topic was in its nascent stages within academic circles. However, there is a noticeable uptick in the number of publications from 2015 onwards, with a gradual increase to three articles, reflecting a budding awareness and interest in the subject matter. The period from 2016 to 2019 shows a more pronounced increase, with the number of articles fluctuating but generally trending upwards. This period could be characterized by a growing recognition of the importance of innovation and green finance in mitigating risks within the textile industry, likely driven by escalating environmental concerns and the industry's impact on sustainability.

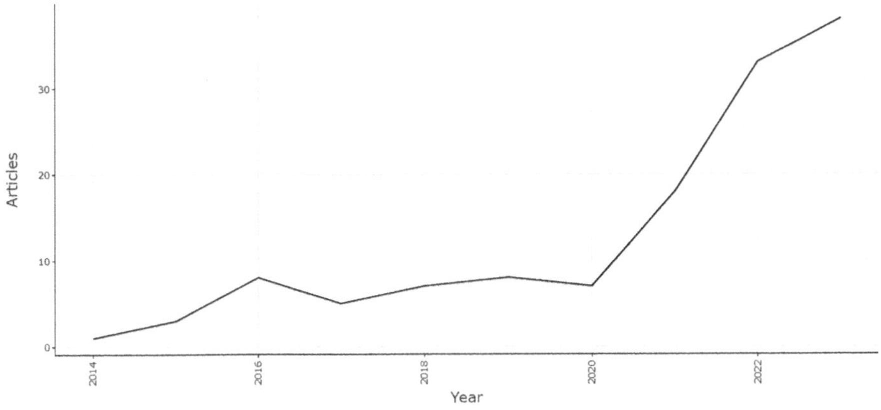

Fig. 5 Annual scientific production trends in innovation and green finance management in the textile industry

The year 2020 marks a slight dip in publications, possibly attributable to global disruptions caused by the COVID-19 pandemic. Despite this, the academic interest did not wane significantly, suggesting a resilient and ongoing commitment to exploring these themes. Post-2020, there is a remarkable surge in publications, with 2021 and 2022 showing 18 and 33 articles, respectively. This surge is indicative of a heightened collective realization of the critical need for innovation and sustainable practices in the textile industry. The sharp increase in 2021 might also suggest a response to the lessons learned during the pandemic, emphasizing the need for resilient and sustainable business models.

The peak in 2023, with 38 articles, underscores the current and pressing nature of this topic. It reflects the culmination of previous years' developments and indicates that the academic community is now deeply engaged in exploring and addressing the challenges at the intersection of risk management, innovation, and green finance in the textile industry. The data from Fig. 5 not only highlight the growing scholarly attention towards innovation and green finance in the textile industry but also mirror the industry's own evolution in response to global sustainability challenges. This trend suggests a promising direction for future research and practical applications in the industry, aiming towards a more sustainable and innovative future.

Analyzing the bibliometric data on innovation and green finance management in the textile industry reveals a nuanced landscape of research interest and impact over time. The data span from 2014 to 2023, tracking the mean total citations per article (Mean TC per Art), the number of articles (N), the mean total citations per year (Mean TC per Year), and the citable years since publication (Table 3).

In 2014, the field seems nascent, with only one article and no citations. This absence of citations could be due to the emerging nature of the topic or limited awareness in the academic community. A gradual increase in interest is observed in 2015, with an average of 0.22 citations per year across three articles. Although still modest, this uptick indicates growing academic recognition. The year 2016 marks a significant surge in impact, with a mean of 44.62 citations per article among eight

Table 3 Average citations per year for the researched publications

Year	Mean TC per art	N	Mean TC per year	Citable years
2014	0	1,00	0,00	10
2015	2	3,00	0,22	9
2016	44,62	8,00	5,58	8
2017	2	5,00	0,29	7
2018	37	7,00	6,17	6
2019	31,25	8,00	6,25	5
2020	13,29	7,00	3,32	4
2021	21,61	18,00	7,20	3
2022	6,03	33,00	3,02	2
2023	2,5	38,00	2,50	1

articles, translating to 5.58 citations per year. This peak suggests a pivotal moment for the field, possibly due to a breakthrough in research or a rising global emphasis on green practices in the textile industry. In 2017, there was a sharp decline, reverting to levels similar to 2015. This fluctuation could be attributed to various factors, including changes in research focus, publication of less impactful studies, or external factors affecting the industry. The interest and impact rebound in 2018 and 2019, with mean citations per year reaching 6.17 and 6.25, respectively. This resurgence indicates a sustained, growing interest in the intersection of innovation, green finance, and the textile industry.

A slight decrease is noted in 2020, with a mean of 13.29 citations per article and 3.32 citations per year. This decline could be influenced by the global events of the time, possibly shifting academic priorities or disrupting research activities. In 2021, there's a substantial increase in both the number of articles (18) and the mean total citations per year (7.20). This uptick might reflect a maturing field with more researchers contributing to the discourse and the increasing relevance of green practices in the textile industry. The year 2022 sees a significant increase in the number of articles (33) but a decrease in the mean total citations per year (3.02). This pattern could suggest that while the field is attracting more research, individual articles are, on average, less cited, possibly due to a saturation of the topic or varying quality of the publications. In 2023, there's a large number of articles (38) but a relatively low mean citation rate (2.5 per year). This trend could indicate that while the topic remains popular among researchers, it may be reaching a plateau in terms of groundbreaking insights or that recent articles have not yet had enough time to accumulate citations.

The bibliometric analysis shows a field that is growing in popularity but experiencing fluctuations in research impact. These trends reflect the evolving nature of innovation and green finance in the textile industry, highlighting periods of significant breakthroughs and potential saturation points. The data underscore the importance of ongoing research to maintain momentum and continue advancing sustainable practices in the industry.

The analysis reveals a noticeable trend in the production of relevant literature across various journals over time, as detailed in Fig. 6. From 2014 to 2015, there was a complete absence of publications in all listed journals, indicating a lack of academic focus on these topics within the textile industry during this period. This zero activity might reflect either a nascent stage of research in these areas or a lack of recognition of their importance.

A gradual inception of interest is observed from 2016 onwards, with sparse contributions in "Environmental Footprints and Eco-Design of Products and Processes" and "Wool Textile Journal." This slow start suggests a growing but still limited recognition of the relevance of innovation and green finance in the textile sector. The year 2019 marks a slight increase, particularly with the "Journal of Cleaner Production" beginning to contribute to the field. This change might indicate a rising awareness and interest in sustainable practices and innovation within the industry. A significant uptick in publications is evident from 2021, with a broader range of

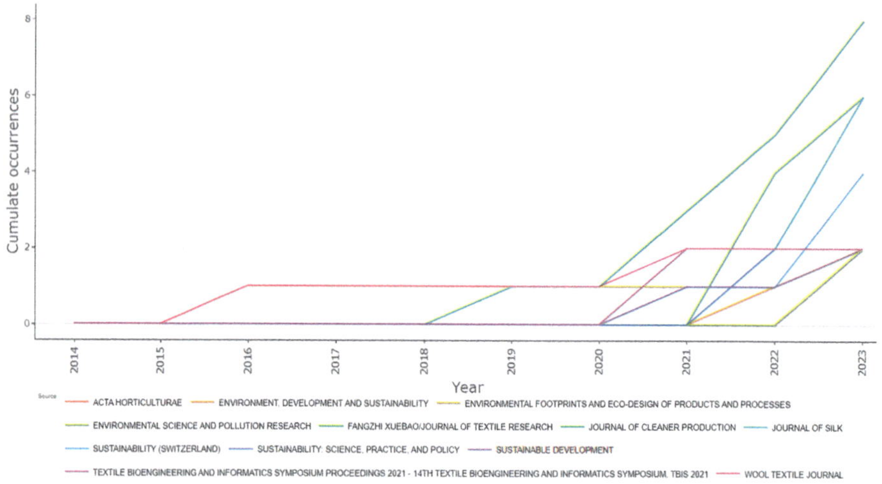

Fig. 6 Sources' production over time

journals like "Sustainability (Switzerland)," "Sustainable Development," and "Textile Bioengineering and Informatics Symposium Proceedings" contributing to the discourse. This surge reflects a heightened academic interest and possibly a response to increasing industry and societal demands for sustainable and innovative practices in the textile industry.

By 2023, there is a robust and diverse output across almost all journals listed, with "Journal of Cleaner Production," "Fangzhi Xuebao/Journal of Textile Research," and "Journal of Silk" leading in publication volume. This diversity and volume indicate that innovation and green finance in the textile industry have become prominent and widely recognized topics of academic inquiry.

Figure 7 reveals interesting patterns in authors' publication frequencies and their impact over time, as measured by total citations (TC) and citations per year (TCpY). This analysis reflects a dynamic field with varying degrees of author influence and impact over time. It underscores the evolving nature of innovation and green finance management in the textile industry, with some authors establishing significant early influence that may wane over time, while others gradually build their academic presence.

Figure 8 shows a ranking of the most cited documents in the field, indicating their relevance and impact. The prominence of these documents suggests they address key aspects of innovation and green finance in textiles, covering sustainable practices, technological advancements, financial models for eco-friendly operations, and policy implications.

The intersection of innovation and green finance management in the textile industry is a burgeoning field of study, as evidenced by a rich tapestry of bibliometric data. This literature review synthesizes insights from key scholarly contributions, offering an integrated perspective on emerging trends and challenges in this

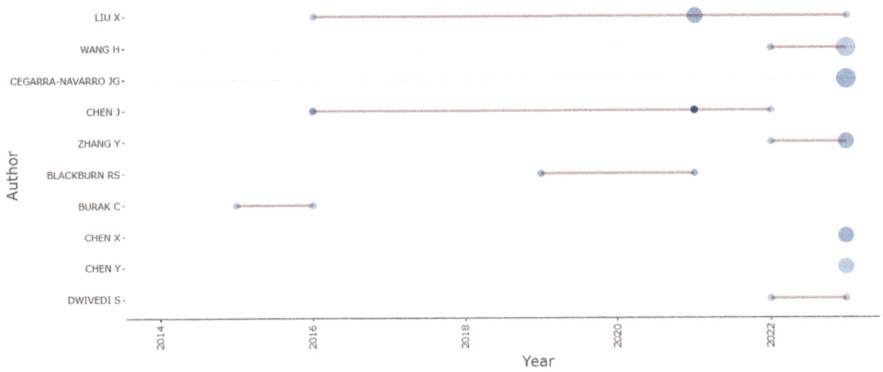

Fig. 7 Authors' production over time

domain. The pioneering work of Liew et al. (2018) marked a significant stride in understanding the role of biotechnology in sustainable textile production. Garnering a high citation count, this research underscores the criticality of biotechnological innovations for green finance management in textiles, highlighting its efficiency and sustainability benefits. Fang et al. (2021) made an investigation into the chemical underpinnings of sustainable textile production processes stands as a testament to the importance of chemical sciences in advancing sustainable practices in the industry. Majumdar and Sinha (2019) presented a nuanced view of sustainable production and consumption patterns in the textile industry. Their research emphasizes the interconnectedness of production processes and consumer behavior, offering insights into how sustainability can be embedded at every level of the textile value chain.

Li et al. (2016) took a novel approach by integrating nanotechnology into textile production. This study's significant citation rate reflects the growing interest in employing cutting-edge technologies to enhance the environmental sustainability of textile manufacturing processes. The research by Nayak and Mishra (2016) provided a pivotal analysis of fashion and textile trends, focusing on sustainable practices. Their work is a crucial link in understanding the consumer side of sustainable textiles, emphasizing the role of fashion trends in driving green practices. Hebeish et al. (2016) delved into the biological aspects of textile production, emphasizing the role of macrobiological processes in creating sustainable textile materials. Their research provides essential insights into the biological innovations that can aid in greening the textile industry. Sharma et al. (2021) explored the broader implications of clean production methodologies in the textile industry. This research is instrumental in highlighting how clean production is not just an environmental concern but also a critical factor in financial and operational efficiency. Ali et al. (2021) brought a unique perspective by linking intellectual capital to green finance in the textile industry. Their study underscores the role of intellectual resources in driving innovation and sustainability in the sector. Velmurugan et al. (2016) and Schellenberger et al. (2019) contributed significant insights into the role of photochemical and photobiological processes in sustainable textile production. These studies highlight the potential of

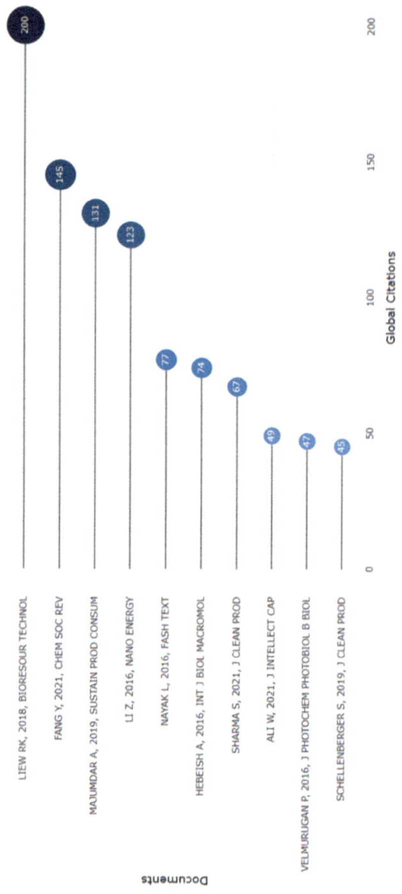

Fig. 8 Most global cited documents

harnessing light-based technologies for environmentally friendly textile manufacturing. The reviewed literature collectively indicates a robust and growing interest in integrating innovation and green finance management within the textile industry. The diverse approaches, ranging from biotechnology and nanotechnology to consumer behavior and clean production methodologies, demonstrate the multidisciplinary nature of the challenges and solutions in this field.

Figure 9, focusing on author productivity through Lotka's law, presents a distinct pattern in the distribution of research contributions. The vast majority of authors, representing 93.7%, have contributed to only a single document. This significant proportion suggests that a large number of researchers are involved in the field, but their engagement might be limited to isolated efforts or singular contributions.

A much smaller fraction of authors, about 5.1%, have written two documents. This indicates a more sustained but still moderate level of involvement in the research area. The number of authors who have contributed three or four documents is even lower, each group accounting for only 0.6% of the total. These figures reflect a highly skewed distribution of author productivity, with a small number of researchers producing multiple works, while the majority contribute only once. This pattern is characteristic of many academic fields, where a large number of researchers engage in a topic but only a few continue to contribute extensively. In the context of innovation and green finance management in the textile industry, this might suggest that while the field attracts a broad range of scholars, sustained and in-depth research is limited to a smaller group. This could pose challenges for the development of a cohesive and comprehensive body of knowledge, as the field may be dominated by a multitude of single-document authors, potentially leading to a fragmented research landscape.

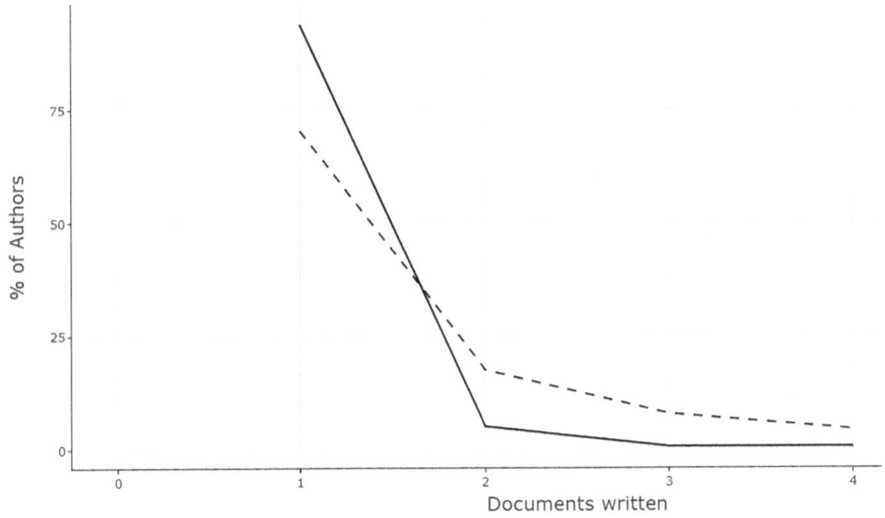

Fig. 9 Author productivity through Lotka's law

Figure 10 reveals some intriguing trends and challenges regarding the academic output of various affiliations over time. A consistent pattern of publication emerges from the Luxury and Fashion Management SKEMA Business School, beginning with a modest output of two articles in 2015 and stabilizing at four articles annually from 2016 to 2023. This demonstrates a sustained interest and steady research output in the field, suggesting a strong, ongoing focus on these topics within the institution. Contrastingly, Donghua University shows a more dynamic trajectory. Initially, there were no publications in 2015, but a gradual increase is observed from 2016 onwards, peaking significantly in 2022 and 2023 with 7 and 13 articles, respectively. This sharp rise indicates a growing emphasis and possibly a strategic focus on the subject matter in recent years, signaling Donghua University's increasing engagement and potential leadership in this research area.

Politecnico di Milano presents a different scenario. There were no articles published in the field until 2018, after which the institution maintained a consistent output of four articles per year. This could indicate a newfound interest in the subject starting in 2018, followed by a steady engagement without significant growth in volume.

The Hong Kong Polytechnic University and Sharda University both depict a late but sudden increase in research publications. Neither institution had any articles until 2021 for the former and 2023 for the latter, after which there is an immediate rise to four articles. This pattern might suggest a recent initiation into the field, possibly driven by emerging trends or new capabilities within these institutions.

The data from Fig. 10 highlight varied approaches and commitments to research in innovation and green finance management in the textile industry among different academic institutions. While some show early and consistent involvement, others

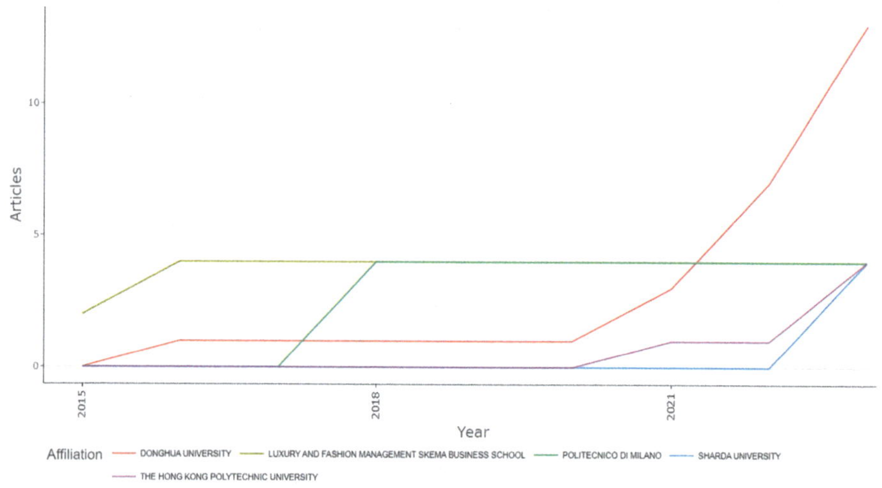

Fig. 10 Affiliations' production over time

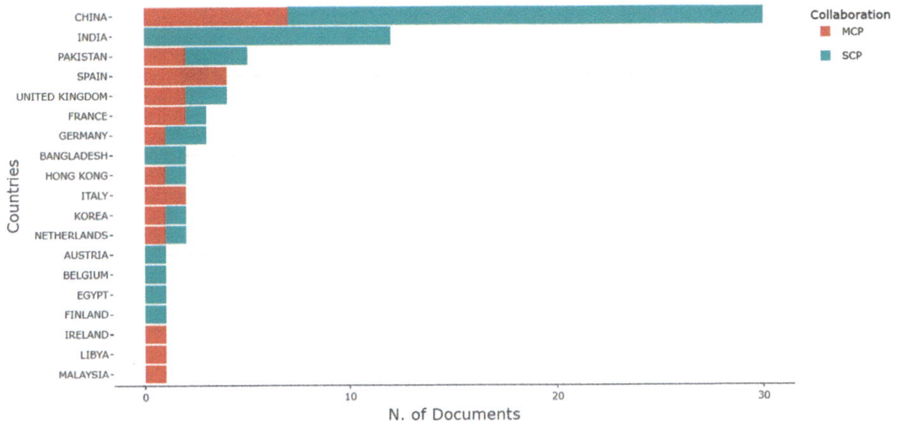

Fig. 11 Corresponding author's countries

display a late but rapid increase in interest, reflecting the dynamic nature of academic research and its responsiveness to emerging trends in the industry.

The data present a clear dominance of China in the field, both in terms of the total number of articles (30) and single-country publications (SCPs, 23), indicating a strong, localized research interest. The relatively lower multi-country publication (MCP) count (7) with an MCP ratio of 0.233 suggests a moderate level of international collaboration (Fig. 11).

India follows as the second most prolific country with 12 articles, all of which are SCPs, indicating a focused, domestic research effort without international collaboration. This is contrasted starkly with countries like Spain and Italy, each having all their publications (four and two, respectively) as MCPs, demonstrating a strong inclination towards international collaboration.

Pakistan, the United Kingdom, France, Germany, Hong Kong, Korea, and the Netherlands present a balanced mix of SCPs and MCPs, reflecting a blend of localized research and international partnerships. The MCP ratios in these countries vary, with Pakistan having a notably high ratio (0.400), suggesting a significant leaning towards international collaboration relative to its total output.

The data collectively reveal a multifaceted picture of research in the textile industry's innovation and green finance management. It underscores China and India's dominant, locally-focused research efforts, a strong European tendency towards international collaborations (especially in Spain and Italy), and a diverse approach in countries with fewer publications, oscillating between domestic focus and international partnerships. This landscape reflects the varying priorities, resources, and collaborative networks across the globe in this research area.

China leads the chart with the highest total citations (TC) of 346, though its average citations per article stand at 11.50 (Fig. 12), suggesting a high volume of research with moderate per-article influence. India follows with a TC of 268 and a

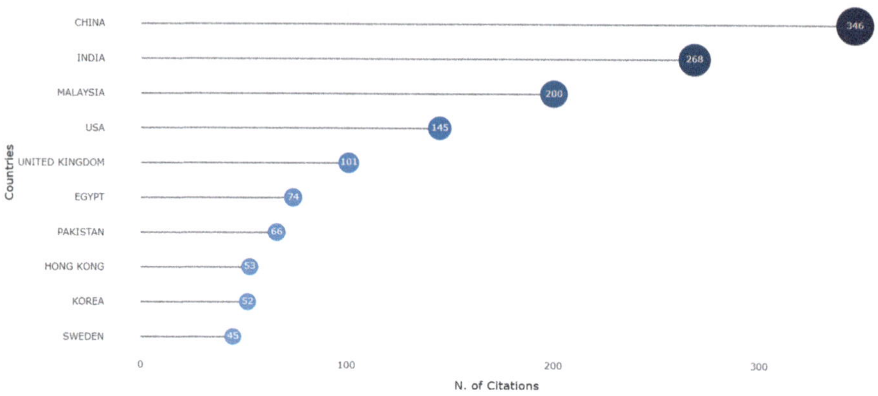

Fig. 12 Most cited countries

notably higher average citation rate of 22.30 per article, indicating fewer but more impactful publications.

Malaysia and the USA demonstrate an extraordinary pattern where their TC and average citations per article are identical (200 and 145, respectively). This unusual coincidence suggests a scenario where each country may have produced a single, highly influential paper or a few papers with uniformly high citation rates. The United Kingdom, with a TC of 101 and average citations of 25.20, along with Egypt (TC 74, average citations 74.00), Pakistan (TC 66, average citations 13.20), and Hong Kong (TC 53, average citations 26.50), show diverse patterns in their research impact, indicating different levels of research output and influence. Korea and Sweden, with TCs of 52 and 45 and average citations per article close to their TCs (26.00 and 45.00, respectively), round out the list. These countries, though contributing lesser in terms of quantity, seem to have produced highly influential individual papers. This analysis underscores the varied nature of research contributions and their impact across different countries in the field of innovation and green finance in the textile industry, highlighting emerging trends and challenges in global research collaboration and knowledge dissemination.

Figure 13 shows the scientific production of various countries in the realms of innovation and green finance management within the textile industry, reveals some intriguing trends and challenges. At the forefront, China stands out with a significant lead, having contributed 88 publications. This dominance could be indicative of China's substantial investment in both the textile industry and green finance initiatives, reflecting its position as a global leader in textile manufacturing and innovation. Following China, India shows a strong presence with 36 publications, possibly highlighting its growing emphasis on sustainable practices in its burgeoning textile sector.

The United Kingdom and the United States are tied with 21 publications each, suggesting a balanced interest and contribution to the field from these nations. Italy and Pakistan, each with 17 publications, indicate their emerging focus in this area,

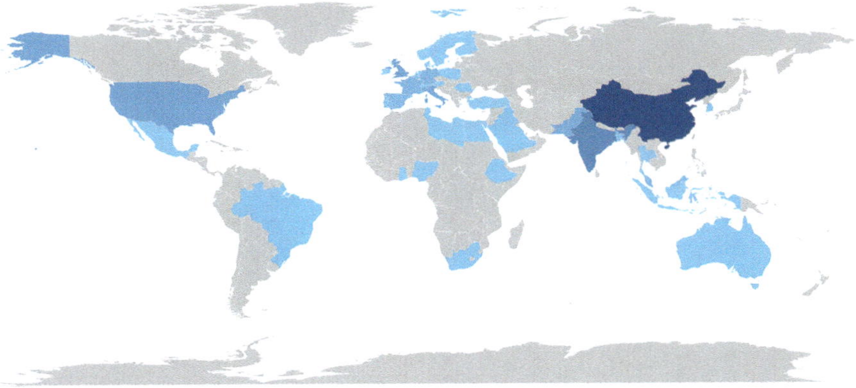

Fig. 13 Countries' scientific production

potentially driven by Italy's renowned fashion industry and Pakistan's growing textile sector. France (13), Germany (12), Malaysia (10), and Spain (10) showcase moderate levels of scientific output, reflecting a consistent but less dominant engagement in this research area. Countries like Bangladesh, Iraq, Brazil, Portugal, and Saudi Arabia, with publications ranging from 5 to 7, demonstrate a budding interest in this field. This interest could be propelled by the need to integrate sustainable practices in their respective textile industries, which are significant contributors to their economies.

The analysis not only highlights the leading contributors to this field but also underscores the widespread global interest in integrating sustainable practices in the textile industry. The varying levels of scientific production across countries reflect the differing stages of development and focus on green finance and innovation in the textile sector. This diversity presents both challenges and opportunities for collaborative research, knowledge sharing, and the advancement of sustainable practices worldwide.

Figures 14, 15, depicting the word map of researched publications in the domain of innovation and green finance management in the textile industry, reveals insightful emerging trends and challenges. This factorial analysis categorizes terms into distinct clusters based on their dimensions (Dim.1 and Dim.2), which represent their relative importance and interrelation in the research landscape.

Key insights include the prominence of terms like "green innovation" and "technological innovation," situated in Cluster 1, indicating a significant focus on integrating environmental considerations with innovation. These terms, with their negative Dim.1 and positive Dim.2 values, suggest a nuanced intersection between technological advancement and green practices. Cluster 2, featuring terms like "sustainability," "innovation," "eco-innovation," and various aspects directly linked to the textile industry (like "textile" and "fashion"), highlights the industry's growing emphasis on sustainable practices. The positioning of these terms, mostly with positive Dim.2 values, underscores their rising importance in the current research narrative. Interestingly, "circular economy," "recycling," and "waste" form the core of

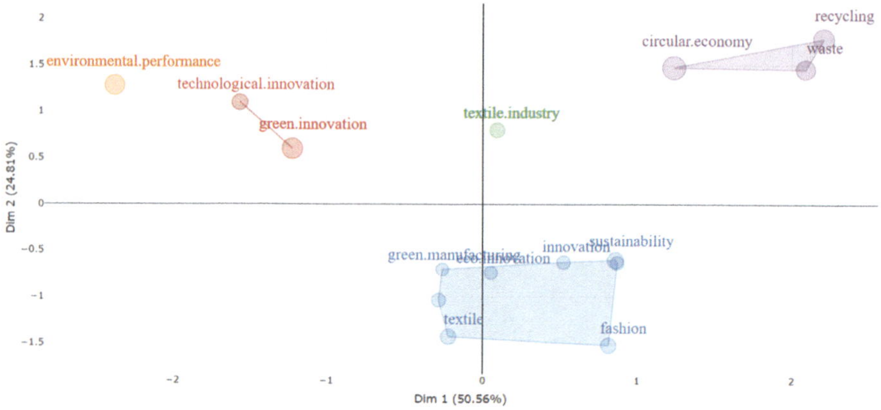

Fig. 14 Word map of the studied publications factorial analysis

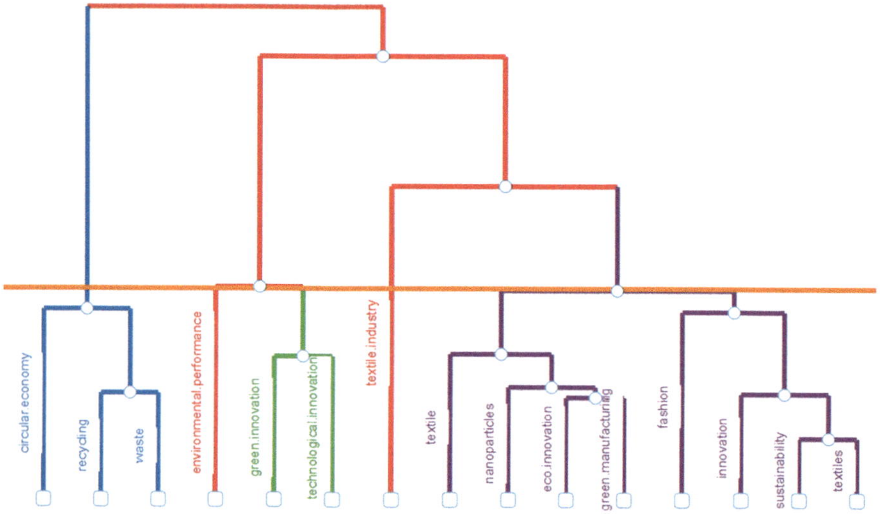

Fig. 15 Factorial analysis dendrogram of the studied publications

Cluster 4, reflecting an increased research focus on sustainable resource management within the industry. Their high positive values in both dimensions signify their critical role in shaping future trends in green finance and innovation within the textile sector. In contrast, "environmental performance" stands out in Cluster 5, with its highly negative Dim.1 value and positive Dim.2 value, indicating it as a pivotal yet challenging aspect of the industry that requires more focused research and innovation efforts.

The word map illustrates a clear trend towards integrating green and sustainable practices within the textile industry, emphasizing recycling and circular economy

Table 4 Factorial analysis indicators of the researched publications

Word	Dim. 1	Dim. 2	Cluster
Green.innovation	−1,22	0,6	1
Sustainability	0,86	−0,61	2
Textile.industry	0,09	0,8	3
Circular.economy	1,24	1,48	4
Innovation	0,52	−0,63	2
Eco.innovation	0,05	−0,74	2
Nanoparticles	−0,29	−1,04	2
Textile	−0,22	−1,43	2
Textiles	0,87	−0,64	2
Environmental.performance	−2,38	1,28	5
Fashion	0,81	−1,53	2
Green.manufacturing	−0,26	−0,71	2
Recycling	2,2	1,79	4
Technological.innovation	−1,56	1,1	1
Waste	2,09	1,46	4

as key components. It also highlights the need for more focused research on environmental performance, suggesting this is an area ripe for innovative solutions and advancements.

The factorial analysis indicators presented in Table 4 offer a multifaceted view of the themes and trends within the realm of innovation and green finance management in the textile industry. This analysis, mapped across two dimensions and clustered into distinct groups, reveals intriguing patterns and focal points in the researched publications.

The first cluster emphasizes green innovation and technological innovation, indicating a strong alignment with progressive environmental approaches integrated with technological advancements. This cluster highlights the evolving nature of green technologies in the textile sector, with a particular focus on innovation that is environmentally friendly. Sustainability, innovation, eco-innovation, nanoparticles, textile, textiles, fashion, and green manufacturing form the second cluster. This group illustrates a comprehensive blend of concepts, ranging from general sustainability and innovation to more specific aspects like nanoparticles and fashion. The presence of terms like eco-innovation and green manufacturing suggests a keen interest in sustainable practices within the textile industry, emphasizing the growing emphasis on environmental responsibility and ethical production methods. The third cluster, represented solely by the term "textile industry," signifies the central role of this sector in the broader context of the study. It acts as a pivot around which other themes revolve, highlighting its importance as a focal point of research in the field of green finance and innovation. Circular economy and recycling, in the fourth cluster, point towards a strong inclination towards sustainable resource management and waste reduction.

This cluster emphasizes the shift towards a circular economy, where resource use is optimized, and recycling is prioritized to minimize environmental impact. The fifth cluster, represented by environmental performance, stands out with its strong dimensional values. This indicates a significant focus on measuring and evaluating the environmental impact of the textile industry. This cluster underscores the importance placed on assessing and improving the environmental footprint of industry practices. These clusters and their associated terms paint a comprehensive picture of the emerging trends and challenges in the textile industry, particularly in the areas of sustainability, innovation, and green finance. The emphasis on circular economy, eco-innovation, and environmental performance suggests a growing awareness and commitment to sustainable practices within the industry.

3 Conclusion

The analysis reveals a clear upward trajectory in academic interest in the intersection of risk and innovation management within the textile industry. Starting with minimal scientific production in 2014, there has been a noticeable increase in publications from 2015 onwards. This trend suggests a growing recognition of the importance of innovation and green finance in addressing environmental challenges and sustainability goals. The slight dip in publications in 2020, likely due to the COVID-19 pandemic, did not significantly dampen academic interest. This resilience indicates a sustained commitment to exploring innovation and sustainability themes, even in the face of global disruptions. The peak in publications in 2023, with 38 articles, underscores the current and pressing nature of this topic. It reflects the culmination of previous years' developments and indicates deep engagement by the academic community in addressing the challenges at the intersection of risk management, innovation, and green finance in the textile industry.

The analysis highlights a diverse range of research topics within the field, including biotechnology, nanotechnology, consumer behavior, and clean production methodologies. This diversity underscores the multidisciplinary nature of the challenges and solutions in the textile industry. The data reveal a highly skewed distribution of author productivity, with a small number of researchers producing multiple works while the majority contribute only once. This pattern may pose challenges for the development of a cohesive body of knowledge in the field.

Different academic institutions exhibit varying levels of commitment to research in innovation and green finance in the textile industry. Some institutions have shown sustained interest, while others have rapidly increased their research output in recent years.

China leads in terms of total publications, reflecting its significant investment in both the textile industry and green finance initiatives. India, the UK, and the USA also demonstrate strong contributions. The varying levels of research production across countries reflect differing stages of development and focus on sustainable practices.

The word map analysis reveals key themes, including green innovation, sustainability, circular economy, and environmental performance. These themes reflect a growing emphasis on integrating green and sustainable practices in the textile industry, with a focus on resource management and environmental impact assessment. Researchers should continue to explore innovative technologies and practices that promote sustainability in textile production. This includes further investigation into eco-friendly materials and processes. A deeper understanding of consumer behavior and its impact on sustainable fashion choices is essential. Future research should delve into consumer preferences, trends, and the role of fashion in driving green practices. Collaboration between academia, industry, and policymakers is crucial to drive the adoption of sustainable practices in the textile industry. Research should focus on policy implications and strategies for industry stakeholders to transition towards sustainability. Investigating the economic implications of green finance and its role in supporting innovation in the textile sector is an avenue worth exploring. Understanding the financial mechanisms for sustainable practices can facilitate their adoption.

The bibliometric analysis provides a comprehensive overview of the current state of research in innovation and green finance management in the textile industry. It illuminates emerging trends, knowledge gaps, and opportunities for future investigations. Researchers, industry stakeholders, and policymakers can use these insights to guide their efforts in advancing sustainable practices within the textile industry.

References

Ali, W., Wen, J., Hussain, H., Khan, N. A., Younas, M. W., & Jamil, I. (2021). Does green intellectual capital matter for green innovation adoption? Evidence from the manufacturing SMEs of Pakistan. *Journal of Intellectual Capital, 22*(5), 868–888.

Bakos, J., Siu, M., Orengo, A., & Kasiri, N. (2020). An analysis of environmental sustainability in small and medium-sized enterprises: Patterns and trends. *Business Strategy and the Environment, 29*(3), 1285–1296.

Bertola, P., & Teunissen, J. (2018). Fashion 4.0. Innovating fashion industry through digital transformation. *Research Journal of Textile and Apparel, 22*(4), 352–369.

Costa, I., Martins, F. G., & Alves, I. (2019). Ecological Footprint as a sustainability indicator to analyze energy consumption in a Portuguese textile facility. *International Journal of Energy and Environmental Engineering, 10*, 523–528. https://doi.org/10.1007/s40095-018-0268-6

Desore, A., & Narula, S. A. (2018). An overview on corporate response towards sustainability issues in textile industry. *Environment, Development and Sustainability, 20*, 1439–1459.

Fang, Y., Chen, G., Bick, M., & Chen, J. (2021). Smart textiles for personalized thermoregulation. *Chemical Society Reviews, 50*(17), 9357–9374. 10.1039/D1CS00003A

Gonçalves, A., & Silva, C. (2021). Looking for sustainability scoring in apparel: A review on environmental footprint, social impacts and transparency. *Energies, 14*(11), 3032.

Guo, L., Zhao, S., Song, Y., Tang, M., & Li, H. (2022). Green finance, chemical fertilizer use and carbon emissions from agricultural production. *Agriculture, 12*(3), 313.

Hebeish, A., Shaheen, T. I., & El-Naggar, M. E. (2016). Solid state synthesis of starch-capped silver nanoparticles. *International Journal of Biological Macromolecules, 87*, 70–76.

Huang, D. (2022). Green finance, environmental regulation, and regional economic growth: From the perspective of low-carbon technological progress. *Environmental Science and Pollution Research, 29*(22), 33698–33712.
Jia, F., Yin, S., Chen, L., & Chen, X. (2020). The circular economy in the textile and apparel industry: A systematic literature review. *Journal of Cleaner Production, 259*, 120728.
Kumar, L., Nadeem, F., Sloan, M., Restle-Steinert, J., Deitch, M. J., Ali Naqvi, S., & Sassanelli, C. (2022). Fostering green finance for sustainable development: A focus on textile and leather small medium enterprises in Pakistan. *Sustainability, 14*(19), 11908.
Lara, L., Cabral, I., & Cunha, J. (2022). Ecological approaches to textile dyeing: A review. *Sustainability, 14*(14), 8353.
Li, Z., Chen, J., Zhou, J., Zheng, L., Pradel, K. C., Fan, X., & Wang, Z. L. (2016). High-efficiency ramie fiber degumming and self-powered degumming wastewater treatment using triboelectric nanogenerator. *Nano Energy, 22*, 548–557.
Liew, R. K., Azwar, E., Yek, P. N. Y., Lim, X. Y., Cheng, C. K., Ng, J. H., & Lam, S. S. (2018). Microwave pyrolysis with KOH/NaOH mixture activation: A new approach to produce micro-mesoporous activated carbon for textile dye adsorption. *Bioresource Technology, 266*, 1–10.
Liu, X., Yang, Y., Jiang, Y., Fu, Y., Zhong, R. Y., Li, M., & Huang, G. Q. (2023). Data-driven ESG assessment for blockchain services: A comparative study in textiles and apparel industry. *Resources, Conservation and Recycling, 190*, 106837.
Lu, Q., Farooq, M. U., Ma, X., & Iram, R. (2022). Assessing the combining role of public-private investment as a green finance and renewable energy in carbon neutrality target. *Renewable Energy, 196*, 1357–1365.
Majumdar, A., & Sinha, S. K. (2019). Analyzing the barriers of green textile supply chain management in Southeast Asia using interpretive structural modeling. *Sustainable Production and Consumption, 17*, 176–187.
Nayak, L., & Mishra, S. P. (2016). Prospect of bamboo as a renewable textile fiber, historical overview, labeling, controversies and regulation. *Fash Text, 3*, 2.
Nouinou, H., Asadollahi-Yazdi, E., Baret, I., Nguyen, N. Q., Terzi, M., Ouazene, Y., & Kelly, R. (2023). Decision-making in the context of Industry 4.0: Evidence from the textile and clothing industry. *Journal of Cleaner Production*, 136184.
Rahman, M., Billah, M. M., & Hack-Polay, D. (2019). What is hindering change? Anticipating the barriers to the adoption of enzyme-based textile processing in a developing country. *Business Strategy & Development, 2*(2), 137–147.
Schellenberger, S., Hill, P. J., Levenstam, O., Gillgard, P., Cousins, I. T., Taylor, M., & Blackburn, R. S. (2019). Highly fluorinated chemicals in functional textiles can be replaced by re-evaluating liquid repellency and end-user requirements. *Journal of Cleaner Production, 217*, 134–143.
Sharma, S., Prakash, G., Kumar, A., Mussada, E. K., Antony, J., & Luthra, S. (2021). Analysing the relationship of adaption of green culture, innovation, green performance for achieving sustainability: Mediating role of employee commitment. *Journal of Cleaner Production, 303*, 127039.
Sierdovski, M., Pilatti, L. A., & Rubbo, P. (2022). Organizational competencies in the development of Environmental, Social, and Governance (ESG) criteria in the industrial sector. *Sustainability, 14*(20), 13463.
Velmurugan, P., Shim, J., Bang, K. S., & Oh, B. T. (2016). Gold nanoparticles mediated coloring of fabrics and leather for antibacterial activity. *Journal of Photochemistry and Photobiology b: Biology, 160*, 102–109.
Wang, F., Cai, W., & Elahi, E. (2021). Do green finance and environmental regulation play a crucial role in the reduction of CO_2 emissions? An empirical analysis of 126 Chinese cities. *Sustainability, 13*(23), 13014.
Wang, M. L. (2023). Effects of the green finance policy on the green innovation efficiency of the manufacturing industry: A difference-in-difference model. *Technological Forecasting and Social Change, 189*, 122333.

Wang, Y., & Yang, Y. (2021). Analyzing the green innovation practices based on sustainability performance indicators: A Chinese manufacturing industry case. *Environmental Science and Pollution Research, 28*, 1181–1203.

Xing, C., Zhang, Y., & Tripe, D. (2021). Green credit policy and corporate access to bank loans in China: The role of environmental disclosure and green innovation. *International Review of Financial Analysis, 77*, 101838.

Zhang, S., Wu, Z., Wang, Y., & Hao, Y. (2021). Fostering green development with green finance: An empirical study on the environmental effect of green credit policy in China. *Journal of Environmental Management, 296*, 113159.

The Innovative Influence of Industry 4.0 on the Textile Industry

Rui Dantas

Abstract This study aims to perform a bibliometric analysis on the burgeoning field of research at the intersection of Industry 4.0 and the textile industry. It seeks to identify and analyze the most influential contributions, chart the growth of literature over time, and discern the core research themes and potential gaps for future exploration. Utilizing bibliometric tools such as VOS viewer and Biblioshiny, this study systematically quantifies the research output, maps co-authorship and citation networks, and examines the thematic landscape through keyword co-occurrence analysis. The findings highlight a significant uptrend in research on the technological integration of Industry 4.0 within the textile sector. Key areas of focus include automation, IoT, and smart manufacturing, reflecting a paradigm shift towards more adaptive, predictive, and self-optimized production processes. The analysis also reveals a geographical concentration of research activity, particularly in Asia and Europe, with China and the UK leading in terms of output and international collaboration, respectively. Despite the technological emphasis, the study uncovers a relative scarcity in research on the socio-economic and environmental implications of Industry 4.0 applications in the textile industry. The article concludes with a call for future research to broaden the scope to encompass sustainability, economic, and policy perspectives, thereby enriching the current understanding of this industrial transformation.

1 Introduction

The evolution of industries over time can be categorized into distinct eras, each characterized by revolutionary advancements in technology and methods of production. Starting from the utilization of water and steam power in the first industrial revolution to the introduction of electricity in the second and the emergence of computer and automation technology in the third, the world has witnessed profound

R. Dantas (✉)
Instituto Piaget—Instituto Superior de Estudos Interculturais e Transdisciplinares de Almada (ISEIT), Avenida Jorge Peixinho, No. 30 - Quinta da Arreinela, 2805-059 Almada, Portugal
e-mail: rui.dantas@ipiaget.pt

Lisbon Accounting and Business School, Polytechnic Institute of Lisbon (ed.), *Industry and Innovation: Textile Industry*, SDGs and Textiles,
https://doi.org/10.1007/978-3-031-57804-5_10

transformations in manufacturing paradigms. Currently, we are amidst the fourth industrial revolution, popularly termed as Industry 4.0. This revolution is typified by the amalgamation of cyber-physical systems, the Internet of Things (IoT), artificial intelligence, and big data analytics. These advancements have radically transformed various industrial sectors, not least among them the textile industry.

The textile industry, one of the oldest and most significant sectors in the global economy, has historically been known for its labor-intensive nature. The integration of Industry 4.0 technologies is promising a seismic shift from traditional modes of production to more automated, efficient, and sustainable methods. Such a transformation could mean unprecedented levels of product customization, reduced lead times, enhanced quality control, and a more sustainable production footprint. Nevertheless, while the potential benefits are vast, they also present a new set of challenges in terms of adaptability, technology adoption, skill-set evolution, and infrastructure overhaul.

In the scholarly realm, a myriad of research has emerged over the last decade elucidating the convergence of Industry 4.0 and the textile sector. These studies range from exploring new technological implementations in textile manufacturing processes to investigating the socio-economic implications of this new industrial age. Given the rapid proliferation of literature on this topic, there exists a crucial need for a comprehensive overview that delineates the trajectories of research, identifies the core themes, and pinpoints knowledge gaps.

Bibliometric analysis stands as a robust tool in fulfilling this need. It aids in the quantitative assessment of research output in a particular domain, offering insights into the most influential authors, institutions, countries, and journals. Moreover, it provides a macroscopic view of the thematic evolution, collaboration networks, and future directions of research. To date, while bibliometric analyses have been conducted on Industry 4.0 in a broader sense, a dedicated examination centered on its influence in the textile industry remains an untouched avenue.

Therefore, the primary objective of this study is to comprehensively map the landscape of research pertaining to the interplay of Industry 4.0 and the textile industry using bibliometric analysis. In doing so, this study aspires to:

- identify the leading authors, institutions, countries, and journals that have significantly contributed to the discourse on Industry 4.0 in the textile sector;
- chart the chronological growth of publications in this domain and predict future research trends;
- uncover the core thematic areas of research and their inter-relations;
- illuminate potential knowledge gaps, offering avenues for future investigations.

The insights gleaned from this study are envisioned to benefit a broad spectrum of stakeholders. Academics and researchers can gain clarity on the state of the art, leading voices, and potential areas of exploration. Industry professionals and policymakers can leverage the findings to understand technological trends, challenges, and opportunities presented by the intersection of Industry 4.0 and textiles. Additionally, educators can glean valuable information on the evolving knowledge areas, aiding in curriculum design and skill development initiatives.

To contextualize the forthcoming analysis, it is vital to understand the essence of Industry 4.0 and its core components. Industry 4.0 is more than just a buzzword; it represents a paradigm shift where machines, products, and systems interconnect and communicate in real time to make decentralized decisions. At the heart of this revolution lies the Internet of Things (IoT), enabling devices to collect and exchange data. When complemented with artificial intelligence and big data analytics, it paves the way for smart manufacturing, where processes become more adaptive, predictive, and self-optimized. Within the textile industry, this translates to smart textiles, wearables with embedded sensors, predictive maintenance of machinery, and real-time supply chain optimization, among other innovations.

As the textile industry stands on the cusp of a transformative phase powered by Industry 4.0, a systematic understanding of the existing body of knowledge becomes imperative. This study, through its bibliometric analysis, aims to offer such a panoramic view, bridging the past and present research trajectories and illuminating the path for future inquiries.

2 Systematic Literature Review

Bibliometric analysis serves as an insightful tool to gauge the magnitude of scholarly output on specific subjects and comprehend its temporal evolution, trends, and dynamics. Table 1 elucidates the publication data pertaining to the influence of Industry 4.0 on the textile sector (a detailed analysis of the main information).

The data span the period from 2011 to 2022. This timeframe suggests that the influence of Industry 4.0 on textiles has been a topic of interest for at least a decade in the Scopus and Web of Science databases. An impressive annual growth rate of 29.89% indicates that there has been a surging interest in this domain, reflecting the escalating impact of Industry 4.0 on the textile sector. A total of 215 sources have been identified, demonstrating a wide-ranging investigation. This number includes journals, books, and other types of media. The dataset consists of 315 documents, with an average age of 3.68 years. This suggests that most research on this topic is relatively recent. The average citations per document stand at 15.73. This is a testament to the relevance and significance of the content, as it indicates that each article or source is being referenced or cited by other works. With 1002 authors contributing to this corpus of work, it's clear that the topic has engaged a sizable scholarly community.

The bibliometric data underscore that the impact of Industry 4.0 on the textile industry has piqued substantial academic interest over the past decade. This interest is not only confined to traditional research articles but also spans conference discussions and reviews. The relatively recent average age of the documents, paired with a high annual growth rate, suggests that the textile sector's intersection with Industry 4.0 remains an evolving and vital area of study. The diverse authorship and blend of collaborative and individual research further enrich this academic landscape. To carry out the study and pinpoint the principal characteristics of the research undertaken,

Table 1 Main information about data

Description	Results
Timespan	2011:2022
Sources (journals, books, etc.)	215
Documents	315
Annual growth rate %	29,89
Document average age	3,68
Average citations per doc	15,73
Authors	1002
Authors of single-authored docs	19
Single-authored docs	59
Co-authors per doc	3,82
International co-authorships %	13,97
Article	166
Book	3
Book chapter	13
Conference paper	87
Conference paper article	3
Conference review	32
Note	2
Review	7
Short survey	2

the VOSviewer software was utilized to execute a cluster analysis on documents tagged with "industry 4.0," "innovation," and "textile industry" from 2011 to 2022. The outcomes of this analysis are depicted in Fig. 1.

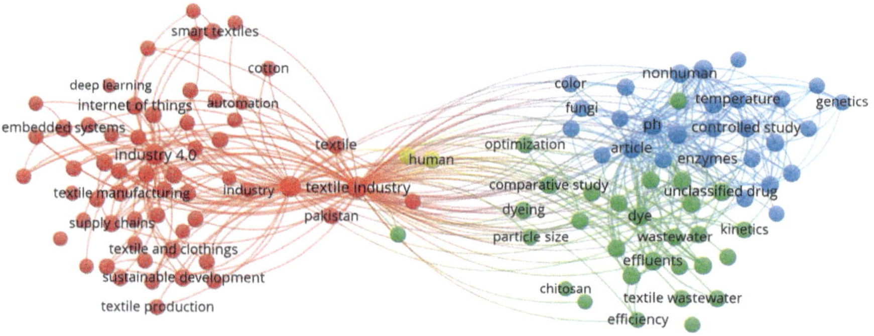

Fig. 1 Visual representation of cluster analysis on "Industry 4.0," "innovation," and "textile industry" publications

The concept of Industry 4.0, often referred to as the "fourth industrial revolution," has captured the attention of scholars and industry experts alike due to its transformative potential across various sectors. One industry that has witnessed significant shifts due to Industry 4.0 is the textile sector (Majumdar et al., 2021). The emergence of Industry 4.0 is deeply rooted in the digitalization of manufacturing processes, integration of cyber-physical systems, and the utilization of advanced data analytics (Nounou et al., 2022; Fromhold-Eisebith et al., 2021). The development of the Internet of Things (IoT) has played a pivotal role in bridging the gap between physical and digital realms, paving the way for enhanced connectivity and real-time data exchange (Ahmad et al., 2020; Manglani et al., 2019).

Advanced manufacturing processes, augmented by digital tools, have redefined traditional textile production methods. The implementation of smart factories facilitates real-time monitoring, predictive maintenance, and autonomous decision-making, ensuring seamless textile production (Imran et al., 2018; Ku et al., 2020). Enhanced automation and the use of robotic technologies have enabled precision in textile manufacturing, leading to reduced material wastage and increased efficiency (Nascimento et al., 2019). The integration of sensors and IoT devices has also made it possible to track and trace materials across the supply chain, ensuring transparency and sustainability (Lalic et al., 2019; Mohiuddin Babu et al., 2022).

The textile industry has long been under scrutiny for its environmental impact. With the integration of Industry 4.0, there's an emphasis on sustainable manufacturing processes, ensuring minimal resource wastage and maximum energy efficiency (Margherita and Braccini, 2020; Ciliberto et al., 2021). Advanced monitoring systems have empowered manufacturers to reduce their carbon footprint by optimizing energy consumption and waste production (Fallahpour et al., 2021). Moreover, real-time data analytics tools facilitate environmental impact assessment, promoting eco-innovation in textile design and manufacturing (Pandey et al., 2020; Salo et al., 2020).

The era of mass production is gradually making way for mass customization in the textile industry. Advanced data analytics, coupled with augmented reality tools, allow manufacturers to customize designs based on consumer preferences, resulting in personalized products (Mourtzis et al., 2019; Akram et al., 2022). The enhanced connectivity also paves the way for improved consumer engagement and feedback, leading to products that truly resonate with consumer needs (De Silva et al., 2019).

While the transformative potential of Industry 4.0 in the textile sector is evident, challenges persist. Issues related to data privacy, security, and the need for significant capital investment are primary concerns for textile manufacturers (Chen, 2019; Wijewardhana et al., 2021). There is also a pressing need for skilled labor that can adeptly navigate the advanced tools and technologies integral to Industry 4.0 (Chen, 2020). However, the opportunities outweigh the challenges. The integration of advanced technologies is expected to drive the global competitiveness of the textile industry, ensuring its sustained growth in the future (Guan et al., 2019; Jiménez-Marín et al., 2021).

As the textile industry delves deeper into the realm of Industry 4.0, it's anticipated that technologies like artificial intelligence, blockchain, and 3D printing will further redefine the sector (Nazam et al., 2022). The convergence of biotechnology and

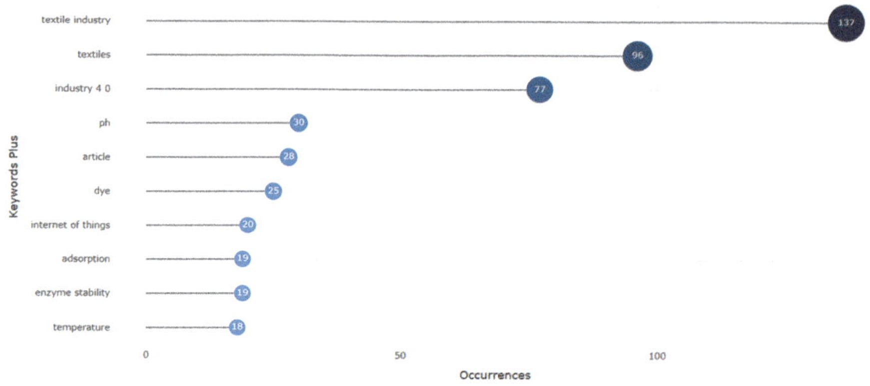

Fig. 2 Most frequently occurring words in the studied documents

textiles, facilitated by Industry 4.0 tools, could pave the way for "smart textiles" with unprecedented functionalities (Barman et al., 2022; Farooq & Zhang, 2022).

The transformative influence of Industry 4.0 on the textile industry is multi-faceted, offering numerous opportunities for innovation, sustainability, and enhanced consumer experiences. As the sector continues to evolve, it's imperative for stakeholders to stay abreast of the latest technological developments, ensuring that the textile industry remains at the forefront of the fourth industrial revolution (Bai et al., 2020; Jimeno-Morenilla et al., 2021).

Figure 2 provides a bibliometric insight into the prevalence of specific terms within the researched publications on the topic of Industry 4.0's influence on the textile industry.

"Textile industry" and "textiles" are the most frequently occurring terms with 137 and 96 occurrences, respectively. This is expected as they are central to the theme of the researched articles. The term "industry 4 0" appeared 77 times, reinforcing the prominence of Industry 4.0 discussions in the textile sector. Its relevance cannot be understated as the fourth industrial revolution brings about digital transformation, impacting various sectors including textiles. Terms like "ph," "dye," "adsorption," "enzyme stability," and "temperature" suggest technical aspects of textile processing or research. For instance, "PH" and "dye" likely relate to the dyeing process in textiles. "Adsorption" might pertain to the process of dye uptake by fibers.

"Enzyme stability" suggests research on enzymes used in textile processing, perhaps for sustainable or eco-friendly methods. "Temperature" can be related to various processes in textile manufacturing, from dyeing to finishing.

"Internet of things" (IoT) is mentioned 20 times, indicating the increasing integration of IoT technologies in the textile industry. This could relate to smart textiles, real-time monitoring of manufacturing processes, or supply chain optimizations using IoT.

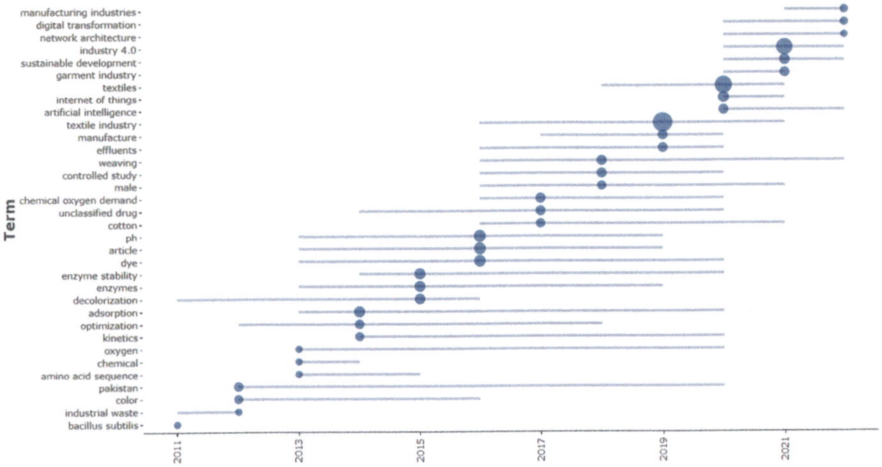

Fig. 3 Trend topics in the studied documents

The term "article" appeared 28 times, which is a meta-reference indicating the nature of the content being analyzed. It might suggest that many publications explicitly state that they are articles or use the term in their methodology or discussions.

Figure 2 gives a snapshot of the convergence of traditional textile terms with modern industrial and technological terms. The presence of both traditional textile processing terms and emerging tech terms like "Internet of things" showcases the evolving landscape of the textile industry in the era of Industry 4.0.

Figure 3 provides a comprehensive overview of the evolving trends in topics within the researched publications. While some themes have been consistent areas of interest over the years, others, especially those related to technological innovations like Industry 4.0 and IoT, are emerging as new focal points in the textile industry's academic landscape.

The topic "textile industry" emerges as the most frequently researched topic with 137 mentions, followed by "textiles" at 96 mentions, and "industry 4.0" at 77 mentions. These three topics hold a significant lead in frequency when compared to the other listed topics, highlighting their central role in the researched publications. The topic "industry 4.0" has seen a rapid rise in research interest in recent years, with its median year of publication being 2021. This suggests that discussions around Industry 4.0 and its relevance to the textile industry have gained momentum recently. Topics like "ph", "article", "dye", "adsorption", "enzyme stability", and "enzymes" have earlier median years (ranging from 2013 to 2016), indicating that they were subjects of interest in earlier years, but might have plateaued or reduced in prominence recently. The topic "internet of things" (IoT) stands out with its narrow range from 2020 to 2021, suggesting that IoT's relevance to the textile industry is a newer area of exploration.

The range between the first quartile (Q1) and the third quartile (Q3) for topics like "textile industry," "textiles," and "dye" spans several years. This indicates sustained interest in these areas over time.

In contrast, topics like "industry 4.0" and "internet of things" have a tighter range, showing that they have become prominent more recently.

"Adsorption," "enzyme stability," and "enzymes" appear to be more specialized topics. Their presence in the list indicates that while they might not be as frequently mentioned as broader themes like "textile industry" or "industry 4.0," they hold specific importance in certain niche areas of textile research.

Topics like "ph," "article," "dye," and "enzymes" have both their Q1 and median years before or around 2016. This might suggest that while they were significant themes in the past, they may not be the central focus of recent research endeavors.

The thematic map for researched publications (Fig. 4) provides a comprehensive overview of the predominant themes and sub-themes related to the influence of Industry 4.0 on the textile industry. Let's delve into the data and elucidate the significance of each keyword based on occurrences, cluster label, betweenness centrality (btw_centrality), closeness centrality (clos_centrality), and PageRank centrality (pagerank_centrality).

The most frequently occurring word is "textile industry" with 128 occurrences. This is followed by "textiles" (94 occurrences) and "industry 4.0" (77 occurrences). Some other keywords that are seen with a moderate number of occurrences include "internet of things," "sustainable development," "artificial intelligence," and "embedded systems." "Weaving," "clothes," and "garment industry" have the least occurrences among the given words but are still significantly mentioned.

All of the provided keywords belong to the same cluster (Table 2) labeled as "textile industry." This suggests a concentrated focus of the researched publications around the overarching theme of Industry 4.0's impact on the textile sector.

"Textiles" has the highest betweenness centrality, indicating its role as a critical connector in the context of the textile industry's transformation under Industry 4.0.

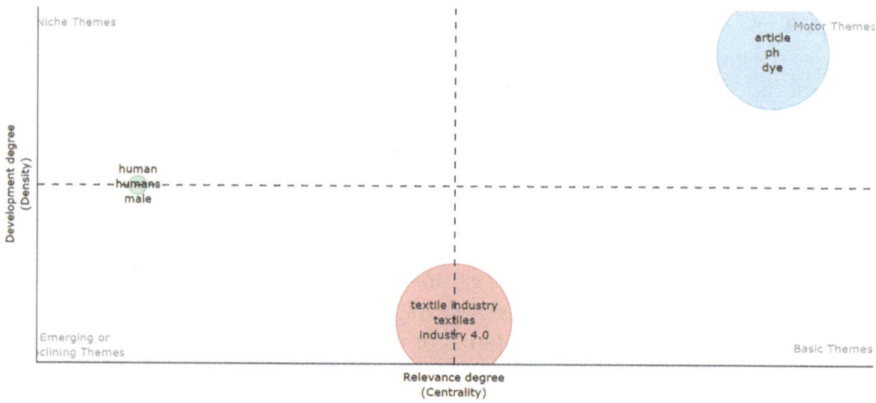

Fig. 4 Thematic map for the studied documents

Table 2 Cluster analysis according to the thematic map

Cluster	CallonCentrality	CallonDensity	RankCentrality	RankDensity	ClusterFrequency
Textile industry	7.1212	54.8776	2	1	781
Article	11.4053	179.2743	3	3	733
Human	3.4620	141.2847	1	2	54

Interestingly, while "textile industry" is the most frequently occurring term, "textiles" and "weaving" have higher closeness centrality values. "Textile industry" has the highest PageRank, followed by "textiles" and "industry 4.0." The prominence of terms like "internet of things," "artificial intelligence," and "embedded systems" in the context of the textile industry signifies the integration of advanced technologies in textile processes and products. The presence of "sustainable development" indicates the increasing emphasis on sustainability in the textile industry as it modernizes. The terms "weaving," "clothes," and "garment industry" suggest specific areas within the textile sector that might be witnessing significant innovations or challenges due to the integration of Industry 4.0 concepts.

2.1 *The Common Research Trends and Meta-analysis*

Figure 5 presents the annual scientific production related to the innovative influence of Industry 4.0 on the textile industry as a number of articles published each year, we can examine the trends, key observations, and potential implications of the data.

There is a clear upward trend in the number of articles published from 2011 to 2022. The growth is not only positive but also accelerating over time, which suggests

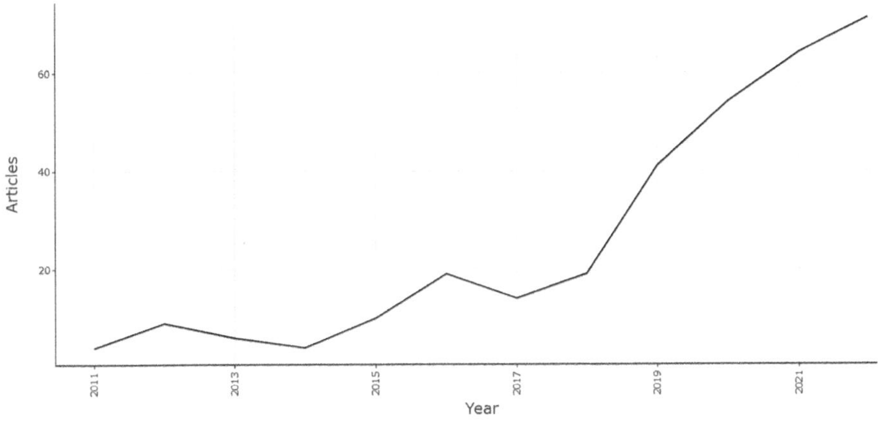

Fig. 5 Annual scientific production on the innovative impact of Industry 4.0 on the textile industry

an increasing interest and focus on the intersection of Industry 4.0 and the textile industry. Between 2011 and 2015, the growth in the number of articles is relatively modest and shows some fluctuations. This period could represent the nascent stage of research where the topic was gradually gaining attention. Starting from 2016, there is a more rapid increase in articles. The years 2016 and 2018 both saw a significant increase, nearly doubling from the previous years, which could indicate key breakthroughs or increased funding and interest in this area of research. The largest jumps occur in the most recent years, from 2019 to 2022, where the numbers rise from 41 to 71. This could be reflective of a maturity phase in the research area, with more developed and comprehensive studies being published.

There is a period of stagnation or modest growth in the early 2010s. This might reflect the early stage of Industry 4.0 concepts being applied to the textile industry. The year 2015 marks an inflection point with a noticeable increase in articles to double digits for the first time. The period from 2016 onwards could be seen as a significant growth phase for Industry 4.0 in the textile sector, with the annual number of articles nearly doubling every few years.

The continuous increase in articles suggests sustained and growing interest and perhaps the ongoing development of new technologies, methods, and applications within the textile industry influenced by Industry 4.0. The increasing trend in publications implies that there is substantial research and development occurring at the intersection of Industry 4.0 and the textile industry, potentially leading to innovative applications and processes.

The growth in academic interest might correlate with the adoption of Industry 4.0 technologies in the textile industry, including automation, data analytics, and the Internet of Things (IoT). The ongoing increase in publications could suggest that the field is still evolving with new areas of interest and research gaps being identified regularly.

Progress in related technologies such as AI, machine learning, and sensor technology could be driving the increased research activity.

The textile industry's need for efficiency, sustainability, and cost reduction could be leading to more research into how Industry 4.0 can address these issues. Increased funding from governments and private sectors for Industry 4.0 initiatives may be facilitating more research. The recent surge in articles might also reflect a response to global events that necessitate innovation in the textile industry, such as the COVID-19 pandemic or supply chain disruptions. An analysis of a single figure without context from the rest of the article might miss out on nuances that explain why these trends are happening. There may be a publication lag that isn't accounted for, where research conducted in 1 year is only published in subsequent years.

Table 3 presents a summary of the average number of citations per article (MeanTCperArt), the number of publications (N), the average citations per year per article (MeanTCperYear), and the number of citable years since publication.

The average citations per article (MeanTCperArt) show fluctuating trends with some years showing high citation rates such as 2015 with 55.2 citations, indicating a possible release of influential research or increased interest in the topic during that year. The average citations per article tend to decrease with the more recent

Table 3 Average citations per year for the studied documents

Year	MeanTCperArt	N	MeanTCperYear	CitableYears
2011	41	4.00	3.15	13
2012	35.22	9.00	2.94	12
2013	23.5	6.00	2.14	11
2014	31.25	4.00	3.12	10
2015	55.2	10.00	6.13	9
2016	22.53	19.00	2.82	8
2017	14.64	14.00	2.09	7
2018	26.74	19.00	4.46	6
2019	10.05	41.00	2.01	5
2020	23.85	54.00	5.96	4
2021	7.58	64.00	2.53	3
2022	4.63	71.00	2.32	2

years (e.g., 2021 and 2022), which is expected as these articles have had less time to accumulate citations.

There is a clear increasing trend in the number of publications from 2011 to 2022, suggesting a growing interest and research activity in the area of Industry 4.0's impact on the textile industry. Despite the increase in the number of publications, the average citations per article are not necessarily increasing, indicating that while quantity is increasing, the impact as measured by citations may not be uniformly high across all papers.

The data indicate that while the total citations per article might be higher for older publications (due to them having more time to accumulate citations), when adjusted for the age of the publication, the picture is slightly different. Notably, 2015 stands out again with the highest MeanTCperYear of 6.13, suggesting that the articles from this year have been particularly impactful on a per-year basis.

Citable years decrease as expected from 2011 to 2022 because the more recent the publication, the less time it has had to be cited. The MeanTCperYear is generally higher for articles with fewer citable years, likely because they have had less time to accumulate citations and their citation rates have not yet stabilized.

The peak in 2015 for both MeanTCperArt and MeanTCperYear could be indicative of seminal works published that year, which had a significant impact on subsequent research. The dips in later years such as 2017, 2019, and particularly 2021 and 2022 suggest either that the works from these years are yet to gain traction, or they may not be as influential, or it could simply be too early to tell given the shorter citation window.

While the interest in Industry 4.0's influence on the textile industry appears to be growing with more publications each year, the impact of these publications as measured by citations seems to vary significantly from year to year. The data from 2015 are particularly notable and may warrant closer examination to understand the

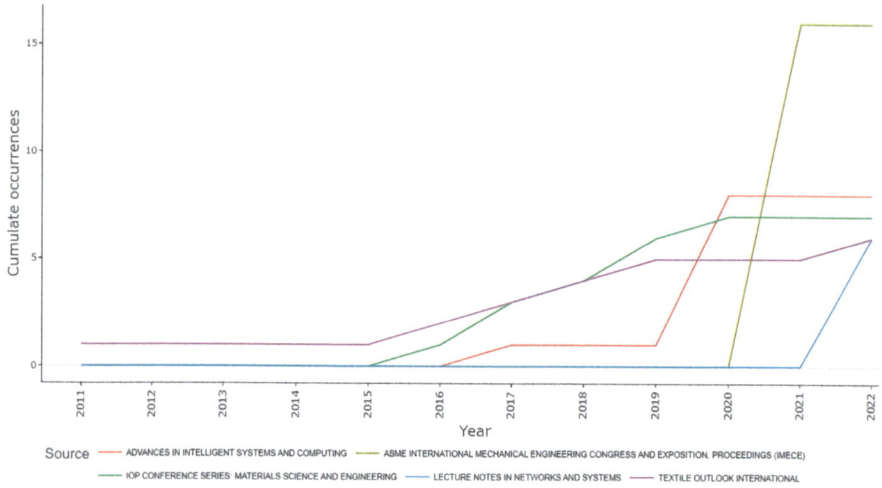

Fig. 6 Sources' production over time

drivers behind the high citation rates for that year. Recent publications have yet to show their full impact in the field, and ongoing citation analysis will be important to track their influence over time.

Textile Outlook International is the only source with a consistent annual output starting from 2011 through 2022, showing a gradual increase from one publication in 2011 to six publications in 2022 (Fig. 6). This indicates a steady interest in the topic over the years within this outlet.

A significant change occurs in 2021, where there is a marked increase in publications from the ASME International Mechanical Engineering Congress and Exposition, Proceedings (IMECE), which went from having no publications to 16 in a single year. This suggests a possible theme or special focus on Industry 4.0 within the conference that year. The Advances in Intelligent Systems and Computing sees its first publication in this field in 2017, and then consistently has one publication annually until 2020 when it increases to eight publications annually for 2020 and 2021, which remains constant into 2022. IOP Conference Series—Materials Science and Engineering shows a trend of increasing output from 2016 onward, peaking at seven publications in 2020, which suggests growing interest from this community in the textile industry's innovations related to Industry 4.0. The Lecture Notes in Networks and Systems series does not register any publications until 2022, where it suddenly presents six publications. This could indicate a new entrant into the field or a new series that has started to focus on the relevant topics.

There is a clear upward trend in the number of publications over time across all sources, which reflects increasing scholarly and industry interest in the impacts of Industry 4.0 on the textile industry. The data suggest that this interest has become more pronounced in the most recent years (2021 and 2022), possibly indicating that

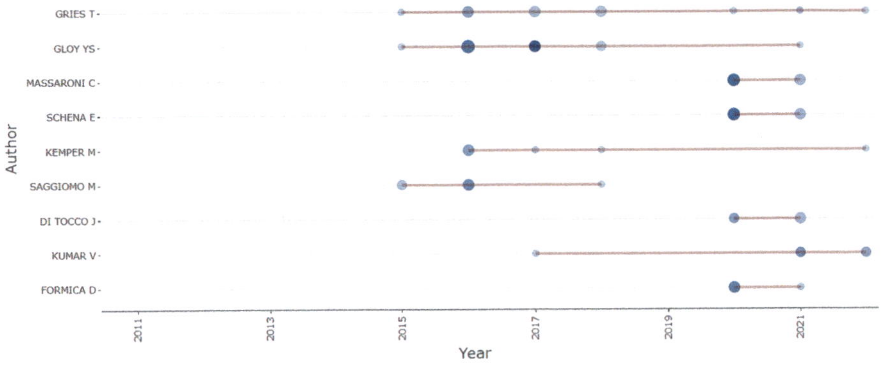

Fig. 7 Authors' production over time

the textile industry is undergoing significant changes due to technological advancements and that these changes are gaining attention in academic and industry circles. The absence of any publications in some sources in the early years could indicate that the intersection of Industry 4.0 and the textile industry was not yet a prominent topic of discussion or that it was being covered in other venues not listed in this data set. The spikes in publications in certain years (particularly in 2021 for IMECE and 2020 for Advances in Intelligent Systems and Computing) may suggest the occurrence of special issues, focused conferences, or a particular surge in research funding or innovation around those times.

Figure 7 represents a bibliometric analysis of the scholarly output and impact of various authors within the context of the textile industry and its intersection with Industry 4.0.

Gloy Y. seems to have had a peak in influence between 2016 and 2017, with a high frequency of publications and a very high citation impact, particularly in 2017, which indicates that the work published during this period was highly regarded and widely referenced in the field. Di Tocco J. and Formica D. showed a burst of activity in 2020 with more publications and citations compared to 2021. Specifically, Di Tocco J. had a higher frequency in 2021 but with fewer total citations, suggesting that their later work might not have had enough time to accumulate citations or was not as impactful as their earlier work. Massaroni C. and Schena E. both had a productive year in 2020, each with four publications and a total of 32 citations, indicating significant contributions to the field that year. Kumar V. shows an increasing trajectory of influence, with a growing number of citations, especially in 2021, indicating ongoing relevance in the field. Gries T. has a consistent record of publishing from 2015 to 2022, with the highest impact in 2016. There seems to be a decline in citations over time, with no citations recorded for 2020 and 2022 as of the data available. Saggiomo M. had their most impactful year in 2016 with three publications and a significant number of citations. However, after that year, there seems to be a drop in both output and impact. Kemper M. had several publications with a peak in 2016 (29

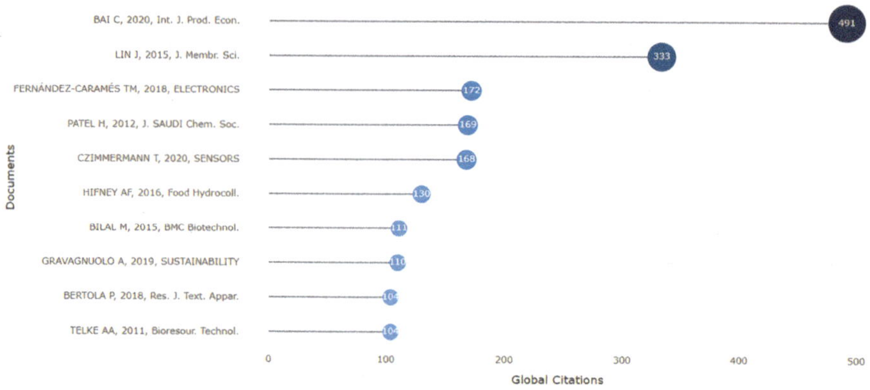

Fig. 8 Most global cited documents

TC), but there's no data on any citations post-2017, indicating a possible decline in academic output or impact after this point.

It can be inferred that authors have varying degrees of output and influence over time, and the impact of their work can be seen as both immediate and long-term. The citation patterns may reflect the evolving nature of the field and the reception of the community to their work. It is important to note that the number of citations is also influenced by the time available for the academic community to reference a work, with older publications having more opportunity to be cited. The data also suggests that key contributions to the field of textile industry in relation to Industry 4.0 have been made during this time frame, with some authors standing out due to their significant contributions.

In the landscape of academic research, the measurement of a paper's impact is often gauged by the frequency with which it is cited. Figure 8 presents a compelling overview of the most globally cited documents that have contributed to shaping the current state of the textile industry within the context of Industry 4.0.

The impact of Industry 4.0 on production economics has been documented, with a significant focus on how industries can utilize these advancements to optimize their operations (Bai et al., 2020). Research has also explored the technological advancements associated with membranes, which are essential components in textile treatments and wastewater management, a pivotal aspect of sustainable textile production (Lin et al., 2015).

Furthermore, the textile industry's integration with modern electronics has been explored, highlighting how this confluence can revolutionize textile manufacturing, especially in the domain of smart textiles (Fernández-Caramés & Fraga-Lamas, 2018). Another dimension of Industry 4.0's influence in the textile domain can be seen in the chemical processes. The exploration of chemical reactions, their optimizations, and the integration of new materials have opened up avenues for innovative textile products (Patel & Vashi, 2012).

Sensors play an integral role in the new industrial paradigm. Their influence is markedly evident in the textile industry, wherein they are utilized for quality checks, process optimizations, and even in the creation of wearable tech (Czimmermann et al., 2020). From the perspective of food hydrocolloids, there's an interesting interplay between food science and textiles, especially in the domain of biotextiles, an emerging area of research (Hifney et al., 2016).

In the realm of biotechnology, advancements have paved the way for biotextiles, emphasizing the synthesis of innovative materials that are environmentally friendly and sustainable (Bilal & Asgher, 2015). The sustainability aspect of the textile industry under the influence of Industry 4.0 is gaining traction. Efforts are made to ensure that the industry adopts sustainable practices, minimizes waste, and reduces its carbon footprint (Gravagnuolo et al., 2019).

A direct insight into the textile and apparel industry's transformation highlights the innovations in design, manufacturing processes, and the overall supply chain management due to Industry 4.0 (Bertola & Teunissen, 2018). Lastly, biotechnological advancements have influenced the utilization of bioresources in the textile industry, leading to eco-friendly and sustainable textile products (Telke et al., 2011). The textile industry's metamorphosis under the aegis of Industry 4.0 is multidimensional. It encompasses technological, environmental, and economic facets, making it an evolving sector poised for sustainable growth in the coming decades.

Given in Fig. 9, here's an analysis of author productivity in the context of the textile industry's innovation due to Industry 4.0:

The vast majority of authors (899) have contributed only a single document. This accounts for approximately 89.7% of the authors, which is a common finding in bibliometric studies and aligns with Lotka's Law. It suggests that there is a large number of researchers who have shown interest in the field but have not contributed

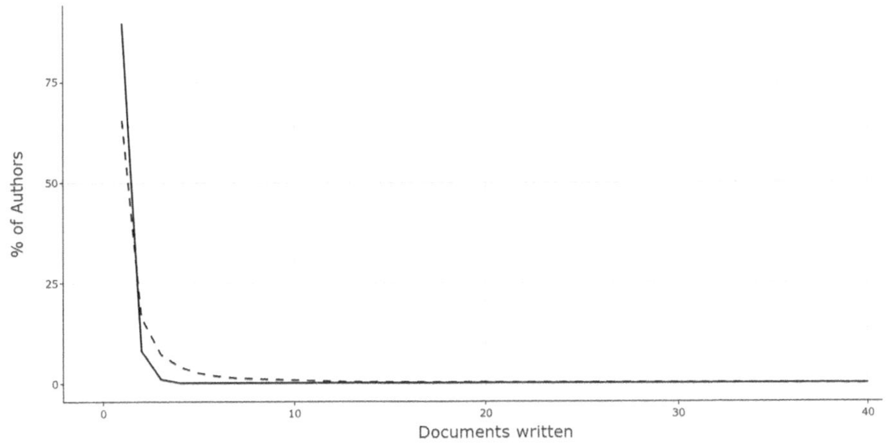

Fig. 9 Author productivity through Lotka's law

extensively, possibly due to the emerging nature of Industry 4.0 in textiles or because they are contributing in a limited scope or as part of a one-off research project.

A much smaller number of authors (81), representing about 8.1% of the total, have written two documents. This significant drop from the number of single-document authors is also consistent with Lotka's Law, indicating that a smaller number of researchers are more actively engaged in ongoing research in this field.

As the number of documents increases, the number of authors decreases dramatically, which is again in line with Lotka's Law. Only 11 authors have written three documents (1.1%), and the number decreases further with only two authors having written four or five documents, representing a mere 0.2% for each of these categories. This trend continues, with two authors having contributed six or seven documents each.

There is an interesting jump in the data where one author has contributed twelve documents and another thirteen, which suggests that there may be key thought leaders or particularly prolific researchers in the area of Industry 4.0's influence on the textile industry.

This bibliometric pattern where most authors publish only one paper and progressively fewer authors publish more papers is characteristic of academic productivity in many fields and is described by Lotka's Law. The data support the idea that the field of Industry 4.0 in the textile industry, like many specialized fields, has a small core of highly productive authors and a large number of less prolific contributors.

Analyzing the affiliations' production over time provides insight into the evolution of research output from various academic institutions regarding the influence of Industry 4.0 on the textile industry (Fig. 10). The University of Agriculture shows a steady increase in publications from 2011, starting with one publication per year, reaching a peak of six publications by 2021 and maintaining that through 2022. This indicates a growing and sustained interest in Industry 4.0 within this institution. RWTH Aachen University had no publications until 2016 but then shows a significant increase, reaching 12 publications by 2021. This could suggest a strategic focus on Industry 4.0 research starting around 2016, which has since resulted in a strong output. Donghua University also shows no output until 2017 but then a quick rise to 7 publications by 2021. This may imply a more recent but aggressive investment in Industry 4.0 research. The University of Minho had no publications until 2018 but shows rapid growth in output from 2019 onwards, doubling its output from four in 2021 to eight in 2022. Universitas Al Azhar Indonesia had a sudden start in 2019 with six publications and has maintained that output consistently. Università Campus Bio-Medico di Roma had no publications until 2020 but showed a remarkable increase to 13 by 2021, suggesting a possible new and focused initiative in this research area. Universiti Teknologi Malaysia remained without publications until 2020 but then increased to 6 by 2021, indicating a nascent interest that might be expanding.

When comparing the institutions, RWTH Aachen University and Università Campus Bio-Medico di Roma have shown the most substantial increase in a relatively short time frame, which could indicate either a large-scale investment in research in this domain or a successful research team generating results. The University of Agriculture represents a consistent and gradual increase, which may reflect a long-term

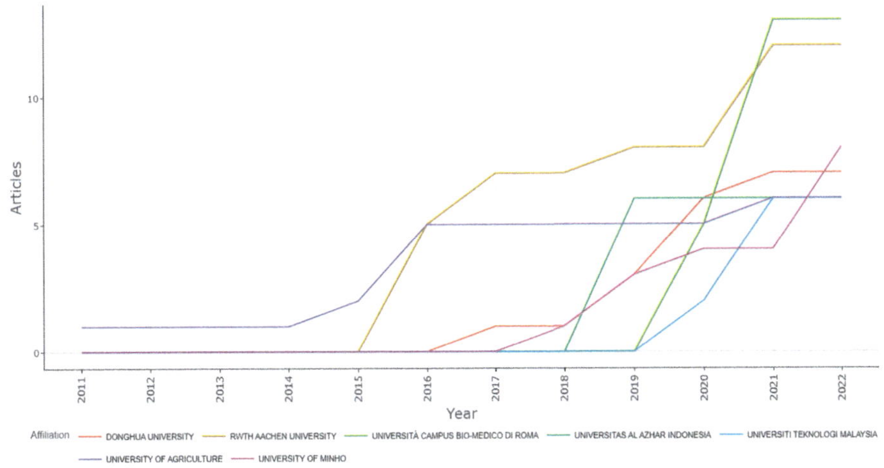

Fig. 10 Affiliations' production over time

commitment to research in the area of Industry 4.0's impact on textiles. The variability among institutions can also be indicative of the different stages of adoption and integration of Industry 4.0 in their respective textile research programs or could reflect varying levels of resources allocated towards such research.

The increasing number of articles from almost all institutions from 2016 onwards suggests a global increase in interest and investment in research related to Industry 4.0 in the textile industry. The late but rapid increase in publications from several institutions may reflect an industry response to technological advancements or a reaction to competitive pressures to innovate within the textile industry.

Figure 11 provides a detailed overview of the geographic distribution of research output concerning the impact of Industry 4.0 on the textile sector. There is a strong presence of Asian countries in the research of Industry 4.0's impact on the textile industry, with China and India leading in article count but not necessarily in international collaboration. European countries like Germany, the UK, and Portugal show significant research output with varying degrees of collaboration. The UK, in particular, stands out for its collaborative efforts.

The ratios of multi-country publications to total publications provide insight into the extent of international collaboration, with the UK and Pakistan standing out for their higher ratios, despite their lower overall publication counts. While certain countries have higher outputs of research articles, the nature of the collaborations (single country vs. multi-country) is crucial in understanding the dynamics of knowledge creation and dissemination in the textile industry's adaptation to Industry 4.0. The UK, despite having fewer total publications, may be seen as playing a pivotal role in driving international collaborations in this research area.

Figure 12 outlines the most cited countries in the context of the impact of Industry 4.0 on the textile industry. China leads with the highest number of total citations, standing at 1104. With an average of 36.80 citations per article, it indicates not only

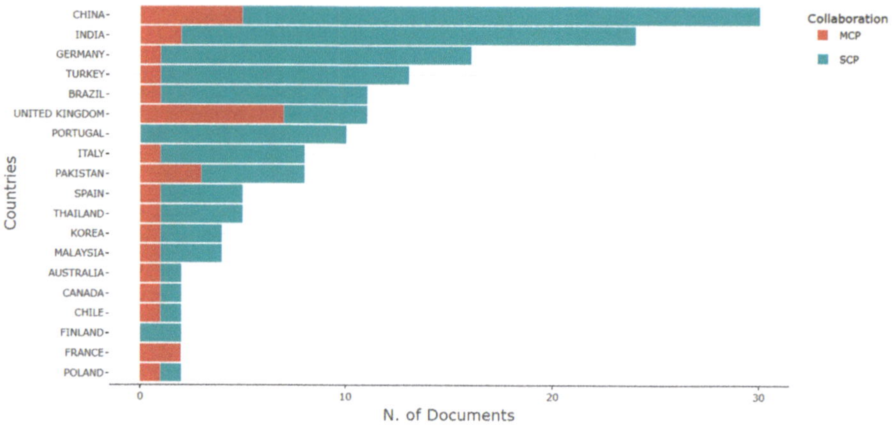

Fig. 11 Corresponding author's countries

a high volume of research output but also a significant impact per publication. This suggests that China is a key player in Industry 4.0 within the textile industry, both in terms of quantity and quality of research. India follows with 554 total citations but with a lower average of 23.10 citations per article. While India's research is influential, it doesn't seem to achieve the same impact per article as China. However, it is still a significant contributor to the field, considering the substantial total citation count. Italy has fewer total citations (439) compared to China and India but has a much higher average citation rate per article at 54.90. This outstanding average suggests that while Italy may produce fewer publications, their research is highly regarded and frequently cited, indicating high quality or seminal work in the field.

The data suggest that there is a wide disparity in the influence of different countries on the research of Industry 4.0 in the textile industry. Countries like China and India

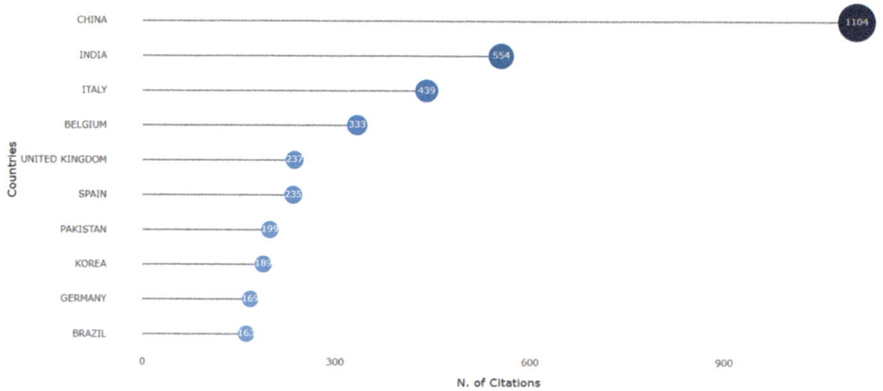

Fig. 12 Most cited countries

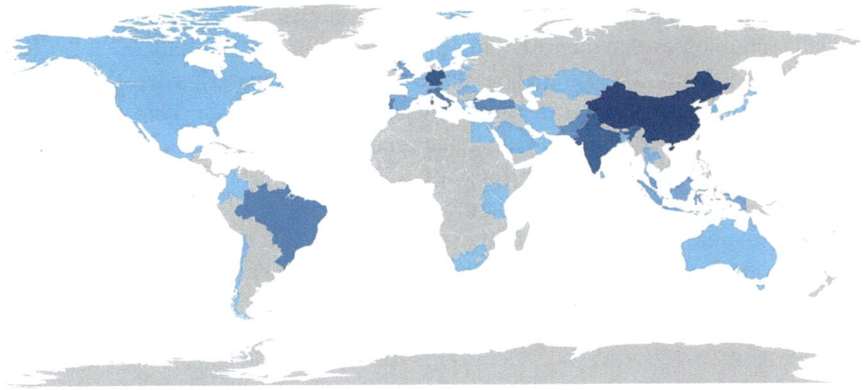

Fig. 13 Countries' scientific production

have a high volume of research output which is also well-cited, indicating both quantity and substantial influence. Nations such as Italy and Belgium may produce less research output but have higher average citations per article, pointing to quality and seminal contributions that have a significant impact on the field. The variability in total citations and average citations per article across these countries can be due to multiple factors, including the size of the research community, the focus of the research, the novelty, and the overall research and development investment in Industry 4.0 technologies within the textile sector.

To conduct a bibliometric analysis on the given data regarding the scientific production by country in the context of the influence of Industry 4.0 on the textile industry, we will analyze the frequency of publications from various countries (Fig. 13). China leads the list with 86 publications. This high number reflects China's prominent role in both textile manufacturing and the adoption of Industry 4.0 technologies. China's significant investment in research and development, as well as its large manufacturing base, likely contribute to this high level of scientific output.

With 65 publications, Germany's strong position underscores its historical leadership in engineering and industrial technology. Germany is the birthplace of the Industry 4.0 concept, and its research community is deeply integrated into furthering this area, particularly in advanced manufacturing sectors like textiles.

India's presence with 57 publications may be indicative of its rapidly growing textile industry and its push towards modernization and adopting new technologies. The country has a significant interest in enhancing manufacturing through digitalization and automation, resonating with the principles of Industry 4.0.

These bibliometric data indicate a strong correlation between a country's industrial base in textiles and its research output related to Industry 4.0. Leading countries are likely investing significantly in research to integrate advanced technologies into their textile industries to enhance competitiveness, efficiency, and sustainability. The presence of countries from Asia, Europe, and South America demonstrates a global recognition of the importance of Industry 4.0 in the textile sector. The focus on

Industry 4.0 in these publications ranges from robotics, data analytics, IoT, to other digital technologies that are reshaping the manufacturing landscape.

To provide a bibliometric analysis of a word map we must consider several factors, such as the placement of terms in the two-dimensional space (Dim.1 and Dim.2). Such map was generated to visualize the relationships between different keywords or subjects within the literature (Figs. 14, 15, Table 4). The analysis identified five distinct themes or categories within the publications.

If we assume that a higher positive value on Dim.1 indicates a greater emphasis on technological aspects, then words like "model," "data," and "system" are most

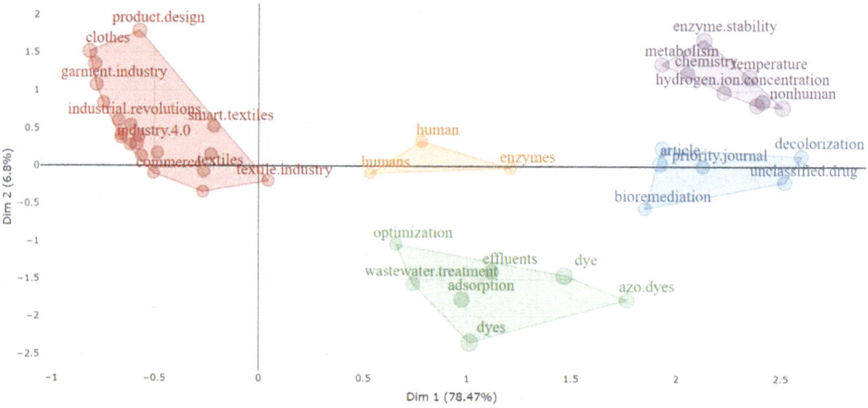

Fig. 14 Word map of the researched publications factorial analysis

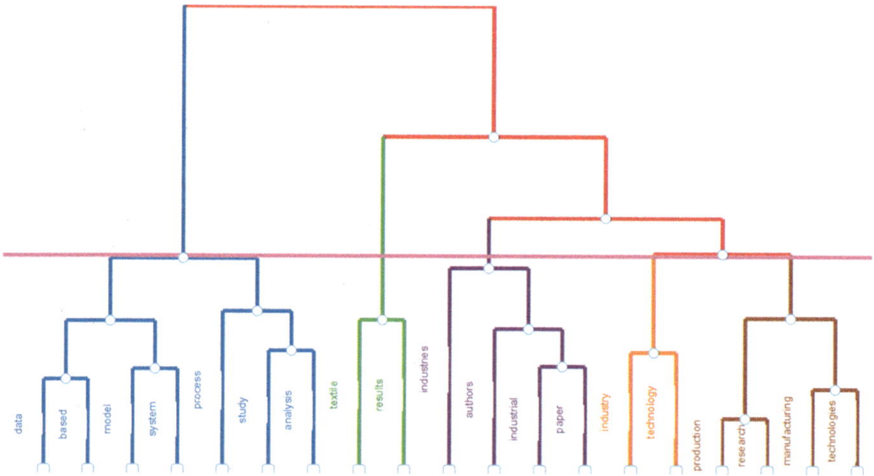

Fig. 15 Factorial analysis dendrogram of the researched publications

Table 4 Factorial analysis indicators of the studied documents

Word	Dim.1	Dim.2	Cluster
Industry	0.2	0.37	1
Textile	0.05	0.05	2
Production	0.01	0.76	3
Study	0.5	−0.2	4
Manufacturing	0.1	1.01	3
Process	0.65	0.09	4
Data	0.93	0.17	4
Model	1.07	−0.06	4
Industrial	−0.24	0.58	5
Analysis	0.71	−0.2	4
Technology	0.17	0.58	1
Based	0.94	0.03	4
Technologies	0.04	1.1	3
System	1.1	−0.22	4
Research	0.04	0.76	3
Results	0.16	−0.21	2
Industries	−0.29	0.12	5
Paper	−0.4	0.52	5
Authors	−0.07	0.46	5

strongly associated with this theme. Conversely, words with negative values such as "industries" and "paper" may represent a more general or foundational approach.

If this axis represents something like practical application or industry relevance, then terms like "manufacturing," "technologies," and "research" with higher values on Dim.2 might signify a focus on practical implications in the field, while terms with lower values such as "study" and "results" may indicate a more theoretical or abstract focus.

"Industry" and "technology" are in Cluster 1 (Fig. 15), suggesting that these publications frequently discuss the technological aspects of Industry 4.0 as it pertains to the industry at large. The term "textile" is in Cluster 2 with a lower position on both dimensions, perhaps indicating that while it is central to the subject matter, it may not be as deeply discussed in the technological or process-oriented context as other terms. "Manufacturing" and "technologies" are in Cluster 3 with high scores on Dim.2, which might reflect a focus on the practical application of Industry 4.0 in the production process.

Terms like "model," "data," "based," and "system" fall in Cluster 4, with high scores on Dim.1 but varying on Dim.2. These could be more related to the theoretical modeling and data analysis aspect of Industry 4.0. Words like "industrial," "industries," "paper," and "authors" are in Cluster 5, which might signify a broader discussion of the industrial context or the nature of the publications themselves rather than a focused technological discussion.

The bibliometric analysis points to a research landscape in the textile industry that is being shaped by Industry 4.0, with a strong focus on data and system modeling, an appreciation for the production and technology advancements, and a diverse approach to research methodologies. This landscape is dynamic and reflects the interplay between technology, industry practices, and the academic study of these elements, while also hinting at potential under-researched areas that could be explored further.

3 Conclusion

The bibliometric analysis reveals the significant integration of Industry 4.0 technologies and concepts in the textile industry. The emphasis is not just on the mechanistic transformation but also on sustainability. The centrality metrics highlight the interconnectedness and relative importance of various terms, revealing that while the broader theme revolves around the textile industry, more specific areas like weaving and garments are also gaining significant attention.

While the interest in Industry 4.0's influence on the textile industry appears to be growing with more publications each year, the impact of these publications as measured by citations seems to vary significantly from year to year. Recent publications have yet to show their full impact in the field, and ongoing citation analysis will be important to track their influence over time.

The increasing number of publications can be seen as a potential indicator of the growing importance of collaborative research and its impact on the industry. Universities ramping up their research output may be seeking collaborative partnerships with other academic institutions or industry players. The data reflects a clear trend of increasing academic interest in the intersection of Industry 4.0 and the textile industry, with several institutions showing rapid increases in publication output after certain points in time. This trend underscores the textile industry's evolving nature and the critical role of technological innovation in driving academic research and potential industrial transformation.

The innovative influence of Industry 4.0 on the textile industry is a multifaceted research area with a burgeoning corpus of academic literature, as reflected in the bibliometric analysis. The progression of research output from a variety of academic institutions showcases a growing engagement with the subject matter. The commitment to the integration of Industry 4.0 into the textile sector is exemplified by the consistent rise in publications, with certain universities signaling a robust focus through heightened publication activity in recent years.

The geographical spread of research, predominantly emanating from Asia and Europe, underscores a regional emphasis on developing advanced technological competencies within the textile industry. China's preeminent role in both research output and impact, as measured by citation analysis, alongside the UK's distinctive position in spearheading international collaborations, suggests that different nations are carving out unique niches within the research ecosystem.

The thematic substance of the research corpus encompasses robotics, data analytics, the Internet of Things (IoT), and other digital innovations, indicating a strong inclination towards transformative technologies that are recalibrating the manufacturing processes in the textile industry. However, the analysis also reveals that while there is a significant amount of research on the technological aspects of Industry 4.0, there may be less emphasis on the socio-economic and environmental implications of these changes.

The bibliometric study illuminates potential knowledge gaps and opens up new avenues for investigation. The nuanced understanding of the impact of Industry 4.0 technologies on workforce skills, job roles, and employment patterns in the textile industry remains less explored. Furthermore, while there is a proliferation of research on efficiency and productivity gains, there is a comparative paucity of insights into the sustainability outcomes of Industry 4.0 adoptions in textiles.

Future research could pivot towards these under-explored dimensions, enriching the understanding of Industry 4.0's influence beyond the technological upgrades. The interrelation between Industry 4.0 technologies and circular economy principles in the textile industry could offer fertile ground for research, addressing a critical gap in literature around sustainable practices. Additionally, the role of policy, both at national and international levels, in facilitating or impeding the adoption of Industry 4.0 in textiles warrants closer scrutiny.

The bibliometric analysis reveals a vigorous academic discourse around the nexus of Industry 4.0 and the textile industry, with a strong technological bias. To foster a comprehensive body of knowledge, future research should broaden the scope to include the social and environmental aspects of Industry 4.0 transformations. There is also an imperative for cross-disciplinary research that bridges technology with sustainability, economics, and policy to holistically address the challenges and opportunities presented by Industry 4.0 in the textile sector.

References

Ahmad, S., Miskon, S., Alabdan, R., & Tlili, I. (2020). Towards sustainable textile and apparel industry: Exploring the role of business intelligence systems in the era of industry 4.0. *Sustainability, 12*(7), 2632.

Akram, S. V., Malik, P. K., Singh, R., Gehlot, A., Juyal, A., Ghafoor, K. Z., & Shrestha, S. (2022). Implementation of digitalized technologies for fashion industry 4.0: Opportunities and challenges. *Scientific Programming*.

Bai, C., Dallasega, P., Orzes, G., & Sarkis, J. (2020). Industry 4.0 technologies assessment: A sustainability perspective. *International Journal of Production Economics, 229*, 107776.

Barman, J., Tirkey, A., Batra, S., Paul, A. A., Panda, K., Deka, R., & Babu, P. J. (2022). The role of nanotechnology based wearable electronic textiles in biomedical and healthcare applications. *Materials Today Communications*, 104055.

Bertola, P., & Teunissen, J. (2018). Fashion 4.0. Innovating fashion industry through digital transformation. *Research Journal of Textile and Apparel, 22*(4), 352–369.

Bilal, M., & Asgher, M. (2015). Dye decolorization and detoxification potential of Ca-alginate beads immobilized manganese peroxidase. *BMC Biotechnology, 15*, 1–14.

Chen, C. L. (2019). Value creation by SMEs participating in global value chains under industry 4.0 trend: Case study of textile industry in Taiwan. *Journal of Global Information Technology Management, 22*(2), 120–145.

Chen, C.-L. (2020). Cross-disciplinary innovations by Taiwanese manufacturing SMEs in the context of Industry 4.0. *Journal of Manufacturing Technology Management, 31*(6), 1145–1168.

Ciliberto, C., Szopik-Depczyńska, K., Tarczyńska-Łuniewska, M., Ruggieri, A., & Ioppolo, G. (2021). Enabling the Circular Economy transition: A sustainable lean manufacturing recipe for Industry 4.0. *Business Strategy and the Environment, 30*(7), 3255–3272.

Czimmermann, T., Ciuti, G., Milazzo, M., Chiurazzi, M., Roccella, S., Oddo, C. M., & Dario, P. (2020). Visual-based defect detection and classification approaches for industrial applications—A survey. *Sensors, 20*(5), 1459.

De Silva, R. K. J., Rupasinghe, T. D., & Apeagyei, P. (2019). A collaborative apparel new product development process model using virtual reality and augmented reality technologies as enablers. *International Journal of Fashion Design, Technology and Education, 12*(1), 1–11.

Fallahpour, A., Wong, K. Y., Rajoo, S., Fathollahi-Fard, A. M., Antucheviciene, J., & Nayeri, S. (2021). An integrated approach for a sustainable supplier selection based on Industry 4.0 concept. *Environmental Science and Pollution Research*, 1–19.

Farooq, A. S., & Zhang, P. (2022). A comprehensive review on the prospects of next-generation wearable electronics for individualized health monitoring, assistive robotics, and communication. *Sensors and Actuators A: Physical*, 113715.

Fernández-Caramés, T., & Fraga-Lamas, P. (2018). Towards the internet-of-smart-clothing: A review on IoT wearables and garments for creating intelligent connected E-textiles. *Electronics, 7*(12), 405.

Fromhold-Eisebith, M., Marschall, P., Peters, R., & Thomes, P. (2021). Torn between digitized future and context dependent past–How implementing 'Industry 4.0'production technologies could transform the German textile industry. *Technological Forecasting and Social Change, 166*, 120620.

Gravagnuolo, A., Angrisano, M., & Fusco Girard, L. (2019). Circular economy strategies in eight historic port cities: Criteria and indicators towards a circular city assessment framework. *Sustainability, 11*(13), 3512.

Guan, Z., Xu, Y., Jiang, H., & Jiang, G. (2019). International competitiveness of Chinese textile and clothing industry—A diamond model approach. *Journal of Chinese Economic and Foreign Trade Studies, 12*(1), 2–19.

Hifney, A. F., Fawzy, M. A., Abdel-Gawad, K. M., & Gomaa, M. (2016). Industrial optimization of fucoidan extraction from Sargassum sp. and its potential antioxidant and emulsifying activities. *Food Hydrocolloids, 54*, 77–88.

Imran, M., Hameed, W. U., & Haque, A. U. (2018). Influence of industry 4.0 on the production and service sectors in Pakistan: Evidence from textile and logistics industries. *Social Sciences, 7*(12), 246.

Jiménez-Marín, G., Elías Zambrano, R., Galiano-Coronil, A., & Ravina-Ripoll, R. (2021). Business and energy efficiency in the age of industry 4.0: The hulten, broweus and Van Dijk sensory marketing model applied to spanish textile stores during the COVID-19 crisis. *Energies, 14*(7), 1966.

Jimeno-Morenilla, A., Azariadis, P., Molina-Carmona, R., Kyratzi, S., & Moulianitis, V. (2021). Technology enablers for the implementation of Industry 4.0 to traditional manufacturing sectors: A review. *Computers in Industry, 125*, 103390.

Lalic, B., Rakic, S., & Marjanovic, U. (2019). Use of industry 4.0 and organisational innovation concepts in the Serbian textile and apparel industry. *Fibres & Textiles in Eastern Europe, 3*(135), 10–18.

Lin, J., Ye, W., Zeng, H., Yang, H., Shen, J., Darvishmanesh, S., & Van der Bruggen, B. (2015). Fractionation of direct dyes and salts in aqueous solution using loose nanofiltration membranes. *Journal of Membrane Science, 477*, 183–193.

Ku, C. C., Chien, C. F., & Ma, K. T. (2020). Digital transformation to empower smart production for Industry 3.5 and an empirical study for textile dyeing. *Computers & Industrial Engineering, 142*, 106297.

Majumdar, A., Garg, H., & Jain, R. (2021). Managing the barriers of Industry 4.0 adoption and implementation in textile and clothing industry: Interpretive structural model and triple helix framework. *Computers in Industry, 125*, 103372.

Manglani, H., Hodge, G. L., & Oxenham, W. (2019). Application of the internet of things in the textile industry. *Textile Progress, 51*(3), 225–297.

Margherita, E. G., & Braccini, A. M. (2020). Industry 4.0 technologies in flexible manufacturing for sustainable organizational value: reflections from a multiple case study of Italian manufacturers. *Information Systems Frontiers*, 995–1016.

Mohiuddin Babu, M., Akter, S., Rahman, M., Billah, M. M., & Hack-Polay, D. (2022). The role of artificial intelligence in shaping the future of Agile fashion industry. *Production Planning & Control*, 1–15.

Mourtzis, D., Zogopoulos, V., & Xanthi, F. (2019). Augmented reality application to support the assembly of highly customized products and to adapt to production re-scheduling. *The International Journal of Advanced Manufacturing Technology, 105*, 3899–3910.

Nascimento, D.L.M., Alencastro, V., Quelhas, O.L.G., Caiado, R.G.G., Garza-Reyes, J.A., Rocha-Lona, L., & Tortorella, G. (2019). Exploring Industry 4.0 technologies to enable circular economy practices in a manufacturing context: A business model proposal, *Journal of Manufacturing Technology Management, 30*(3), 607–627.

Nazam, M., Hashim, M., Nuță, F. M., Yao, L., Zia, M. A., Malik, M. Y., & Dimen, L. (2022). Devising a mechanism for analyzing the barriers of blockchain adoption in the textile supply chain: A sustainable business perspective. *Sustainability, 14*(23), 16159.

Nounou, A., Jaber, H., & Aydin, R. (2022). A cyber-physical system architecture based on lean principles for managing industry 4.0 setups. *International Journal of Computer Integrated Manufacturing, 35*(8), 890–908.

Pandey, R., Pandit, P., Pandey, S., & Mishra, S. (2020). Solutions for sustainable fashion and textile industry. *Recycling from Waste in Fashion and Textiles: A Sustainable and Circular Economic Approach*, 33–72.

Patel, H., & Vashi, R. T. (2012). Removal of Congo Red dye from its aqueous solution using natural coagulants. *Journal of Saudi Chemical Society, 16*(2), 131–136.

Salo, H. H., Suikkanen, J., & Nissinen, A. (2020). Eco-innovation motivations and ecodesign tool implementation in companies in the Nordic textile and information technology sectors. *Business Strategy and the Environment, 29*(6), 2654–2667.

Telke, A. A., Ghodake, G. S., Kalyani, D. C., Dhanve, R. S., & Govindwar, S. P. (2011). Biochemical characteristics of a textile dye degrading extracellular laccase from a Bacillus sp. ADR. *Bioresource Technology, 102*(2), 1752–1756.

Wijewardhana, G. E. H., Weerabahu, S. K., Nanayakkara, J. L. D., & Samaranayake, P. (2021). New product development process in apparel industry using Industry 4.0 technologies, *International Journal of Productivity and Performance Management, 70*(8), 2352–2373.

Textile Innovation, Sustainability, Survival, and the Future

Tânia Alves de Jesus

Abstract The textile industry was undoubtedly one of the first sectors to be affected by the effects of globalization. Today the great and main strength of the textile industry is and will be its capacity for resilience and to recovery. But for the future, it is not enough to innovate and evolve because success is limited in time. It seems textiles are on a path that has no turning back towards sustainability. This research describes and discusses some perspectives on the future of the textile sector in Portugal in terms of its survival and sustainability. In conclusion, the future of textiles in the world, especially in Portugal, is undoubtedly challenging. It seems that the market of the future will face significant concerns related to the environmental and social impact of fashion. For humans and the textile industry to survive, the mission must demonstrate economic viability and combine truly environmentally sustainable innovation with a better life.

1 Introduction

The textile industry was undoubtedly one of the first sectors to be affected by the effects of globalization. Even though the textile industry is one of the most industrialized sectors, it is still an activity where the workforce is still very present since an important and significant part of the production work continues to be done manually. European industry, and the Portuguese cannot be excluded, inevitably suffered from the entry of low-wage cost countries into the labor force sector. A good number of the textile companies simply disappeared. To face competition from low-cost countries like China, textile producers had to know how to innovate and reinvent themselves. Some companies, Portuguese and European, were able to reposition themselves and change their focus to high added value products. They became more functional and

T. A. de Jesus (✉)
Department of Accounting and Auditing, Lisbon Accounting and Business School, Avenida Miguel Bombarda, 20, 1069-035 Lisboa, Portugal
e-mail: tajesus@iscal.ipl.pt

Lisbon Accounting and Business School, Polytechnic Institute of Lisbon (ed.), *Industry and Innovation: Textile Industry*, SDGs and Textiles,
https://doi.org/10.1007/978-3-031-57804-5_11

were able to survive, managing, in certain aspects, the development of themselves in new markets, instead of producing low-cost artifacts.

Portugal has adopted, and still is adopting, a strategy of qualitative nature based on a specialized and professionalized industry, showing that its products meet the most demanding criteria of quality as well as sustainability. The diversification of customers, thus, became fundamental. Some factories have stopped working solely with large surfaces or textile giants, to include working with smaller companies and orders for smaller details, but who seek the quality of the product or the design, and others will follow. These adaptations, which imply a higher price on acquisitions, caused and continue to cause a lower volume of orders, however, as a result, the total turnover was not significantly affected. Most importantly factories gain independence as they are no longer dependent on big orders because become more functional. This functionalization, in textiles, came to disturb the relations between manufacturers a little. They had to learn to relate and communicate with each other. Traditional textile manufacturers insisted on preserving their know-how, keeping it reserved, almost secret. But is there a better way to explore and expand the historical Portuguese know-how in textile production than to bet on inter-company relationships and interconnected knowledge?

Today, in a certain way, anyone who wants to be at the forefront of the market is forced to share their skills and knowledge in joint projects. Inter-company relations must no longer be closed in microcompartments. The great and main strength of the textile industry is and will be its capacity for resilience and to recovery.

The capacities of renewal and reinvention of European entrepreneurs are and have been remarkable, including, already with some relevance, the Portuguese textiles that are ubiquitous in our environment. They are often associated with clothing but also found in furniture and in other sectors such as health or transport, in these cases with technical textiles. This research describes and discusses the Portuguese textile industry way in innovation in terms of its survival, sustainability, and future.

2 Innovative Textiles and Main Challenges

The main innovations in technical textiles relate to the domains of sports, leisure, health, and personal protection (personal safety). The main trends are the development of technical textiles, less polluting finish, and consumers who like personalized clothing.

The development of technical textiles, whether they be ropes or nets, or cosmetotextiles has a direct impact on well-being and health (moisturizing clothing, better washability, or providing massages, etc.). Another example of smart textiles is smart sensing, a piece of clothing with connections that allow monitoring of physical activities (Bunsell, 2018). The sports equipment used today, whether in football, cycling, triathlon, or surfing, is already highly technical and has nothing to do with what was produced in the past.

A less polluting finish is an operation that improves textiles by dyeing, printing, or imparting specific properties (functionalization). It is a challenge for the industry because it hopes to use less polluting treatment techniques that respect the environment more (plasma treatment, for example).

Likewise, consumers who like made-to-measure are also a main trend. It is a niche market in a phase of growth with a strong demand for personalized clothing. The consumers themselves are setting this trend. Yes, the consumers!

There are three main challenges in this sector, the technological and ecological transition, the development of smart textiles, and creating the factory of the future.

Coupled with the use of fibers of biological origin that allow them a long life and recycling, such as flax for example, the technological and ecological transition is, if not the first, one of the most relevant challenges. In Portugal, quality linen has always been produced, and this knowledge, passed down from ancestors, should not be disregarded. The linen plant has a very long historical past in the Iberian Peninsula. As evidenced by traces found in archaeological findings its use dates to prehistoric times, and when the Romans invaded the Iberian Peninsula, the cultivation was revitalized, and seeds and linen fabrics were taken to Rome (Lanca & Baptista, 1993).

In addition to linen, other good suppliers of plant fibers include hemp, thistle, corn, banana leaf, and nettles (Khan et al., 2024). It is necessary to maintain the good quality of these biological fibers that have been passed down, as well as their great recycling capacity, using this good quality to produce new textiles or other products, such as thermal insulation. Recycling capacity is indeed important; the number of textiles thrown in landfills every year is simply absurd.

Moreover, with the use of new technologies, including digital, artificial intelligence, and nanotechnology, the development of smart textiles will certainly occupy a considerable space, probably essential, in the future of the fiber and textile industry. Since textiles, even with all the automation, will continue to require a considerable amount of labor, it will also be important to find new assembly methods in the factory of the future, such as 3D knitting for example, methods that allow for lower costs in terms of space and perhaps even allow for production reallocation.

2.1 How Will the Market Evolve?

The market will naturally evolve. The first step to be taken by textile manufacturers is to start/continue working in collaboration with other industries, given that textiles, as highly flexible and adaptable materials, can be, and are, used in a wide range of applications.

The maintenance of technical textiles can be delicate, as it becomes necessary to perpetuate their conferred properties; the challenge for the near future will therefore be to make them more resistant, to use and washing for example, giving them greater durability and reusability, all at a reasonable cost.

The path lies in perpetuating excellent know-how. New (many) creative and design teams must be supported to prevent manipulation by major sponsors, and all new creators must be given the opportunity to showcase their prototypes. It is necessary to support the creation of new businesses within textiles, and, notably, support young people by providing and sustaining the emergence of new horizons and new brands.

We are in a time of emerging from a pandemic, war in Europe, and inflationary crisis, but also a time of slight growth in average wages, if measured comparatively to the inflation rate. The crisis, we know, shakes people and people shake companies, and companies force themselves to invest in new technologies, diversify their activities, and invest in more technological products, or rather, more technical products; technical in the sense of having a deep knowledge of an art. Betting on technical textiles, new technologies, and more sophisticated products, and above all, greater international exposure, have changed/are changing the face of a sector that was once almost considered lost.

The textile and clothing industry is one of the most important industries in Portugal and in the European Union. Currently, the Portuguese textile and clothing industry represents 8% of national exports of goods and accounts for 8% of turnover and 8% of production of the Portuguese manufacturing industry. In addition, in Portugal, the sector represents 17% of employment. The main foreign customers of Portuguese textiles are Spain, France, Germany, and the United States (Dinis & Guimarães, 2023). So, it is important to keep up with what is happening in these countries… that's right, update and deepen knowledge about the habits, characteristics, and ways of being of the people who inhabit them, because it is people, and not these countries, who are/will be customers!

Regarding technological and ecological transition, it is worth remembering that synthetic fibers have made it possible to develop fabrics with various technical properties, and the future certainly holds many surprises for us; in this field, important advances have occurred; advances that have been used to develop natural fabrics, simple or incorporated, which have shown the same, and even better, technical performance. We already know about Merino wool, wonderfully soft, high quality, pleasantly soft on the skin, breathable, and temperature-regulating, or Tencel, an environmentally friendly fiber with antibacterial properties and highly technological, which is not extracted from an animal but rather from eucalyptus pulp dissolved in a non-toxic solvent, and eucalyptus is a tree that requires very little water for its growth and no pesticides are used throughout its life. The elasticity of Tencel makes it a material of excellent resistance, enduring regular wear and passing the washing machine test with distinction (Khan et al., 2024). The lifespan of Tencel clothing is thus much longer than that of common fabrics. Tencel is also known for its great moisture absorption capacity, as it has the same operating mechanism as merino wool; its fiber can absorb large amounts of moisture in its core while remaining dry on the surface, enough to combine an active life with style, in all circumstances.

Evolution/innovation has naturally occurred, and fortunately, Merino wools and Tencel fibers are no longer alone in the market (Bustamante et al., 2021). It is truly important to keep up with these innovations that combine a positive environmental approach with technicity in the service of human well-being. As previously mentioned, some of these innovations are already found in various clothing pieces and are frequently used in fields other than ready-to-wear, such as sports equipment, military uniforms, and even astronaut suits. It is only a matter of time until these innovations reach the whole world and all people; if not before, it will be when they acquire a true environmental consciousness. It is therefore time for the textile industry to confirm its pioneering spirit, stop being a promise, and openly embark on the path of environmentally sustainable innovation, aware that this procedure can and will keep it on a path of social interaction and demonstrate economic viability.

2.2 *What Innovations Are Possible?*

The first innovation may come from the sea, more specifically, from phaeophyte algae. The brown algae that sometimes appear during walks by the sea are, in fact, aquatic plants, combined with cellulose to form a textile fiber, which will, in fact, allow the manufacture of clothes and other textile accessories (Baghel et al., 2021). This combination has as its first advantage the saving of water, one of the main sources of criticism in the textile industry, and, in addition, it also allows the recovery of some specific properties of algae such as anti-odor, antibacterial protection, and moisture rejection.

A second environmentally responsible fiber that arrives via the ocean is seaqual fiber. Its basic idea is to go against the grain by recovering plastics from the oceans to create polyester fiber and obtain yarns that will then be mixed with natural fibers to create ready-to-wear or sportswear and simultaneously cause the disappearance of the infamous seventh continent that annually kills thousands of marine animals between California and Hawaii.

According to Fashionnetwork (2017), seaqual manufacturing also has the advantage of consuming 92% less water in the dyeing process, 40% less energy during the dyeing process and releasing 33% less carbon dioxide emissions into the atmosphere. The result is recycled fabrics with controlled traceability manufactured responsibly and complying with all international quality standards.

The second possible innovation comes close to the dish because it concerns two well-known fruits: coconut and pineapple. Fiber and vegetables are good for health and clothing. Coconut fiber is obtained from the skin of the fruit, after its interior has been consumed by the food industry. Instead of these husks being thrown in the trash, they are sent to factories that recover the coconut fibers by grinding and processing them until they obtain yarn that will be combined with polyester and produce a new fiber. This fiber, in addition to having notable anti-UV properties, is also antibacterial and deodorizing, which allows to create fabrics, perfect for hot days, for example (Ali et al., 2022).

Also derived from an essential fruit, pineapple fiber is also beginning to be increasingly used in the design of textiles, particularly thanks to the leather that can be extracted from it. First used in the Philippines for traditional clothing, this material called piñatex has become popular as a substitute for animal leather in shoes or accessories. This natural and vegetable alternative is resistant and allows dyeing and printing (Hijosa, 2015).

We point to the pair of lotus and eucalyptus as the big winner of interesting combinations. So, another possible innovation may come from the vegetable fiber category. Considering that the most used resources today are decreasing exponentially, lotus fiber appears as a great alternative. Exotic at first glance, given its Asian origin, lotus fiber offers the textile in which it is integrated a light, silky and airy finish, also having anti-stain properties. The lotus seems to align with the goals of sustainability, ecology, and practicality, for an increasingly sustainable fashion that, it seems, will accompany everyday life in the future (Aishwariya & Thamimam, 2019).

In a similar biodegradable approach, eucalyptus fiber makes it possible to obtain a soft, silky, and resistant material that perfectly absorbs moisture (Morais & Curto, 2022). This behavior prevents or at least reduces the development of bacteria, when compared to synthetic materials, and offers excellent breathability to the fabric (ability to eliminate water vapor generated by perspiration during effort) thus managing, in a certain way, to regulate the perspiration. Eucalyptus is a source of other interesting fibers that are better known today in the ready-to-wear sector, such as the previously mentioned Tencel or Lyocell. Particularly pleasant to wear, these fibers give us a real feeling of freshness with a no less pleasant bonus, a soft touch, very similar to silk.

There are other fabrics that are more technical and resistant, such as armalith fabric. This fabric is obtained by combining cotton fibers, lycra fibers and ultra-high molecular weight polyethylene fibers (UHMWPE), so it is not a completely natural fabric (Solino et al., 2014). However, it is undoubtedly a possible innovation because it is no longer reserved for space or military applications and may be purchased on the market. The principle used is simple; it is about integrating technology into the fabric itself, leaving the technology to be used as the lining of a garment. As a material with a high level of compression resistance and designed to resist abrasion and with a low coefficient of friction, this fabric manages to combine the authenticity of denim, with the reinforcement, comfort, and appearance, for example, of a real jeans. Thus, protection and durability, in addition to being the keywords of this fabric, make it the fabric for bulletproof clothing, the technical jeans of tomorrow or the most efficient protective gloves. These properties make it the obvious choice for motorcyclists, cyclists, skaters, skateboarders, and all those who practice activities subject to falls or other activities that cause or may cause abrasion.

One last possible revolution, space technology, but down to earth, it is as if we were taken to space. Space tecnology uses phase change materials. It is, in fact, a microencapsulated natural wax that can be applied to practically any textile material. These capsules can absorb and store body heat, functioning as a natural heat absorber. The overheating that could trigger sweating can thus be avoided/contained quite efficiently. This technology works when the body produces heat, causing the microcapsules to capture excess heat and keep it in the tissue. When the temperature drops, that same heat will be returned to the user's body. These thermal materials begin to act in advance, even before moisture forms. This way, the amount of sweat can be significantly reduced (Ahmad et al., 2023).

In conclusion, it is not enough to innovate and evolve because success is limited in time. It is becoming increasingly necessary to monitor events that indicate future trends in society and markets in an informed manner to prevent crystallization over time.

2.3 The Immediate Future

It is essential to find out what the trends in the use of times, materials, and fashion will be, which cannot be excluded at all, and from now on, for the near future. What will sell more and better in the immediate future?

The answer can be obtained by going to fairs, conferences, and presentations where it is possible to see them. As a rule, Première Vision[1] reveals to professionals the first major fashion trends for the year following its event; The one held in February shows the spring–summer trends, and the one held in September, in turn, shows the autumn–winter trends. The exhibition that was held in February 2023 left the textile and fashion industries with indications of what is expected to happen in the immediate future. According to Fashionunited (2024), the influence of plants on fabrics is highlighted, with a special focus on natural; natural, here in the sense of what is in a state of nature. The origin of the fabrics, however, alone cannot guarantee the success of the textile industry, so it will necessarily have to be associated with fashion, and, equally, fashion will certainly not occur without the intervention of the human imagination. Here we can see the abovementioned communication and cooperation between companies, studios, and sector specialists. In addition, this communication includes tellurism, and within it the search to understand the influence of the soil on social customs and people's character.

It seems that it will be celebrated the return to good graces of leather after green-bashing and the return of the notion of legitimate naturalness, of the use of leather compared to synthetic alternatives. However, there are issues concerning traceability, sustainability, and animal welfare. According to Omoloso et al. (2020), findings confirm that companies in the leather supply chain are embarking on sustainability efforts in terms of energy efficiency, waste management reduction of greenhouse gas emissions, and health and safety social sustainability activities.

2.4 And What About the Future, Beyond the Immediate?

The first observable trend for the textile market of the future is, clearly, the rise of ethical fashion. Consumers are increasingly concerned about the environmental and social impact of fashion and are increasingly looking for brands and/or producers

[1] Première Vision is an international materials sourcing fair which takes place twice a year since 1973 in Paris (Premièrevision, 2024).

that respect not only the environment, but also workers' rights (Rese et al., 2022). Ethical clothing brands, which use sustainable materials, are produced responsibly and which, for example, donate part of their profits to social causes, will, tendentially, have increased growth potential (Abbate et al., 2023).

Another trend, although this is not mandatory, is the personalization of clothing. More and more consumers want unique clothes. Clothing brands that offer these customization options also indicate that they will have strong growth potential (Nobile & Cantoni, 2023). Personalization of clothes can reach the point where customers can be offered the possibility of choosing, be it materials, cuts, colors, and even some of their details. This customization will, in principle, allow customers, in a certain way, to create clothes that could be classified as made-to-measure, and as such unique pieces that, increasingly, resemble them and, at least in part, reflect your personality.

Another trend that can be observed is the sale of clothes online. Consumers are increasingly comfortable shopping online and appreciate not only the convenience but also the variety of offers typically available on the internet (Agrawal, 2023). It is difficult, however, to know the exact percentage of online clothing sales in Portugal. Through artificial intelligence, augmented reality, and virtual reality technologies, consumers can enjoy multi-sensory and rich entertainment experiences. But the results show that these technologies are difficult to implement and develop in Portuguese businesses and that Portuguese consumers have little knowledge of them (Carvalho & Abreu, 2023). Findings differ depending on the source and methodology used, but it is possible to verify that clothing brands that have a strong online presence are already providing very satisfactory online shopping experiences and, everything indicates that this field still has strong growth potential. There is no doubt that online has shown significant growth compared to the years before the pandemic, but it is important not to forget that the textile market in Portugal remains largely focused on physical stores (it is that Portuguese taste of touching/feeling), but everything indicates that this focus will change.

Regarding quality and durability, Portuguese consumers are increasingly paying attention to the useful life of the clothes they wear and are increasingly looking for brands or establishments that offer more durable and higher quality clothes. The sustainability trend is also growing in Portugal due to increased consumer awareness about the environment and sustainable production (Carvalho et al., 2020). Therefore, entrepreneurs who invest in quality materials and offer guarantees of durability have a strong potential for growth, attracting consumers to purchase their products.

Consumers are increasingly being asked by the most varied brands to purchase their clothes, being more convinced through communication that is especially and specifically aimed at them, precise, clear, and powerful. An effective communication strategy is thus necessary, one that can reach the intended target audience and that uses the appropriate communication channels if the aim is to have a strong growth potential. This involves using original advertising campaigns and selecting the most relevant social networks, aimed at reaching a certain target audience (independent and creative women, for example). Also, it can be used, in conjunction with the most relevant social networks, the collaboration of blogs to promote specific projects, such as ethical fashion and solidarity.

The slow fashion trend is clearly growing, especially in Europe, due to increased consumer awareness of issues in the fashion industry, such as excessive consumption, pollution, working conditions, etc. (Centobelli et al., 2022). This slow fashion is prioritizing quality over quantity, and offering durable, good quality, comfortable, and ethical clothing, and this allows consumers to feel responsible and choose clothes that fit their needs and values, and as a result, make purchases more responsibly. Therefore, slow fashion can allow a brand to differentiate itself from the competition by offering exclusive, high-quality clothing that meets consumer expectations, increasing customer awareness and retention, generating financial benefits in the medium/long term.

2.5 *The Challenges of Future Textiles*

With new synthetic fibers and the combination of numerous textiles or related capacities (non-wovens, composites), textiles present themselves, today, with a renewed spirit and appear together with new geometries. Today's textile activities are undergoing an impressive revolution, a source of new social and economic realities. Textile material is undergoing an extremely rapid revolution that reminds us of the tremendous explosion of Information and Communication Technologies.

What, then, will be the main challenges facing textiles of the future?

Today, textiles interact with almost all industrial sectors because they are already used both for their technical performance and for their functional properties that interact with various disciplines, such as chemistry, to design new fibers and/or graft all types of molecules onto the surface of a fabric; physics and mechanics, to study the properties of fibers; mathematics and information technology, to imagine and simulate new intertwining's, and biology, to genetically optimize natural fibers or even copy them. Technical textiles result, therefore, from the ability, which the industry has had and will have, to combine textile fibers indefinitely, possibly with other fibers, such as ceramics or quartz, or even other organic or mineral components, to make threads, surfaces, and volumes, according to structures that allow for the most varied formulations. There are countless textile innovations and each innovation that emerges is more interesting than the last discovery. Above all, innovations will tell us what tomorrow's textiles and clothing will be like.

The twenty-first century will be, at least, the century of so-called intelligent textiles and they will, by themselves, be capable of protecting, caring for and facilitating life, humans and otherwise, in various ways. Smart textiles may introduce new modes of sensitive communication since the senses of vision and hearing have been exposed to a coding theory that communications technology has been extending, at least in social networks, in the emerging concept of wearable computer. Smart textiles will detect how their users are moving and besides being comfortable and cozy fabric, adapt perfectly to the body and manage to recognize posture and the activities of the wearer as well, such as walking, running, and jumping. Using a new manufacturing process, shoes and mats will become intelligent, by integrating hardware and software systems to measure and interpret data from pressure sensors in real time. This technique may have many applications, especially in health and rehabilitation, for example, it could be used to produce smart shoes that track the gait of those who, by accident or other cause, have lost the ability to walk and are in the phase recovery trying to walk again. This same technique could also be applied to socks that monitor the foot of a diabetic patient, with a view to predicting, preventing, and avoiding the formation of ulcers, which we know are unfortunately common in this type of patient.

In conclusion, the future of textiles in the world, especially in Portugal, is undoubtedly challenging. The Portuguese market is visibly recovering but is still faced with significant challenges, particularly in terms of costs, resulting from the impact of the war in Ukraine, energy, and fuel, which have increased several times since last year, in addition to the excessive burden of taxes. Fortunately, there is already a slight reduction in the cost of energy at this moment. Even with these constraints, Portugal, with its ancestry, seems to have the conditions to increasingly assert itself as an important hub of influence and knowledge in the textile sector.

It appears that textiles are on a path towards sustainability and there are already many companies investing in alternative fibers to make their pieces. Of course, validation based on certifications is advisable and it will continue to be recommended since it ends up being a validation tool that adds a guarantee to the user who purchased a properly checked and approved part.[2] It is incredible to see technological advancement through the combination of technicality, ingenuity, elegance, and style and to realize that humanity is particularly sensitive to these assumptions.

3 Conclusion

Even with many constraints, it seems that Portuguese textiles are on a path that has no turning back, clearly towards sustainability. On the one hand, there are already many companies investing in alternative fibers to make their pieces. On the other hand, we must be prepared and sensitive to new and future industrial advances that enable the production of high-performance fibers that guarantee a reduction in environmental impact. Recycled synthetic fibers, such as polyester and nylon, among others, as well as artificial fibers from plant components, natural fibers, and recycled natural fibers will have to be at the center of the future project.

All innovations that aim to improve citizens' daily lives, through clothing and beyond, and that, at the same time, reduce environmental impact, will have to receive all our sympathy and attention. It's a matter of survival. For humans and the textile

[2] Standard 100 by OEKO-TEX certification is, for example, a guarantee that a product does not contain substances harmful to human health (Oeko-tex, 2024). The same happens with the Global Recycled Standard (GRS) certification for recycled materials.

industry to survive, the mission must combine truly sustainable progress with a better life.

References

Abbate, S., Centobelli, P., Cerchione, R., Nadeem, S.P., & Riccio, E. (2023). Sustainability trends and gaps in the textile, apparel, and fashion industries. *Environment, Development and Sustainability*, 1–28.

Agrawal, D. K. (2023). Increasing digital dissemination and online apparel shopping behaviour of Generation Y. *Journal of Fashion Marketing and Management, 28*(1), 28–44.

Ahmad, F., Akhtar, K. S., Anam, W., Mushtaq, B., Rasheed, A., Ahmad, S., Farroq, A., & Nawab, Y. (2023). Recent developments in materials and manufacturing techniques used for sports textiles. *International Journal of Polymer Science*. https://doi.org/10.1155/2023/2021622

Aishwariya, S., & Thamima, S. (2019). Sustainable textiles from lotus. *Asian Textile, 28*(10), 56–59.

Ali, B., Hawreen, A., Kahla, N. B., Amir, M. T., Azab, M., & Raza, A. (2022). A critical review on the utilization of coir (coconut fiber) in cementitious materials. *Construction and Building Materials, 351*, 128957.

Baghel, R. S., Reddy, C. R. K., & Singh, R. P. (2021). Seaweed-based cellulose: Applications, and future perspectives. *Carbohydrate Polymers, 267*, 118241.

Bunsell, A. R. (2018). *Handbook of properties of textile and technical fibers* (2nd ed.). Woodhead Publishing.

de Bustamante, M. S., da Silva Mello, A., & Pimenta, C. A. M. (2021). Culture and local development: Felting as an alternative for the use of sheep wool from the Serra da Mantqueira. *Research, Society and Development*, [S. l.], *10*(8), e50310817562.

Carvalho, C. F. M., & Abreu, M. J. (2023). A Perspective on the future of the fashion and textile sector in Portugal with the use of AI, AR and VR technological tools. In J. Cunha, A. C. Broega, H. Carvalho, & B. Providência (Eds.), *Advances in fashion and design research II* (pp. 17–29). Springer Cham.

Carvalho, L. C., Moreira, S. B., Dias, R., Rodrigues, S., & Costa, B. (2020). Circular economy principles and their influence on attitudes to consume green products in the fashion industry: A study about perceptions of Portuguese students, Rodrigues. In S. S. Almeida, P. J. Almeida, N. M. C., (Eds.), *Mapping, managing, and crafting sustainable business strategies for the circular economy* (pp. 248–275). IGI Global Published.

Centobelli, P., Abbate, S., Nadeem, S. P., & Garza-Reyes, J. A. (2022). Slowing the fast fashion industry: An all-round perspective. *Current Opinion in Green and Sustainable Chemistry*, 100684.

Dinis, A. P., & Guimarães, C. (2023). Yearbook: Portugal textile & clothing industry statistics 2022, Associação Têxtil e Vestuário de Portugal. Retrieved 19 January, 2024, from https://cdn.weasy.io/users/sustainable-ffp/yearbook.pdf

Fashionnetwork. (2017). Seaqual, novo tecido reciclado feito de plásticos retirados do oceano. Retrieved 19 January, 2024, from https://pt.fashionnetwork.com/news/Seaqual-novo-tecido-reciclado-feito-de-plasticos-retirados-do-oceano,788051.html

Fashionunited. (2024). What will be the spring-summer 2024 material trends according to Première Vision? Retrieved 19 January, 2024, from https://fashionunited.com/news/fashion/what-will-be-the-ss24-material-trends-according-to-premiere-vision/2023041353302

Hijosa, C. A. A. (2015). Piñatex, the design development of a new sustainable material, Royal College of Art, United Kingdom.

Khan, A., Awais, M., Mohsin, M., Khan, A., & Cheema, K. (2024). Sustainable yarns and fabrics from tri-blends of banana, cotton and tencel fibres for textile applications. *Journal of Cleaner Production*, 140545.

Lanca, J. C. E., & Baptista, J. M. F. (1993). A cultura do linho, Instituto de Estruturas Agrárias e Desenvolvimento Rural, Portugal.

Morais, F. P., & Curto, J. M. (2022). Design and engineering of natural cellulose fiber-based biomaterials with eucalyptus essential oil retention to replace non-biodegradable delivery systems. *Polymers, 14*(17), 3621.

Nobile, T. H., & Cantoni, L. (2023). Personalization and customization in fashion: Searching for a definition. *Journal of Fashion Marketing and Management: An International Journal, 27*(4), 665–682.

Oeko-tex. (2024). Standard OEKO-TEX® standard 100. Retrieved 19 January, 2024, from https://www.oeko-tex.com/importedmedia/downloadfiles/OEKO-TEX_STANDARD100_Standard_EN_DE.pdf

Omoloso, O., Wise, W., Mortimer, K., Jraisat, L., & Omoloso, S. (2020). Corporate sustainability disclosure: A leather industry perspective. *Emerging Science Journal, 4*(1), 1–11.

Premièrevision. (2024) For 50 years, Première Vision has been the source of inspiration and creation. Retrieved from 19 January, 2024, from https://www.premierevision.com

Rese, A., Baier, D., & Rausch, T. M. (2022). Success factors in sustainable textile product innovation: An empirical investigation. *Journal of Cleaner Production, 331*, 129829.

Solino, L. J. S. D. S., Falani, S. Y. A., Campos, M. C., González, M. O. A., & Almeida, M. R. (2014). Open innovation na cadeia têxtil: estudo exploratório de inovações tecnológicas. *Revista Espacios, 35*(12), 9.

The Role of Financing in the Development of Innovation Processes in Textile Companies

Tânia Saraiva

Abstract This article delves into the intricate relationship between financing and innovation in the textile industry, exploring its historical evolution, current practices, and future trends. The textile and clothing industry is currently undergoing a major transformation, transitioning from incremental changes to facing multiple disruptive organizational shifts due to the evolving global business environment. This change, largely impacting small and medium enterprises (SMEs), is driven by factors such as technological advancements, shifts in global labor costs, regional industry protectionism, and fluctuating consumer preferences. As the industry deals with these complexities, the need for new managerial skills and effective change management becomes crucial for survival and growth. This sector, predominantly composed of SMEs, faces challenges including limited resource access, lower negotiation power, and higher interest rates. Despite their readiness for change, these enterprises often lack the necessary information for effective transition management. In this rapidly evolving landscape, financing plays a critical role in the development of innovation processes within textile companies. For these enterprises, securing financing is essential not just for maintaining operations but also for investing in innovation and staying competitive. Consequently, the textile and clothing industry's evolution is closely tied to its financing mechanisms. The shift towards diverse and sophisticated financing strategies reflects the industry's need to adapt to changing market demands and foster innovation, ensuring sustainability and technological progression in the highly competitive global textile market.

T. Saraiva (✉)
Department of Accounting and Auditing, Lisbon Accounting and Business School, Avenida Miguel Bombarda, 20, 1069-035 Lisboa, Portugal
e-mail: tmsaraiva@iscal.ipl.pt

Lisbon Accounting and Business School, Polytechnic Institute of Lisbon (ed.), *Industry and Innovation: Textile Industry*, SDGs and Textiles,
https://doi.org/10.1007/978-3-031-57804-5_12

1 Introduction

The textile and clothing industry is undergoing a substantial transformation, characterized by a shift from its traditional gradual changes to addressing multiple disruptive organizational shifts, as noted by Prakash et al. (2020) and Melnychuk et al. (2020). This change, driven by the evolving global business environment, is compelling small and medium enterprises (SMEs) to adapt to new challenges such as significant investments in new technologies, labor cost adjustments, protectionism, and fluctuating consumer preferences. This sector's transformation is fueled by technological advancements and a focus on sustainability, demanding innovative approaches in both production and financing.

This evolution towards high-tech and eco-friendly practices in the textile industry heavily relies on innovative financial strategies and robust investment support. The industry's transition from traditional financing methods, such as family-owned capital and bank loans, to more diversified mechanisms like venture capital, corporate funding, and government grants, underscores the changing landscape of financing (Fashion for Good + BCG, 2020). These new funding sources not only provide essential capital for research and development but also bring market expertise and strategic partnerships, critical for commercializing innovative textile technologies.

The industry's global and regional financing dynamics are shaped by various economic, political, and cultural factors, significantly influencing how financial support for innovation is accessed and utilized. Trends like globalization have facilitated foreign direct investment and cross-border financial flows, supporting textile innovation in various regions (Findlay & O'Rourke, 2007; and the European Commission, 2020). However, these dynamics also expose the industry to global economic downturns, impacting investment in innovation.

Different regions exhibit distinct patterns in financing textile innovation, reflecting their economic structures, government policies, and market demands. Developed countries, for example, focus on sustainable and high-tech textile innovations with support from well-established financial markets and government incentives. In contrast, developing countries may emphasize manufacturing efficiency and cost-effectiveness, relying on foreign investments and government subsidies.

Modern financial instruments, such as the digitization of financial services and blockchain technology, play a crucial role in overcoming technological barriers and fostering sustainable practices within the industry. Platforms like crowdfunding and fintech solutions are democratizing access to capital, offering flexible financing routes for SMEs and startups (Mollick, 2014; Tapscott & Tapscott, 2016).

The financial stability and support mechanisms of textile enterprises significantly influence the development of innovation processes. A comprehensive approach to assessing financial stability, which includes modeling the financial condition of enterprises and identifying significant factors influencing their financial status, is crucial. This approach, as discussed by Shim (2017), Acharya et al. (2019), Wagner (2017), and Liu and Arunkumar (2018), supports innovation processes by considering both internal and external factors and understanding the dynamics affecting the enterprise.

Furthermore, the assessment of financial and investment activity in textile enterprises is essential in the modern textile industry. Companies focusing on business administration and analysis can make key managerial decisions, including securing investors and sponsors for expansion. This financial scrutiny, essential for business success and market expansion, involves identifying key indicators such as liquidity, solvency, and financial stability, as emphasized by Raheman et al. (2010), Rafique (2011), Muhammad et al. (2011), and Drobyazko et al. (2020).

Strategic financial management and robust risk management systems are crucial for the long-term financial stability of these enterprises. Effective risk management involves balancing the costs of optimizing financial stability against potential losses and adapting the enterprise to changing economic environments and market uncertainties.

The future of financing in the textile industry is set to be shaped by sustainability-focused models, digital and decentralized platforms, regional diversification, and collaborative approaches. These trends indicate a more inclusive, responsible, and innovative financing landscape, offering exciting prospects for the future of textile innovation. As the industry adapts to these changes, its ability to harness these new financing models will be crucial in driving the next wave of growth and innovation (Tidd & Bessant, 2021).

The pursuit of this article is to present the state-of-art about innovation in the textile sector and the role of financing in the development of this sector, particularly in terms of the intricate relationship between finance and innovation within the textile industry. It aims to provide a comprehensive understanding of how financial mechanisms influence the innovation processes in textile companies, what challenges and opportunities lie therein, and how these dynamics are shaped by global and regional economic trends. Furthermore, this discussion will extend to future projections, contemplating emerging financial models and their potential impact on the textile industry's innovation landscape.

2 Literature Review

The textile and clothing industry, traditionally known for its gradual changes, is currently undergoing multiple disruptive organizational changes due to the evolving global business environment. This shift necessitates a focus on change management to help individuals and enterprises transition through these changes effectively and efficiently. As this sector is mostly made up of small and medium enterprises (SMEs), this industry faces unique challenges such as the need for significant investments in new technologies, relocation arbitrage on labor costs, regional industry protectionism, and volatile consumer preferences.

The textile industry is undergoing a significant transformation. This transformation is propelled by technological advancement and sustainability, demanding innovative approaches not only in production but also in financing. The industry's move towards high-tech and eco-friendly practices is heavily reliant on innovative financial strategies and robust investment support. Such financial support plays a vital role in enhancing the industry's flexibility, responsiveness, and competitiveness in the global market (Prakash et al., 2020; Melnychuk et al., 2020).

Historically, the textile industry's reliance on traditional financing methods, such as family-owned capital and bank loans, has gradually shifted to more diversified and sophisticated mechanisms, including venture capital, corporate funding, and government grants. The use of venture capital and corporate funding is a notable change in the financing landscape. These sources not only provide essential capital for research and development but also bring market expertise and strategic partnerships, which are crucial for the commercialization of innovative products and technologies in textiles (Fashion for Good + BCG, 2020). Alongside private investment, government grants and public funding play a significant role, particularly in supporting foundational research and early-stage development in textile innovation. This support often focuses on advancing technologies that align with broader societal and environmental objectives (Fashion for Good + BCG, 2020).

The global and regional dynamics of financing in the textile industry present a complex and multifaceted picture, shaped by various economic, political, and cultural factors. These dynamics significantly affect the access, utilization, and overall impact of financial support for innovation within the textile sector across various regions.

The textile industry is heavily influenced by broader economic trends such as globalization, trade policies, and international market fluctuations. These factors play a crucial role in determining the flow of investment and the availability of financial resources for innovation. For instance, globalization has facilitated increased foreign direct investment and cross-border financial flows, which have been crucial in supporting textile innovation in various countries. However, this also means that the industry is susceptible to global economic downturns, which can tighten financing options and affect the overall investment in innovation (Findlay & O'Rourke, 2007).

It is relevant to mention that existing different patterns in financing textile innovation by regions of the world, often reflecting their economic structures, government policies, and market demands. For example, in developed countries like those in Western Europe and North America, there is a significant focus on sustainable and high-tech textile innovations. These regions often benefit from well-established financial markets, government grants, and incentives that support such innovations. In contrast, in developing countries, particularly in Asia and Africa, the focus may be more on improving manufacturing efficiencies and cost-effectiveness. Here, financing often comes from foreign investments, government subsidies, and in some cases, international aid and development programs aimed at industrial growth (European Commission, 2020).

In line with the financing sources already mentioned, new modern financial instruments are crucial in overcoming technological barriers and adapting to changing market demands, thereby fostering innovation and sustainable practices within the industry (Fashion for Good + BCG, 2020; Prakash et al., 2020). For example, the digitization of financial services is a trend that is expected to influence the future of financing in the textile industry. Platforms like crowdfunding, peer-to-peer lending, and fintech solutions are democratizing access to capital, especially for SMEs and startups. These digital platforms offer a more flexible and accessible financing route compared to traditional banking and venture capital, enabling a wider range of companies to fund their innovative projects (Mollick, 2014).

The integration of blockchain technology in financial transactions is also anticipated to gain traction in the next decade. Blockchain's ability to offer secure, transparent, and efficient financial transactions makes it an attractive tool for managing supply chain finances and facilitating international trade in the textile industry. This technology could revolutionize how financial transactions are conducted, tracked, and verified, leading to more streamlined and trustworthy financing mechanisms (Tapscott & Tapscott, 2016).

In the context of the textile industry, the role of financing in the development of innovation processes is significantly influenced by the financial stability and support mechanisms of the enterprises. This relationship is underscored by various methodological approaches and practical recommendations aimed at assessing and ensuring financial stability, which is crucial for driving innovation.

The financial stability of service sector enterprises, including those in the textile industry, requires a comprehensive approach that encompasses modeling the financial condition of enterprises and identifying significant factors influencing their financial status. This includes both internal and external factors, establishing normative values for internal indicators, and understanding the dynamics of seasonality and regional factors affecting the enterprise (Shim, 2017; Acharya et al., 2019; Liu and Arunkumar, 2018; Wagner, 2017). Such analysis is vital for assessing the financial stability of the enterprises, thereby supporting innovation processes.

Then, the assessment of financial and investment activity in textile enterprises is also a crucial issue in the modern textile industry, as companies that focus on business administration and analysis have significant opportunities for data operation and making key managerial decisions, including finding investors and sponsors to expand production and sales trends. Rafique (2011) asserts that growth, productivity, capital, operating expenses, and credit risk are determinants of a company's profitability. Muhammad et al. (2011) further elucidates that efficient management of working capital is integral to this process, boosting profitability. Raheman et al.'s (2010) study, which posits that liquidity management influences both risk and profitability.

Strategic financial management and a robust risk management system are crucial for the long-term financial stability of these enterprises (Drobyazko et al, 2020). Effective risk management involves balancing the costs of optimizing financial stability against potential losses and adapting the enterprise to changing economic environments and market uncertainties (Shim, 2017; Acharya et al., 2019; Liu and Arunkumar, 2018; Wagner, 2017).

The composition of a firm's capital structure, especially the balance between loans or debts and equity, critically impacts its financial health. The financial support indicators of a textile company, such as the level of authorized capital, equity, income, expenses, and company assets, play a significant role in influencing investor decisions and, in turn, act as a form of advertising for the company itself. These indicators, comprehensively analyzed, reflect the company's financial support and its position in the market. Jarrow (2013) points out that increased debt in the capital structure escalates the risk of default, thereby diminishing the value of the firm's equity. Khan et al. (2013) delve into the significance of equity sequence and role, emphasizing that the way a firm is financed is a subject of paramount importance. This suggests that while debt can facilitate growth, it becomes a liability when it reaches a critical level, affecting the firm's financial stability. This indicates that while leverage can amplify profitability, it also increases the risk, particularly when firms rely heavily on external financing.

In the context of Pakistan's textile industry, the study of Samo and Murad (2019) reveals that if the debt part in a firm's capital structure exceeds certain limits, it can negatively impact the firm's performance. Factors such as insolvency costs, inability to manage risks, and increased operational costs due to inappropriate financial strategies can impede a textile firm's performance. This aligns with the insightful contribution presented by Guedes et al. (2021), who explore how the capital structure of new ventures adapts in response to economic downturns. Their study, based on a panel of 75,826 Portuguese new ventures established between 2006 and 2015 and followed until 2017, reveals that new ventures' capital structures are indeed sensitive to economic cycles. Ventures founded during economic crises exhibited higher debt ratios (both total and short-term), greater profitability, and more robust growth, yet lower tangibility. Additionally, ventures predominantly financed through equity tend to utilize less short-term debt, exhibit higher profitability and liquidity, but often experience slower growth. This research indicates that the capital structure of new ventures not only reacts to economic downturns but also carries significant implications for establishing ventures during recessionary times.

Therefore, the financial and investment activities of textile enterprises are more crucial than ever in today's competitive landscape. Effective financial management, including a thorough analysis of financial statements and understanding of the company's financial health, is essential for making informed management decisions, attracting investors, and securing sponsorship for expansion. This financial scrutiny allows companies to identify key indicators such as liquidity, solvency, and financial stability, which are fundamental to business success and market expansion (Melnychuk et al., 2020; Prakash et al., 2020). This approach allows for strategic planning, effective risk management, and adaptation to market changes (Shim, 2017; Wagner, 2017; Acharya et al., 2019; Liu & Arunkumar, 2019).

Melnychuk et al. (2020) emphasize the importance of assessing the financial status and investment activity of textile enterprises, proving that such an evaluation is crucial in organizing a successful textile business. The authors clarify that the financial statements provide information about the financial health of the enterprise but also influence investor decisions, thereby impacting the company's ability to secure funding for expansion and innovation.

The study by Boichenko et al. (2021) emphasizes the critical importance of financial support in the development of textile enterprises, highlighting the significant influence of capital structure and current financing on enterprise growth. However, in their study, the financial efficiency factors were not found to be significantly impactful. This study also underscored the necessity of tailored financial support for the specific needs of textile enterprises. In contrast, the report of BCG (2020) on the fashion industry shifts focus to sustainability and innovation. It identifies the urgent need for transformative change in the fashion industry driven by consumer demand and regulatory pressures. The industry, despite its size and potential for disruption, has not yet undertaken systemic change at a scale necessary to meet global climate goals. A major challenge identified is the slow commercialization of new technologies due to unmet investment needs.

Shim (2017), addressing the impact of market seasonality, affirmed that it is particularly crucial for the textile sector. The author suggests strategies to mitigate this issue, including offering diverse services that cater to seasonal demands, adopting pricing policies that vary with the seasons, leveraging mass media to attract customers, and enhancing customer service to maintain a competitive edge. These measures stabilize the financial condition of enterprises and create an environment conducive to innovation by ensuring a steady flow of resources and investment.

Innovators in the fashion industry have identified numerous impactful venture opportunities, developing solutions that range from new materials to digital platforms. However, the report categorizes innovations into "soft tech" (mainly digital solutions) and "hard tech" (requiring physical, capital-intensive assets). Startups focused on new raw materials and more sustainable processes often remain less mature and scale more slowly compared to those developing circular business models. This is primarily due to the longer development cycles and greater capital intensity of hard tech innovations.

The report further highlights two critical gaps in innovation development and scaling: the technology gap and the commercialization gap. These gaps represent phases where capital is less accessible, posing significant challenges, especially for hard-tech innovators. The fashion industry's historical focus on cost efficiency has often overlooked disruptive innovation, creating an environment not conducive to investing in R&D and innovation projects.

Additionally, the fashion industry has been slow to adopt technology and innovation-driven approaches. The recent demand for new solutions has only just begun to shape the innovation landscape, with few startups reaching maturity. The absence of clear commercialization paths and structured innovation processes in the fashion industry creates greater uncertainty and misjudgment of risk, hindering stakeholder involvement at various stages of development.

Furthermore, the study by Wadho and Chaudhry (2018) presents a detailed analysis of how product innovation impacts firm performance, concludes that product innovation leads to increased labor productivity and growth. This is particularly relevant for larger firms, which are more likely to engage in innovation, and those that export, especially to Europe and America. These authors observed that exporting firms are more inclined to innovate, with investments in innovation amounting to around 9% of total turnover in 2015. A notable aspect is that most of the innovation investment is directed towards the acquisition of machinery and hardware, which is a dominant innovation strategy in developing countries. In-house R&D, while significant, is not the primary focus. This result aligns with the earlier findings that financial support, particularly in terms of capital structure and current financing, is crucial for the growth of textile enterprises.

The study by Wadho and Chaudhry (2018) also brings attention to that the financial support is more dependent on subsidies and market constraints, which influence the level of investment in innovation. Subsidies generally have a positive impact on private R&D, although this can vary based on how constraints are treated in the innovation process. The study suggests that firms employing innovative strategies often need to adopt multiple approaches, such as new production methods, workplace management changes, and new marketing strategies, to successfully introduce new products.

Conversely, in line with global trends, there has been a marked shift in the textile industry's financing landscape towards sustainability and ethical manufacturing. Investors and financial institutions are increasingly aligning their strategies with environmental sustainability and social responsibility, reflecting a broader commitment to reducing the textile industry's environmental footprint and enhancing its social impact (Fashion for Good + BCG, 2020). The focus on reducing the use of toxic products, simplifying production processes, and enhancing recycling efforts in this industry, also contributes significantly to more financial support.

The role of financing in driving innovation in the textile industry is best illustrated through real-world case studies:

Case Study 1: Patagonia's Investment in Sustainable Textiles Patagonia, a leading outdoor clothing brand, has long been dedicated to sustainability. Their investment in developing and using recycled polyester and organic cotton is a testament to how financial commitment can drive eco-friendly innovations. By allocating significant resources to R&D, Patagonia was able to pioneer new sustainable materials and production processes, which have set a benchmark in the industry for environmental responsibility (Chouinard & Stanley, 2012).

Case Study 2: Adidas and the Boost Technology Adidas's development of the Boost running shoe technology is another example of successful innovation driven by financial investment. By channeling funds into research and collaborating with BASF, a leading chemical company, Adidas developed revolutionary cushioning technology. This investment not only led to a highly successful product line but also significantly boosted the company's market position and revenue (Adidas Group, 2013).

Case Study 3: Levi Strauss & Co.'s Water < Less™ Technology Levi Strauss & Co. introduced the Water < Less™ technology, which significantly reduces the amount of water used in the denim finishing process. This innovation was a result of targeted financial investment in sustainable practices and has led to saving more than 1 billion liters of water since its inception. It's an example of how funding directed towards sustainability can lead to both ecological and economic benefits (Levi Strauss & Co., 2016).

These instances provide concrete evidence of how financial strategies and support have directly impacted the development and implementation of new technologies, processes, and products in the textile sector.

The study of Cesar da Silva et al. (2021) strengthens that the adoption of cleaner production practices was found to significantly improve the operational, environmental, and economic performance of large textile industries. These improvements align with the need to meet the supply demands of the global market and demonstrate the effectiveness of environmental interventions in decision-making processes. Strategic planning, focused on environmental responsibility and green marketing, emerged as a crucial factor in gaining competitive advantages. This approach also encourages stakeholder communication about the benefits of cleaner production practices adoption, including increased productivity, enhanced product quality, improved worker health and safety, water reuse, minimization of waste, reduction of environmental accidents, and replacement of toxic materials.

The emphasis on cleaner production practices adoption aligns with the previously identified need for transformative investments in the textile and fashion industry to embrace technological advancements and sustainability goals. The studies collectively illustrate that while targeted financial strategies are essential for the growth of textile enterprises, broader industry innovation requires strategic investments in cleaner and more sustainable practices.

Historically, the main factors influencing change in the Textiles and Clothing Industry have been the changing habits and needs of consumers and the short life cycles due to fashion trends. However, current trends in fashion are also influenced by macro and micro factors such as the volatility of the global economy, China's trading and production policies, and new technologies in the supply chain (Tudor, 2018).

In the European textile and clothing industry, 99% of enterprises are SMEs. These SMEs face specific environmental threats such as limited material, financial, informational, human, and managerial resources, lower negotiation power, and higher interest rates. Despite their readiness for change to foster growth, SMEs often lack information on how to approach change effectively. Implementing change management can increase the efficiency of change projects, fostering employee acceptance and involvement (Tudor, 2018).

Therefore, in the context of financing, the development of innovation processes in networks of textile companies, understanding, and managing these dynamic factors is crucial. This involves not only adapting to consumer trends and global market shifts but also efficiently managing resources and navigating the complex financial landscape. Managers in this sector must prioritize strategic planning, change management, and investment in new technologies to drive innovation while adapting to these rapid changes in the industry.

The future of financing in the textile industry is poised at the cusp of significant transformations, driven by evolving economic landscapes, technological advancements, and changing consumer preferences. As the industry looks forward, several key trends and projections are expected to shape the way innovation is financed.

One of the most prominent future trends in the textile industry is the shift towards sustainable and ethical financing models. This change is in response to the growing consumer demand for environmentally friendly and socially responsible products. Investors and financiers are increasingly aligning their portfolios with sustainable practices, leading to a rise in green bonds, impact investing, and ESG (Environmental, Social, and Governance) compliant financing. These models not only provide the necessary capital for innovation but also ensure that such innovations align with broader sustainability goals (European Commission, 2020).

Finally, the future of financing in the textile industry is likely to see more collaborative and cross-industry initiatives. Partnerships between textile companies, technology firms, research institutions, and even non-traditional players like tech startups may become more common. These collaborations can pool resources and expertise, leading to innovative solutions that cut across industry boundaries. Such partnerships not only share the financial burden but also foster a culture of co-creation and open innovation (Tidd & Bessant, 2021).

In summary, there is a complex landscape where the development of textile enterprises and the broader fashion industry's innovation process are heavily influenced by the availability and structure of financial support. The future of financing in the textile industry is set to be shaped by a mix of sustainability-focused models, digital and decentralized platforms, regional diversification, and collaborative approaches. These trends point towards a more inclusive, responsible, and innovative financing landscape, offering exciting prospects for the future of textile innovation. As the industry adapts to these changes, its ability to harness these new financing models will be crucial in driving the next wave of growth and innovation. It was also evident that there be must not overlooked that there is a clear need for tailored financial strategies to foster growth in textile enterprises.

3 Conclusion

In conclusion, the textile industry is undergoing a significant transformation, driven by global business environment changes, as detailed by Prakash et al. (2020) and Melnychuk et al. (2020). This shift necessitates a focus on change management, particularly for SMEs, which comprise a large portion of this sector. These companies face challenges such as the need for significant investments in new technologies, labor cost adjustments, regional industry protectionism, and fluctuating consumer preferences.

The transformation is propelled by technological advancement and sustainability, demanding innovative approaches in both production and financing. The industry's move towards high-tech and eco-friendly practices relies heavily on innovative financial strategies and robust investment support, as highlighted in the works of Prakash et al. (2020) and Melnychuk et al. (2020). This support is critical in enhancing the industry's flexibility, responsiveness, and competitiveness in the global market.

The traditional reliance of the textile industry on family-owned capital and bank loans has shifted towards more diversified financing mechanisms, including venture capital, corporate funding, and government grants, as discussed in the report by Fashion for Good + BCG (2020). These sources provide capital for research and development, bringing market expertise and strategic partnerships crucial for commercializing innovative products and technologies in textiles.

Global and regional financing dynamics in the textile industry are shaped by economic, political, and cultural factors, influencing financial support for innovation. Globalization, trade policies, and market fluctuations play a key role in investment flow and resource availability for innovation, as evidenced by the findings of Findlay and O'Rourke (2007) and the European Commission (2020). Developed countries often focus on sustainable and high-tech textile innovations, supported by financial markets, government grants, and incentives. In contrast, developing countries may prioritize manufacturing efficiency and cost-effectiveness, with financing often coming from foreign investments and government subsidies.

Modern financial instruments, including digitization of financial services and blockchain technology, are crucial in overcoming barriers and fostering innovation and sustainable practices. These instruments democratize capital access and offer secure, efficient financial transactions, essential for managing supply chain finances and facilitating international trade, as discussed by Mollick (2014) and Tapscott and Tapscott (2016).

The financial stability and support mechanisms of textile enterprises significantly influence innovation development. Assessing financial stability requires a comprehensive approach, considering internal and external factors and understanding the dynamics affecting the enterprise, as outlined by Shim (2017), Acharya et al. (2019), Wagner (2017), and Liu and Arunkumar (2018). Strategic financial management and robust risk management are crucial for long-term stability, balancing costs against potential losses, and adapting to market changes.

The assessment of financial and investment activity in textile enterprises is crucial, as companies focusing on business administration and analysis have opportunities to make key managerial decisions, including securing investors and sponsors for expansion. Effective financial management allows companies to identify key indicators like liquidity, solvency, and financial stability, fundamental for business success and market expansion, as noted by Raheman et al. (2010), Rafique (2011), and Muhammad et al. (2011).

The future of financing in the textile industry will likely see sustainable and ethical financing models, collaborative initiatives, and a mix of digital and decentralized platforms. These trends indicate a more inclusive, responsible, and innovative financing landscape, essential for driving the next wave of growth and innovation in textile enterprises, as suggested by the research of Samo and Murad (2019) and Guedes et al. (2021). As the industry adapts to these changes, its ability to harness these new financing models will be crucial in driving the next wave of growth and innovation. The combined insights from these studies reveal a complex landscape where the development of textile enterprises and the broader fashion industry's innovation process are heavily influenced by the availability and structure of financial support.

References

Acharya, V. V., Björn, I., Steffen, S., & Teichmann, D. (2019). Does the lack of financial stability impair the transmission of monetary policy? SSRN. Electronic copy. Retrieved from https://ssrn.com/abstract=2691049

Adidas Group. (2013). For the Love of sport. In *Adidas group annual report*. Retrieved from https://www.adidas-group.com/media/filer_public/2014/03/05/adidas-group_gb_2013_en.pdf

Boichenko, K., Mata, M. N., Mata, P. N., & Martins, J. N. (2021). Impact of financial support on textile enterprises' development. *Journal of Risk and Financial Management, 14*(3), 135.

Cesar da Silva, P., Neto, G. C. O., Correia, J. M. F., & Tucci, H. N. P. (2021). Evaluation of economic, environmental and operational performance of the adoption of cleaner production: Survey in large textile industries. *Journal of Cleaner Production*, 278.

Chouinard, Y., & Stanley, V. (2012). The responsible company: What we've learned from Patagonia's first 40 years. Patagonia.

Drobyazko, S., Barwinska-Malajowicz, A., Slusarczyk, B., Chubukova, O., & Bielialov, T. (2020). Risk management in the system of financial stability of the service enterprise. *Journal of Risk and Financial Management, 13*(12), 300.

European Commission. (2020). Horizon 2020: The EU framework programme for research and innovation. Retrieved from https://ec.europa.eu/programmes/horizon2020/

Findlay, R., & O'Rourke, K. H. (2007). *Power and plenty: Trade, war, and the world economy in the second millennium*. Princeton University Press.

Guedes, M. J., Saraiva, T. M., & Felício, T. (2021). Does the capital structure of new ventures differ as a response to a financial crisis? Capital structure responsiveness of portuguese new ventures to an economic downturn. In N. Teixeira & I. Lisboa (Eds.), *Handbook of research on financial management during economic downturn and recovery* (pp. 77–93). IGI Global.

Jarrow, R. (2013). A leverage ratio rule for capital adequacy. *Journal of Banking and Finance, 37*(3), 973–976.

Khan, A., Muttakin, M. B., & Siddiqui, J. (2013). Corporate governance and corporate social responsibility disclosures: Evidence from an emerging economy. *Journal of Business Ethics, 114*, 207–223.

Levi Strauss & Co. (2016). Open source: Water innovation. Retrieved from https://levistrauss.com/wp-content/uploads/2016/03/Open-Source-Water-Innovations.pdf

Ley, K., van Mazijk, R., Boger, S., Watten, D., Pardo, C. M., & Zuo, C. (2020). Financing the transformation in the fashion industry: Unlocking investment to scale innovation. *Fashion for Good + BCG*. Retrieved from https://fashionforgood.com/wp-content/uploads/2020/01/FinancingTheTransformation_Report_FINAL_Digital-1.pdf

Liu, C., & Arunkumar, N. (2019). Risk prediction and evaluation of transnational transmission of financial crisis based on complex network. *Cluster Computing, 22*(Suppl 2), 4307–4313.

Melnychuk, Y., Slatvinskyi, M., Tkachuk, I., Vinnytska, O., Chvertko, L., & Korniienko, T. (2020). Assessment of financial and investment activity of textile enterprises as a guarantee of successful business in the region. *Industria Textila, 71*(3).

Mollick, E. (2014). The dynamics of crowdfunding: An exploratory study. *Journal of Business Venturing, 29*(1), 1–16.

Muhammad, M., Jan, W. U., & Ullah, K. (2011). Working capital management and profitability an analysis of firms of textile industry of Pakistan. *Journal of Managerial Sciences, 6*(2), 155–165.

Prakash, Y., Charwak, B., & Kumar P. V. (2020). Textile industry in new India; challenges and opportunities. *International Journal of Indian Culture and Business Management 21*(4), 435–458. https://doi.org/10.1504/IJICBM.2020.111349

Rafique, M. (2011). Effect of profitability & financial leverage on capital structure: A case of pakistan's automobile industry. SSRN. https://doi.org/10.2139/ssrn.1911395.

Raheman, A., Afza, T., Qayym, A., & Mahmood, A. (2010). Working capital management and corporate performance of manufacturing sector in Pakistan. *International Research Journal of Finance and Economics, 47*, 151–163.

Samo, A. H., & Murad, H. (2019). Impact of liquidity and financial leverage on firm's profitability—an empirical analysis of the textile industry of Pakistan. *Research Journal of Textile and Apparel, 23*(4), 291–305.

Shim, J. (2017). An investigation of market concentration and financial stability in property–liability insurance industry. *Journal of Risk and Insurance, 84*, 567–597.

Tapscott, D., & Tapscott, A. (2016). Blockchain revolution: How the technology behind bitcoin is changing money, business, and the world. *Portfolio.*

Tidd, J., & Bessant, J. (2021). *Managing innovation: Integrating technological, market and organizational change.* Wiley.

Tudor, L. (2018). Change in textile and clothing industry. *Industria Textila, 69*(1), 37–43.

Wadho, W., & Chaudhry, A. (2018). Innovation and firm performance in developing countries: The case of Pakistani textile and apparel manufacturers. *Research Policy, 4787*, 1283–1294.

Wagner, W. P. (2017). Trends in expert system development: A longitudinal content analysis of over thirty years of expert system case studies. *Expert Systems with Applications, 76*, 85–96.

Management Control Systems and Innovation in Textile Industry

Jéssica Nunes Martins

Abstract This study delves into the symbiotic relationship between Management Control Systems (MCS) and innovation in the textile industry through a comprehensive bibliometric analysis, aiming to bridge the current knowledge gap. Utilizing a systematic literature review, the research scrutinizes scholarly articles from 2004 to 2023, encompassing various documents from databases like Scopus and Web of Science. The methodology includes publication pattern analysis, citation analysis, co-word analysis, and authorship trends, offering a quantitative overview of the research landscape in this field. Findings highlight a growing focus on technological innovation, sustainable practices, and the strategic role of MCS in decision-making, while underscoring the lack of understanding in the cultural and organizational aspects of innovation management. The study reveals the essential role of MCS in shaping organizational culture and driving innovation, emphasizing the need for integrating sustainability and technological advancements. It identifies gaps and potential areas for future research, stressing the importance of international collaboration and exploring under-represented topics to enrich the textile industry's innovation and adaptability.

1 Introduction

In the dynamic world of the textile industry, innovation emerges as a cornerstone for sustainable growth and competitive advantage. As global markets evolve rapidly, the need for effective management control systems (MCS) to foster and manage innovation becomes increasingly evident. This study aims to bridge the gap between MCS and innovation in the textile sector through a comprehensive bibliometric analysis, offering insights into the current state of research and identifying future research

J. Nunes Martins (✉)
NOVA Information Management School (NOVA IMS), Universidade Nova de Lisboa, Campus de Campolide, 1070-312 Lisboa, Portugal
e-mail: jessicanunesmartins@gmail.com

Insight: Piaget Research Center for Ecological Human Development, Lisboa, Portugal

Lisbon Accounting and Business School, Polytechnic Institute of Lisbon (ed.),
Industry and Innovation: Textile Industry, SDGs and Textiles,
https://doi.org/10.1007/978-3-031-57804-5_13

trajectories. The textile industry, characterized by its rich history and cultural significance, is a major contributor to the global economy. However, it faces numerous challenges including environmental concerns, rapidly changing consumer preferences, and intense global competition. In this context, innovation is not just a strategy but a necessity for survival and growth. MCS play a crucial role in this regard, providing the framework for organizations to nurture, manage, and sustain innovation. Despite its importance, scholarly understanding of how MCS facilitate innovation in the textile industry remains fragmented.

This study recognizes the critical need to synthesize existing literature to understand the interplay between MCS and innovation in the textile industry. The primary objective is to conduct a bibliometric analysis to map the existing academic terrain, identify key themes, trends, and gaps in the literature, and suggest directions for future research. This approach not only aids in comprehending the current state of research but also in discerning the evolution of themes and collaborations in this domain.

Adopting a bibliometric analysis methodology, this study analyzes scholarly articles from databases recognized by Scopus and Web of Science. The analysis involves examining publication patterns, citation analysis, co-word analysis, and authorship patterns. Such a method provides a quantitative overview of the research landscape, allowing for the identification of influential studies, authors, and journals in the field. This approach helps in uncovering the most discussed topics and emerging trends in the literature.

The preliminary analysis reveals a growing interest in the study of MCS and innovation in the textile industry over the last two decades. The literature predominantly focuses on technological innovation, sustainable practices, and the role of MCS in facilitating strategic decision-making. There exists a notable gap in understanding the cultural and organizational aspects of innovation management within the industry.

This study is expected to offer valuable insights for academics, industry practitioners, and policymakers. For academics, it provides a consolidated view of the field, highlighting areas ripe for further research. Industry practitioners can benefit from understanding the key drivers and barriers to innovation in the textile industry, while policymakers may find the implications for regulatory frameworks and support mechanisms valuable. This bibliometric analysis aims to serve as a foundational reference point for future explorations into how management control systems can effectively drive innovation in the textile industry.

2 Systematic Literature Review

The textile industry, with its dynamic and innovative character, plays a pivotal role in the global economy. To understand the interplay between management control systems and innovation within this sector, a bibliometric analysis offers valuable insights. This approach quantifies the impact of scholarly work and highlights trends and patterns in academic research. The current study focuses on the period from

Table 1 Input data for the study

Description	Results
Timespan	2004:2023
Sources (journals, books, etc)	44
Documents	54
Document average age	10,8
Average citations per doc	5,5
Keywords Plus	497
Author's keywords	124
Authors	112
Authors of single-authored docs	14
Single-authored docs	26
Co-authors per document	2,33
International co-authorships %	5,56
Article	28
Book chapter	3
Conference paper	13
Conference review	4
Editorial	1
Review	5

2004 to 2023, examining a range of documents including articles, book chapters, and conference papers to unravel the complex relationship between management controls and innovation in the textile industry (Table 1).

Timespan (2004–2023): The 20-year period is significant, covering multiple phases of technological advancements and market changes in the textile industry. This timeframe allows for a comprehensive understanding of the evolving nature of management control systems and their role in fostering innovation. The study encompasses 44 different sources, including journals and books, leading to a collection of 54 documents. This diverse range of sources ensures a broad perspective on the topic. The average age of the documents is relatively high, indicating a well-established field of research with roots in earlier studies. This could also suggest a need for more recent studies to capture the latest trends. The metric signifies the impact and relevance of the research in the academic community. An average of 5.5 citations per document reflects a moderate level of influence and recognition among scholars. The substantial number of keywords, especially in the 'Keywords Plus' category, points to a rich and diverse research landscape. It suggests the interdisciplinary nature of studies linking management control systems with innovation in the textile industry. The involvement of 112 authors, including those who have single-authored documents, highlights the collaborative nature of research in this field. However, the presence of single-authored documents also indicates individual contributions and perspectives. This average suggests a moderate level of collaboration per document, which is essential for cross-pollinating ideas and approaches in research. The relatively low percentage of international co-authorships could point to a more localized research focus or the need for broader international collaboration to enrich the research perspectives.

The variety in document types (articles, book chapters, conference papers, etc.) indicates a multi-faceted research approach. Articles dominate the dataset, which is typical for academic research dissemination, followed by conference papers and reviews. This diversity supports a comprehensive understanding of the topic from different academic and practical viewpoints. The data provides a foundational understanding of the academic landscape concerning management control systems and innovation in the textile industry. The diverse sources, documents, and author contributions indicate a rich, albeit moderately collaborative and internationally connected, research area. The bibliometric analysis sets the stage for deeper exploration into how management strategies influence innovation in this vital industry sector.

The co-occurrence network, which maps the relationships between different nodes (or themes), is primarily clustered around several core areas. The most central node, in terms of betweenness (a measure of how often a node appears on the shortest path between two other nodes), is the "textile industry" itself (Fig. 1). This is unsurprising given the focus of the study, and it highlights the industry's role as a hub that connects various concepts. Innovation, another key node, shows significant betweenness, although it's less central than the textile industry. This suggests that while innovation is a critical theme, it's more of a connecting element between different areas rather than a central focus. Quality control and information management also emerge as important nodes, indicating their pivotal roles in the textile industry. They act as bridges in the network, linking various concepts together. Other concepts like commerce, design, resource allocation, and technology, while connected to the central theme, exhibit lower betweenness and closeness values. This suggests that they are more peripheral in the current discourse. Interestingly, the network reveals clusters around themes like textiles, sustainable development, and software, each presenting a distinct but interconnected area of focus. Textiles, for instance, forms a secondary hub, indicating its specific importance within the broader textile industry context. The presence of nodes like dyeing, color, production management, and energy conservation in distinct clusters highlights specialized areas within the textile industry. These areas, while less central than the core themes, are crucial for understanding the nuanced aspects of the industry.

The bibliometric analysis paints a picture of a network where the textile industry and innovation stand as central themes, interconnected with various other crucial concepts like quality control and information management. The presence of distinct clusters around specific themes like textiles and software suggests specialized areas of focus within the broader context of the textile industry. Management control systems (MCS) play a crucial role in the innovation processes within the textile industry. These systems, which encompass various tools and practices for directing and controlling organizational activities, are pivotal in fostering an environment conducive to innovation. The textile industry, characterized by rapid technological

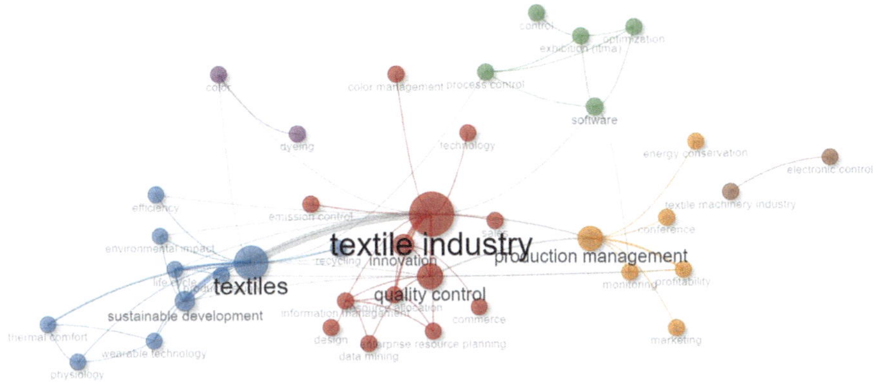

Fig. 1 Cluster analysis of publications on management control systems and innovation in textile industry

advancements and market dynamics, particularly benefits from effective MCS that support strategic decision-making and resource allocation (Akroyd & Kober, 2020; Lee et al., 2014).

The integration of MCS in the textile industry has been studied from various perspectives. One key area of focus is the alignment of MCS with organizational strategy to enhance innovation capabilities. Studies have shown that the strategic use of MCS, tailored to the specific needs of the textile sector, significantly impacts the industry's capacity to innovate and adapt (Laguir et al., 2022; Rehman et al., 2021). Moreover, the role of MCS in facilitating organizational learning and knowledge management, which are critical for innovation, has been emphasized in recent research (Chen et al., 2022; Do et al., 2022).

Another significant aspect is the influence of MCS on the organizational culture within textile firms. A culture that supports experimentation, risk-taking, and creative problem-solving is essential for innovation. Research indicates that MCS can either foster or hinder the development of such a culture, depending on their design and implementation (Quesado et al., 2022; Tran et al., 2022). For instance, performance measurement systems, a component of MCS, can be designed to encourage innovative behaviors by incorporating metrics that reward creativity and innovation outcomes (Chen et al., 2021; Mangenda Tshiaba et al., 2021).

Bibliometric analysis has been employed to explore the evolution of research in this field. Such analyses reveal the most influential studies, authors, and research trends over time. They provide insights into how the understanding of MCS's role in fostering innovation in the textile industry has evolved, highlighting key theories and models that have shaped this research area (Bresciani et al., 2022; Laguir et al., 2022).

The application of MCS in the context of sustainability and environmental innovation in the textile industry has also gained attention. With increasing emphasis on sustainable practices, MCS are being leveraged to drive eco-innovation, ensuring that environmental considerations are integrated into the innovation process (Henri & Journeault, 2010; Khan et al., 2019). This shift reflects a broader trend towards sustainability in business practices and the role of MCS in supporting this transition (Guo et al., 2011; Ngai et al., 2014).

In terms of methodology, both qualitative and quantitative approaches have been used to study the impact of MCS on innovation in the textile industry. Case studies, surveys, and empirical analyses provide a comprehensive understanding of the practical implications of MCS in different contexts within the industry (Andersen & Lueg, 2017; Rahman et al., 2022).

Challenges in implementing MCS for innovation in the textile industry have also been a focus of research. Issues such as resistance to change, the complexity of integrating MCS with existing processes, and the need for alignment with organizational goals are recurrent themes (Rehman et al., 2019; Barros & Ferreira, 2019). Studies suggest that addressing these challenges requires a nuanced understanding of the organizational context and the specific needs of the textile industry (Wickramasinghe & Hopper, 2005; Franco-Santos & Otley, 2018).

In the realm of management control systems (MCS) and innovation within the textile industry, institutional theory provides a profound lens through which the evolution and adoption of these practices can be examined. This theory, fundamentally concerned with the deeper, often implicit rules and norms that shape organizational behavior and change, offers valuable insights into how MCS and innovation are not merely technical or economic choices but are deeply embedded in the institutional fabric of the industry (Kalyar et al., 2020). The textile sector, characterized by its rapid adaptation to technological changes and market demands, exemplifies an institutional environment where traditional practices and modern innovations intersect. Research in this area often highlights how institutional pressures, such as regulatory demands, normative standards from industry associations, and mimetic processes driven by competition play a pivotal role in shaping the adoption and implementation of MCS and innovative practices (Farrukh et al., 2022; Peters & Simaens, 2020). These institutional forces not only dictate the "rules of the game" but also influence how textile companies perceive and respond to technological advancements, sustainability challenges, and global market trends. Thus, institutional theory enriches the understanding of the dynamics between MCS and innovation in the textile industry, providing a nuanced perspective that transcends purely functional or strategic interpretations.

The literature on MCS and innovation in the textile industry is rich and diverse, encompassing various theoretical and practical aspects. The bibliometric analysis indicates a growing interest in this field, with an increasing number of studies exploring the multifaceted role of MCS in driving innovation. Future research could further explore the interplay between MCS and emerging technologies, such as artificial intelligence and the Internet of Things, which are set to revolutionize the textile industry.

The term "textile industry", with 18 occurrences, stands out as the most prevalent, underscoring the primary focus of the analyzed literature. This is closely followed by "textiles" with 16 mentions, which reinforces the emphasis on the industry's

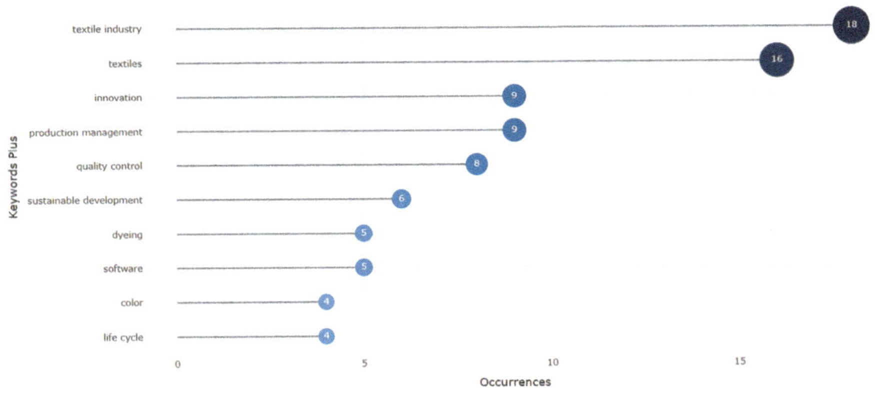

Fig. 2 The most frequently occurring words in the studied publications

core product. Interestingly, the words "innovation" and "production management" both appear 9 times, highlighting their significant roles in the textile industry. These terms suggest a dual focus on novel approaches and efficient management practices within the sector. Furthermore, "quality control", occurring 8 times, indicates a strong emphasis on maintaining standards and ensuring the reliability of textile products. "Sustainable development", mentioned 6 times, reflects an increasing awareness and integration of environmental and ethical considerations in the textile industry. This aligns with global trends towards sustainability and corporate responsibility. The term "dyeing", with 5 occurrences, points to a specific area of focus within the textile production process, likely linked to both innovation and sustainable practices (Fig. 2).

The word "software", also appearing 5 times, might be indicative of the growing role of technology and digital tools in enhancing various aspects of textile production and management. The lower frequency terms, "color" and "life cycle", each mentioned 4 times, could be related to product development and environmental impact assessments, respectively.

The analysis of the most frequently occurring words in the studied publications paints a comprehensive picture of the textile industry's current academic and research focuses. These include a strong emphasis on the core elements of the industry, such as production and textiles themselves, along with a balanced attention to innovation, quality control, and sustainability. The presence of terms like software and life cycle also suggests a forward-looking perspective, considering the role of technology and environmental impact in the industry's future.

The bibliometric analysis of publications in the context of management control systems and innovation in the textile industry reveals several key trends and focal topics over the years (Fig. 3).

Starting with "production management", it saw a concentrated interest around 2007, with a frequency of 9 publications. This suggests a significant focus on optimizing production processes in the early 2000s, with the interest tapering off slightly

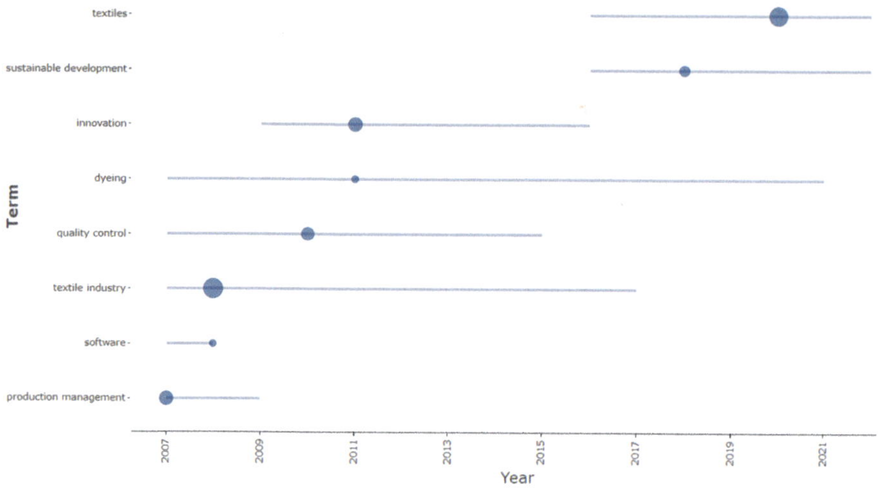

Fig. 3 Trend topics in the studied publications

by 2009. In contrast, the "textile industry" as a topic maintained a more extended period of interest, with 18 publications spread from 2007 to 2017. This indicates a sustained research interest in the industry, perhaps reflecting ongoing changes and challenges in this sector over a decade. The topic of "software" had a brief but intense period of focus around 2007–2008 (5 publications), possibly correlating with technological advancements in textile manufacturing and management systems during that period. "Quality control" remained a relevant topic from 2007 to 2015, with 8 publications. This enduring interest might reflect the industry's ongoing challenges in maintaining product quality amidst evolving production techniques and market demands. "Innovation" as a topic emerged slightly later, gaining traction in 2009 and maintaining research interest until 2016 (9 publications). This could signify a shift in the industry's focus towards novel approaches and technologies to stay competitive and sustainable. The topic of "dyeing" had a unique trajectory, beginning in 2007 and extending until 2021 (5 publications), indicating a long-term interest in this specific aspect of textile production, possibly linked to environmental and efficiency concerns.

"Sustainable development" and "textiles" as separate topics saw increasing attention from 2016 onwards, peaking in 2022. This reflects a more recent shift in the industry towards sustainability and innovation in textile materials and processes, aligning with global environmental concerns and market demands for sustainable products. This analysis reveals a dynamic and evolving landscape in textile industry research, with shifts from traditional production and quality control to more recent focuses on innovation, sustainability, and advanced textiles.

The thematic map in Fig. 4 of the bibliometric analysis on "Management Control Systems and Innovation in Textile Industry" reveals a complex interplay of themes and concepts. This map organizes key terms into clusters, highlighting their relevance

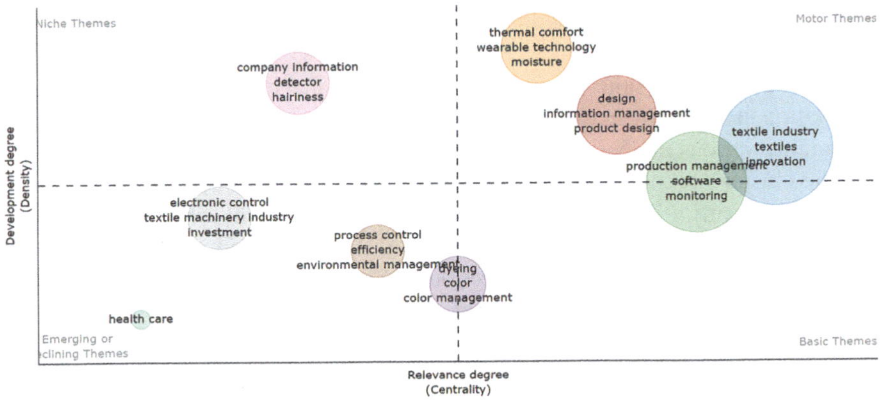

Fig. 4 Thematic map for the studied publications

through metrics like occurrences, betweenness centrality, closeness centrality, and Page Rank centrality (Fig. 4).

The cluster labeled "Design" encompasses terms like "design", "information management", "product design", "carbon footprint", "data mining", and "fabrics." This suggests a strong focus on the design aspect within the textile industry, integrating environmental considerations (carbon footprint) and technological tools (data mining, information management). The centrality measures indicate these terms are not only frequent but also pivotal in connecting various research themes. In the "Textile Industry" cluster, terms like "textile industry", "textiles", "innovation", "quality control", and "sustainable development" are prominent. This cluster is distinguished by its high PageRank centrality, especially for the term "textile industry", suggesting a central focus of the publications. The inclusion of "innovation" and "sustainable development" indicates a trend towards modernizing industry practices and prioritizing sustainability. The "Production Management" cluster, featuring terms such as "production management", "software", "monitoring", and "energy conservation", highlights the importance of efficient production processes. The presence of "software" and "monitoring" points towards an increasing reliance on technology for production optimization. "Dyeing" as a separate cluster, with terms like "dyeing", "color", and "color management", reflects a specialized focus within the textile manufacturing process, possibly linked to both technological advancements and environmental considerations. "Thermal Comfort" and "Process Control" are smaller clusters. "Thermal Comfort" includes terms like "moisture", "physiology", and "wearable technology", indicating research in textile applications for personal comfort and health. "Process Control" emphasizes efficiency and environmental management, underscoring the industry's focus on sustainable practices. Clusters like "Company Information" and "Electronic Control" reveal a focus on technological advancement in machinery and control systems, while "Health Care" suggests an intersection of textiles and medical applications. The thematic map portrays a textile industry at the crossroads of traditional practices and modern innovation, with a pronounced emphasis on sustainability, technology integration, and efficiency in production management.

Table 2 focuses on the interrelation between management control systems and innovation in the textile industry. This analysis uses a thematic map cluster approach, examining various dimensions and the frequency of each cluster.

The cluster "Textile Industry" emerges as the most prominent, with the highest Callon Centrality (9.6627) and a significant Callon Density (109.1046). Its high centrality and density suggest that it is a well-developed and central theme in the literature, with extensive interconnections and importance. This cluster also has the highest frequency (100), reinforcing its dominance in the field. "Design" and "Production Management" follow in terms of importance. The design cluster, while having a lower centrality (4.3056) compared to the textile industry, has a high density (112.7778), indicating a well-developed and focused research area within the field. Its relatively high frequency (23) further underscores its significance. Production management, with a centrality of 6.2269 and a density of 106.6472, also plays a crucial role, showing a strong presence in the literature with considerable development and focus. The clusters "Dyeing", "Thermal Comfort", "Process Control", "Company Information", and "Electronic Control" demonstrate varying degrees of centrality and density. Dyeing and thermal comfort, for example, show lower centrality but relatively high density, indicating specialized, focused areas within the broader textile industry context. The low frequency of these clusters (9 for dyeing and 16 for thermal comfort) suggests they are niche areas within the broader field. "Process Control", "Company Information", and "Electronic Control" have lower centrality and density values, pointing to their emerging or less developed status in the literature. Their frequency also reflects this, with relatively low numbers, indicating they are less explored areas within the field. The "Health Care" cluster is the

Table 2 Cluster analysis according to the thematic map

Cluster	Callon centrality	Callon density	Rank centrality	Rank density	Cluster frequency
Design	4,3056	112,7778	7	7	23
Textile industry	9,6627	109,1046	9	6	100
Production management	6,2269	106,6472	8	5	55
Dyeing	1,9583	51,3889	5	2	9
Thermal comfort	2,8333	123,4127	6	9	16
Process control	1,6667	66,6667	4	3	8
Company information	1,25	120,8333	3	8	12
Electronic control	0,9444	89,4444	2	4	12
Health care	0,25	50	1	1	2

least developed, with the lowest centrality (0.25) and density (50). Its frequency of 2 indicates it is a highly niche area within the context of this analysis, potentially representing an emerging or peripheral area of research in the intersection of management control systems, innovation, and the textile industry.

This analysis reveals a diverse landscape of research within the field, with the textile industry and design as dominant clusters, followed by production management. Other areas like dyeing, thermal comfort, process control, company information, and electronic control, while less central, represent significant specialized fields. Health care stands out as a nascent and niche area, suggesting the potential for future research development.

2.1 The Common Research Trends and Meta-Analysis

The annual scientific production trends in the field of management control systems and innovation in the textile industry, as illustrated in Fig. 5, reveals a fluctuating yet insightful trajectory over the years 2004 to 2023. In the early years from 2004 to 2006, the research output in this area was relatively low, peaking modestly at 3 articles in 2005. This period reflects the nascent stage of academic interest in the intersection of management control systems and innovation within the textile industry. A significant surge is observed in 2007, with the publication of 9 articles. This spike could indicate a growing recognition of the importance of management control systems in fostering innovation in the textile sector, possibly triggered by industry shifts or technological advancements during that period. Following this peak, there was a notable decline, with the years 2008 to 2011 showing a reduced and somewhat inconsistent output, dipping to no publications in 2012. This decline might suggest a temporary shift in research focus or challenges in the textile industry impacting the relevance or viability of such studies.

The period from 2013 to 2017 again shows fluctuating activity, but with a generally higher output than the early years, suggesting a renewed and sustained interest in the topic. This could be attributed to evolving industry practices or emerging challenges in the textile sector that highlighted the role of management control systems in driving innovation. The year 2019 marks another peak with 7 publications, possibly reflecting a response to specific industry needs or advancements in management strategies pertinent to the textile industry. The final years, from 2020 to 2023, show a pattern of moderate and steady research output, with a slight increase in 2021 and 2022, and stabilizing in 2023. This trend could indicate a maturation in the field with established research themes and a consistent recognition of the importance of management control systems in fostering innovation in textiles.

This analysis underscores the dynamic nature of academic research in this domain, influenced by industry trends, technological advancements, and evolving management practices. The varying number of publications over the years reflects the changing focus and interest of researchers and practitioners in understanding and

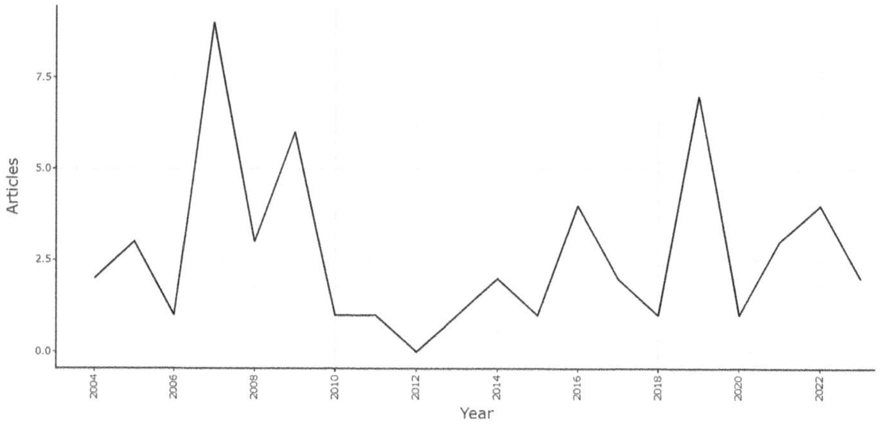

Fig. 5 Annual scientific production trends in management control systems and innovation in textile industry

improving the role of management control systems in driving innovation in the textile industry.

The analysis revealed an intriguing trend in average citations per year for researched publications between 2004 and 2023 (Fig. 6). A noticeable feature in the early years (2004-2010) was the minimal to zero average citations per article, indicating a relatively low impact or interest in this specific research area during that period.

From 2016 onwards, there was a significant increase in the average citations per article, peaking notably in 2019 with an average of 25.14 citations for 7 publications. This sharp increase suggests a growing interest and relevance of the subject matter within the academic and industrial communities. The year 2016, in particular, stands

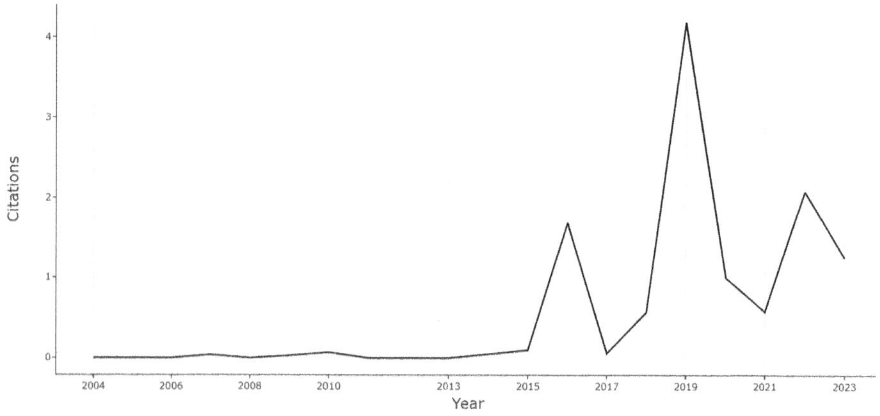

Fig. 6 Average citations per year for the researched publications

out as a turning point with an average of 15.25 citations for 4 publications, marking a substantial leap from the preceding years. Post-2019, while the average citations per article remained relatively high compared to the early years, there was a fluctuation, indicating a dynamic interest in the topic. For instance, the average citations decreased in 2020 and 2021 but rose again in 2022, reflecting ongoing research developments and possibly emerging sub-themes in this field. The data also highlights the number of publications (denoted as "N") which varied each year, reaching its peak in 2019 with 7 publications.

Figure 7 offers a comprehensive overview of the publication trends over time across various sources within the context of the textile industry. The focus is on five key sources: Textile Rental, Pakistan Textile Journal, Lecture Notes in Computer Science (including subseries in Artificial Intelligence and Bioinformatics), Melliand International, and Sustainability (Switzerland).

From 2004 to 2007, there's a discernible increase in publications from Textile Rental, marking its emergence as a significant contributor to textile industry literature. The Pakistan Textile Journal also begins its contribution in 2005, displaying a consistent yet modest increase in output through the years. This steady growth in publications from the Pakistan Textile Journal reflects a sustained interest and possibly a developing research base in the textile sector within Pakistan. Lecture Notes in Computer Science, a source encompassing advanced fields like AI and Bioinformatics, shows a delayed entry into textile-related research, with its first publication recorded in 2016. This could indicate a relatively recent intersection of advanced computational sciences with textile industry research, a trend that could be reflective of the increasing digitization and technological integration in the textile sector. Melliand International presents an interesting case, starting its contributions in 2008 and maintaining a stable output since then. This consistency suggests a focused

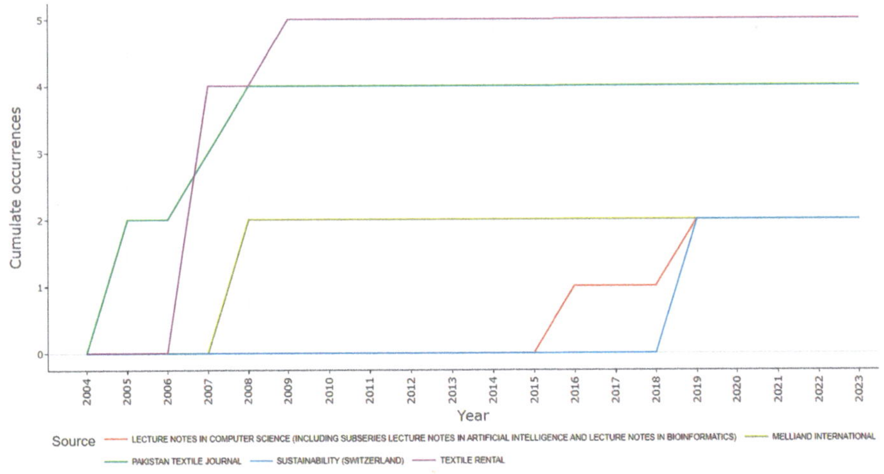

Fig. 7 Sources' production over time

and continuous interest in the subjects covered by this source, possibly catering to a niche but dedicated audience within the textile industry. Sustainability (Switzerland) makes its first appearance in 2019, indicating a recent but timely focus on sustainable practices within the textile industry. This aligns with the global shift towards sustainability and eco-friendly practices, highlighting the textile industry's response to these evolving paradigms. The analysis reveals a gradual but steady increase in the volume of publications across most sources, suggesting a growing academic and industrial interest in the textile sector. The diversity of sources–from industry-specific journals to broader scientific publications–underscores the multifaceted nature of research in the textile industry, encompassing technological, economic, and sustainable aspects.

The analysis of authors' production over time in the context of management control systems and innovation in the textile industry reveals interesting trends and patterns. The data spans from 2006 to 2023, covering a range of authors who have contributed to the field (Fig. 8).

The varied citation patterns among these authors reflect the dynamic nature of research impact. Some works gain immediate recognition, while others may take time to be acknowledged or may serve specific niche areas within the broader field of textile industry innovation and management control systems. This analysis underscores the importance of both immediate and long-term impact in scholarly research.

The analysis of the most globally cited documents in the context of management control systems and innovation in the textile industry reveals a diverse range of publications, each contributing significantly to the field in different ways (Fig. 9).

The integration of management control systems and innovation in the textile industry has been a subject of significant research, as evidenced by a variety of impactful studies. This literature review explores the key findings and methodologies of some of the most globally cited documents in this field, analyzing their contributions to our understanding of how management practices can drive innovation in the textile sector. In "ACS Applied Materials Interfaces", Pakdel et al. (2019) present a highly influential study, demonstrating an impressive rate of citations per year. This work delves into the application of advanced materials in the textile

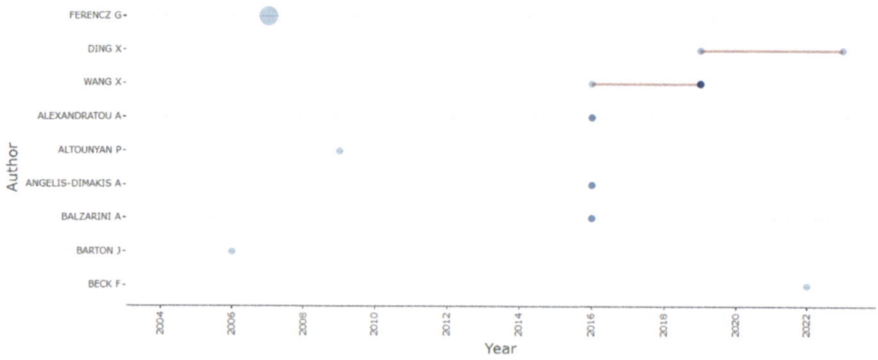

Fig. 8 Authors' production over time

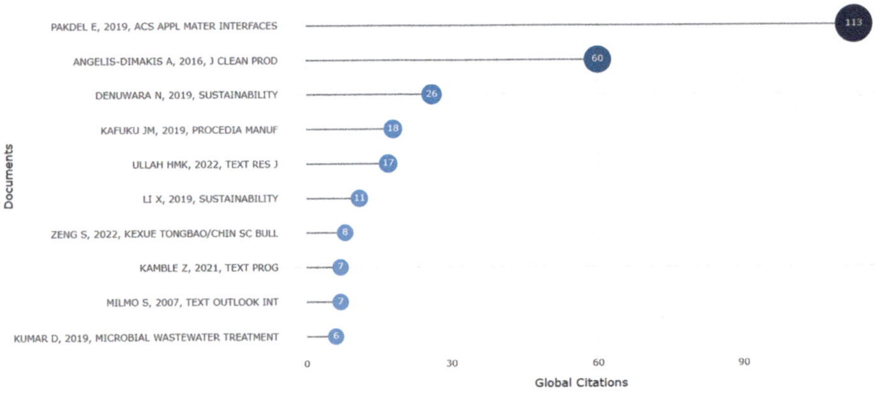

Fig. 9 Most global cited documents

industry, highlighting how innovative materials can revolutionize product quality and sustainability. The study's significant impact reflects its relevance in bridging the gap between material science and textile applications. Angelis-Dimakis et al. (2016) in "Journal of Cleaner Production" explore the environmental aspects of textile production, emphasizing the need for cleaner production methods. The study's high citation count underscores the growing importance of sustainability in the textile industry and the role of management in implementing eco-friendly practices. Denuwara et al. (2019) in their contribution to "Sustainability", examine sustainable practices in the textile industry, focusing on the economic, environmental, and social dimensions. This research emphasizes the holistic approach required for true sustainability in the industry. Kafuku's (2019) article in "Procedia Manufacturing" discusses the integration of technological innovation in textile manufacturing processes. The paper, while not as highly cited as others, provides valuable insights into how technological advancements can be effectively managed and controlled for enhanced productivity and innovation. Ullah et al. (2022) in "Textile Research Journal" provide a more recent perspective on the topic, reflecting the ongoing evolution of management control systems in the textile industry. This paper sheds light on the dynamic nature of management practices in the context of rapidly advancing technological landscapes. The work of Li et al. (2019) in "Sustainability" offers another angle on sustainable practices within the textile industry, specifically focusing on the role of management in fostering a culture of innovation and environmental consciousness. Articles of Zeng et al. (2022) in "Kexue Tongbao/Chinese Science Bulletin" and Kamble and Behera (2021) in "Textile Progress" contribute valuable insights into the nuanced interplay between management control systems and innovation, each from a different cultural and industrial perspective. Kumar et al. (2019) contribute to "Microbial Wastewater Treatment" and expand the conversation to include aspects of wastewater treatment and its significance in the textile industry, highlighting the broader environmental impacts of textile production and the role of innovative management practices in mitigating these impacts. These studies collectively underscore the critical role of management control systems in fostering innovation within the textile industry. They highlight the necessity of integrating environmental sustainability, advanced materials, and technological advancements into management practices to ensure the industry's long-term viability and success.

The analysis of author productivity in the textile industry, as depicted in Fig. 10, is based on Lotka's law. This law is a method used to analyze scientific productivity and is applied here to understand the distribution of document authorship among researchers in the textile industry. In this analysis, the most striking observation is the high proportion of authors who have contributed only one document. A total of 108 authors, constituting 96.4% of the sample, fall into this category. This dominant percentage indicates a field where a vast majority of contributors have limited engagement in terms of published research, possibly suggesting a field with a wide range of transient or occasional researchers.

On the other end of the spectrum, there are significantly fewer authors who have written a higher number of documents. Only 2 authors have written 2 documents each, representing 1.8% of the total author population. Furthermore, a solitary author has contributed 4 documents, and another singular author has contributed as many as 10 documents, each representing 0.9% of the total. This trend shows a steep decline in the number of authors as the number of contributions increases, which is typical in academic research fields as per Lotka's law.

This pattern demonstrates a significant disparity in research contributions within the field. A small number of authors are highly productive, contributing multiple papers, while the vast majority are much less prolific. This could reflect a variety of factors, such as varying levels of expertise, access to resources, or academic focus. Additionally, it could indicate a field that is either highly specialized, where only a

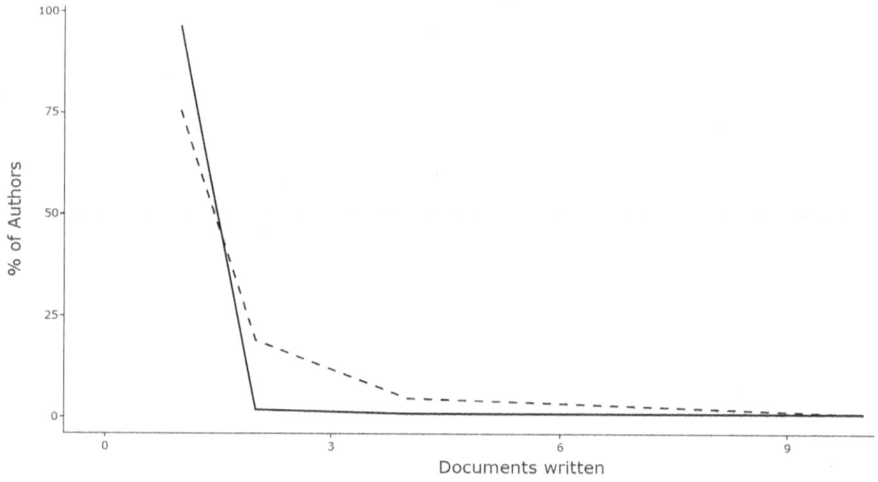

Fig. 10 Author productivity through Lotka's law

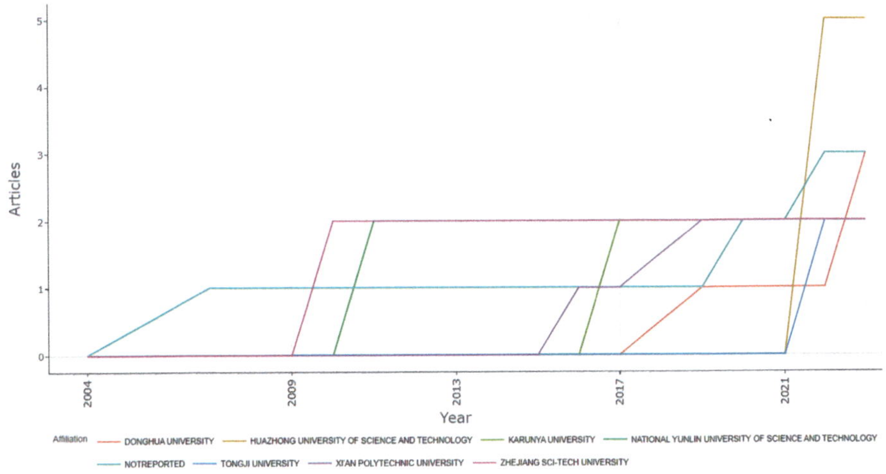

Fig. 11 Affiliations' production over time

few researchers have developed extensive expertise, or one that attracts contributions from a broader range of disciplines, leading to many authors with single contributions.

Starting with Zhejiang Sci-Tech University, it is evident that their contribution to textile-related research began in 2010 (Fig. 11), maintaining a consistent output of two articles per year from then on. This steady production indicates a sustained interest and expertise in the field, likely reflecting the university's commitment to research in textile innovation and management systems. National Yunlin University of Science and Technology shows a similar pattern, but their academic contributions began a year later, in 2011. Like Zhejiang Sci-Tech, they have consistently published two articles per year, suggesting a parallel dedication to the field.

Xi'an Polytechnic University's journey is slightly different. Their research output began in 2016, and it took them three years to reach the consistent output of two articles per year, achieved by 2019. This gradual increase might reflect a developing interest or growing expertise in the textile industry's management and innovation aspects. Karunya University and Donghua University both had a late start in contributing to this field, with their academic outputs beginning in 2017 and 2019, respectively. Karunya University quickly ramped up to a steady output of two articles per year, whereas Donghua University showed a more gradual increase, peaking at three articles in 2023.

The most notable pattern emerges with Huazhong University of Science and Technology and Tongji University. Both universities had no academic output in this field until the very end of the observed period, with Huazhong University producing a significant number of five articles in both 2022 and 2023 and Tongji University contributing two articles in these years. This sudden surge in publications could be indicative of a newfound focus on textile industry research or the culmination of long-term research projects reaching fruition. The data reflects a growing academic interest in the textile industry, particularly in the fields of management control systems and innovation. The varying timelines and outputs of these universities suggest a diverse range of involvement and focus in the textile sector, potentially influenced by institutional priorities, resource allocation, and the evolution of the textile industry itself.

Figure 12 presents an intriguing overview of the global distribution of citations in this field. At the forefront, Australia stands out remarkably, boasting a total citation count (TC) of 113 with an average of 113.00 citations per article. This suggests a highly influential and possibly a pioneering role of Australian research in this sector. Following Australia, the United Kingdom secures a significant position with a TC of 62 and an average of 31.00 citations per article. This indicates a strong, influential academic presence in the textile industry's management and innovation studies. The USA, while lower in total citations (26), still maintains a notable presence in the field, averaging 13.00 citations per article. This could imply a focused but impactful contribution to the research area. China's contribution, with a TC of 23 and an average of 2.90 citations per article, suggests a growing influence in this sector, potentially indicative of the country's expanding role in textile industry research and innovation. France and India, with total citations of 17 and 7, and average citations of 8.50 and 3.50 respectively, indicate a moderate but consistent engagement in this research area. This reflects their ongoing commitment to contributing to academic discourse in this field. Germany, Poland, and Italy, with lower total citations and averages (ranging from 5 TC and 2.50 average citations for Germany to 1 TC and 0.50 for Italy), suggest a more niche or emerging interest in this area of research within these countries.

This distribution of citations highlights the varied levels of engagement and impact different countries have in the realm of management control systems and innovation in the textile industry. It underscores the global nature of academic research and the diverse contributions from different regions to this evolving field (Fig. 13).

India follows, albeit with a considerably lower frequency of 7, suggesting a strong but not as dominant engagement in this research area. The United Kingdom, with 6

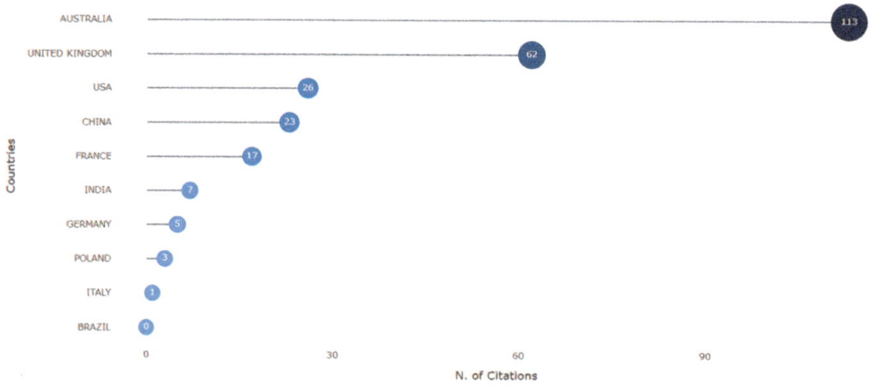

Fig. 12 Most cited countries in management control systems and innovation in textile industry

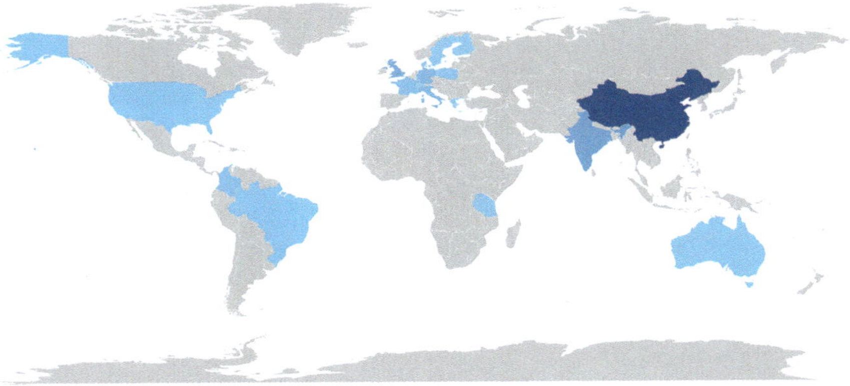

Fig. 13 Countries' scientific production

contributions, indicates a noteworthy involvement, potentially reflecting its historical and contemporary significance in textile innovation and management. Germany and Italy, both with 4 contributions, demonstrate their established presence in the textile sector, possibly due to their focus on high-quality production and innovative practices. Similarly, Colombia, France, Brazil, Poland, and the USA, each with 2 to 3 contributions, show a moderate level of engagement in this research area. This could be reflective of these countries' growing interest or developing capabilities in the textile industry's management and innovation aspects. The remaining countries, namely Australia, Denmark, Finland, Greece, Sweden, Switzerland, and Tanzania, each with a single contribution, indicate a nascent or limited involvement in this specific research domain. This could be due to a variety of factors, including the size of their textile industry, research priorities, or the relative novelty of management control systems and innovation focus in their textile sectors. It provides insightful perspectives on how different countries contribute to the evolving discourse on management control systems and innovation in the textile industry, highlighting a blend of established leaders and emerging contributors in this global field.

The word map from the analysis reveals a complex interplay of various concepts, distributed across different clusters (Fig. 14). The study seems to focus on the intersection of innovation, technology, and management within the textile industry.

Cluster 1 predominantly contains terms directly related to the textile industry and its various facets, such as "textile industry", "textiles", "innovation", "quality control", "dyeing", "color", "commerce", "recycling", "resource allocation", "sales", and "thermal comfort". These terms suggest a strong emphasis on the core aspects of the textile business, including product development, market dynamics, and sustainable practices.

"Sustainable development" and "life cycle" are classified in cluster 3, along with "design", "information management", "product design", and "wearable technology". This indicates a trend towards integrating sustainability and advanced technologies

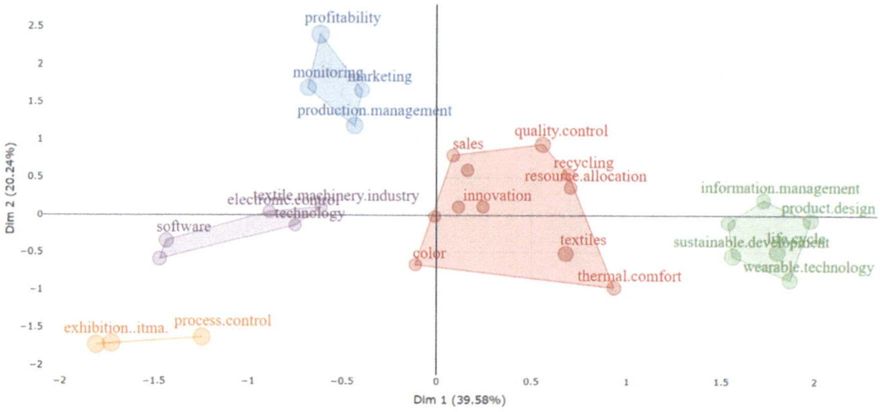

Fig. 14 Word map of the studied publications factorial analysis

into the textile industry. It reflects the industry's shift towards more environmentally friendly practices and the incorporation of new technologies in textile products.

Clusters 2 and 4 seem to represent the management and technological control aspects, respectively. In cluster 2, terms like "production management", "monitoring", "process control", "marketing", and "profitability" indicate a focus on the operational, financial, and strategic management of textile enterprises. Cluster 4, with terms like "software", "electronic control", "energy conservation", and "textile machinery industry", points towards the technological advancements and automation in the industry.

The most intriguing aspect is cluster 5, which includes highly specialized terms such as "process control", "optimization", and "profitability." These terms suggest a focus on the cutting-edge aspects of textile manufacturing and management, perhaps indicating research into the most advanced and efficient production techniques.

The word map indicates a multifaceted approach to studying the textile industry, emphasizing innovation, sustainability, and the integration of new technologies in both product design and manufacturing processes. The analysis reflects the industry's evolving nature, adapting to new challenges and opportunities in a globalized market.

The dendrogram in Fig. 15 illustrates a multifaceted view of the textile industry, where innovation is not only about new products or technologies but also encompasses improvements in process control, sustainability, and management practices. The analysis suggests a shift towards a more integrated and holistic approach in the industry, combining technological, environmental, and operational aspects.

The factorial analysis dendrogram from the study provides a comprehensive overview of how various themes and concepts are interconnected within the scope of the research (Table 3). This dendrogram is structured around two dimensions and clusters that group related concepts.

Cluster 1 includes "textile industry", "innovation", "dyeing", and "process control". These terms are closely associated, indicating a strong relationship between the textile industry and innovation, particularly in the areas of dyeing and process

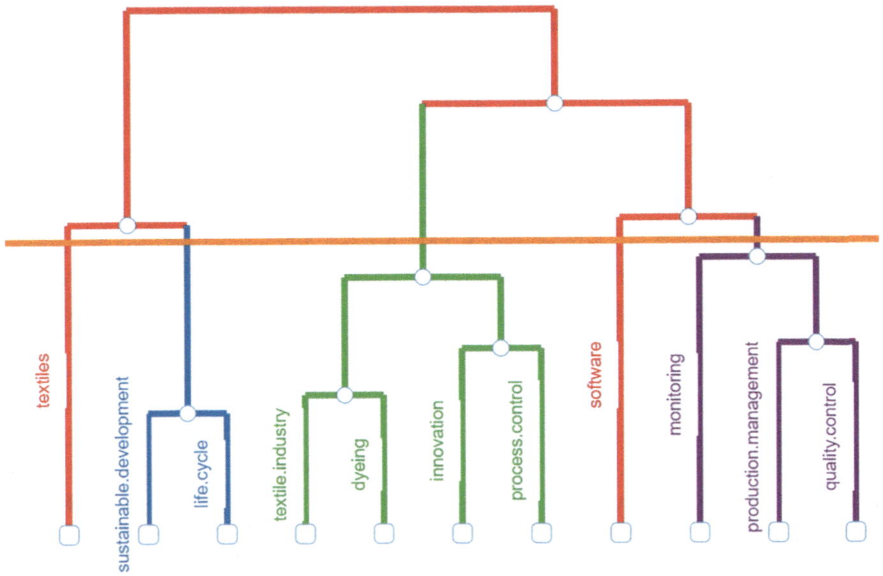

Fig. 15 Factorial analysis dendrogram of the studied publications

Table 3 Factorial analysis indicators of the researched publications

Word	Dim 1	Dim 2	Cluster
Textile.industry	0,14	−0,18	1
Textiles	1,01	0,01	2
Innovation	0,11	−0,94	1
Production.management	−0,87	0,88	3
Quality.control	−0,22	1,09	3
Sustainable.development	1,75	0,89	4
Dyeing	−0,27	−0,27	1
Software	−1,26	−0,03	5
Life.cycle	1,96	1,14	4
Monitoring	−1,55	1,48	3
Process.control	−0,27	−1,49	1

control. The proximity of these words suggests that innovations in dyeing and process control are crucial for the textile industry.

"Textiles" stands alone in cluster 2. Its positioning on the first dimension and near the center on the second dimension could imply its foundational role in the study, serving as a central theme that connects various aspects of the industry.

Cluster 3 groups "production management", "quality control", and "monitoring". These concepts are interlinked, highlighting the importance of management and quality control in the production process within the textile industry. The emphasis here is on the operational side, focusing on efficiency and quality assurance in production.

In cluster 4 "sustainable development" and "life cycle" are grouped together, indicating a strong emphasis on sustainability and environmental considerations in the textile industry. This cluster suggests an emerging trend in the industry towards more sustainable practices and the importance of considering the entire life cycle of textile products.

"Software" stands alone in cluster 5, possibly indicating its emerging yet distinct role in the textile industry. This could reflect the growing importance of technology and software solutions in innovating and streamlining various processes in the textile industry.

The close association between traditional aspects of the textile industry (like dyeing and textiles) and modern concepts (like innovation and software) underscores a trend towards integrating cutting-edge techniques with established practices. This integration is pivotal for the industry's growth and adaptation to contemporary market demands. The strong link between sustainable development and life cycle analysis indicates a growing emphasis on environmental concerns within the textile industry. This shift towards sustainability suggests that the industry is increasingly recognizing the importance of eco-friendly practices and their long-term benefits.

The grouping of production management, quality control, and monitoring highlights the critical role of effective management systems in ensuring efficiency and product quality. This focus reflects an industry-wide acknowledgment that robust control systems are essential for maintaining competitive advantage. The isolated positioning of software in the analysis points to the emerging significance of digital solutions in the textile industry. This trend suggests that there is an increasing reliance on technology for optimizing processes, enhancing productivity, and fostering innovation. The clustering of terms related to production management and process control emphasizes operational efficiency as a major area of focus. This indicates that the textile industry is keenly aware of the need for streamlined operations to stay competitive in a rapidly evolving market.

The bibliometric analysis reveals a textile industry that is actively evolving, integrating innovation with traditional practices, and placing a strong emphasis on sustainability, quality management, and technological advancement. These trends reflect the industry's response to global challenges and its commitment to continuous improvement and adaptation.

3 Conclusion

Central to this study is the role of management control systems (MCS) in driving innovation within the textile industry. The strategic implementation of MCS, aligned with organizational goals and specific sector needs, has shown a significant impact on the industry's ability to innovate and adapt. This alignment has been crucial in facilitating organizational learning and knowledge management, key drivers for innovation. The role of MCS in shaping organizational culture towards fostering creativity and risk-taking has been emphasized, indicating the nuanced influence of these systems on innovation.

The bibliometric analysis underscores the importance of a holistic approach to MCS, integrating sustainability and technological advancements. The evolving nature of MCS, in response to emerging technologies like artificial intelligence and the Internet of Things, presents new opportunities and challenges. The increasing focus on sustainability, reflected in the growing attention to eco-innovation, highlights the industry's shift towards integrating environmental considerations into innovation processes.

The study also identifies potential knowledge gaps and areas for future investigation. The relatively low percentage of international co-authorships suggests a need for broader collaboration, potentially enriching research with diverse perspectives. The dominance of single-authored documents points to the opportunity for more collaborative research endeavors. Moreover, the varied citation patterns among authors suggest that while some research areas receive immediate recognition, others may serve specific niche areas, indicating room for exploring under-represented topics.

The thematic analysis reveals clusters around key areas like design, production management, and sustainable development, each presenting distinct yet interconnected fields of focus. The integration of advanced technologies and digital tools in production management, the emphasis on sustainable practices, and the exploration of specialized areas like dyeing and thermal comfort, all indicate a dynamic research landscape.

The textile industry's academic research is characterized by a blend of traditional practices and modern innovation, with a pronounced emphasis on sustainability, technology integration, and efficiency in production management. Future research should delve deeper into the interplay between MCS and emerging technologies, explore the integration of sustainability in management practices, and foster international collaboration to enrich the research landscape. This approach will ensure the textile industry remains at the forefront of innovation and adapts effectively to the ever-changing global economic and environmental landscape.

References

Akroyd, C., & Kober, R. (2020). Imprinting founders' blueprints on management control systems. *Management Accounting Research, 46*, 100645.

Andersen, C. V., & Lueg, R. (2017). Management control systems, culture and upper echelons–A systematic literature review on their interactions. *Corporate Ownership and Control, 14*(2), 312–325.

Angelis-Dimakis, A., Alexandratou, A., & Balzarini, A. (2016). Value chain upgrading in a textile dyeing industry. *Journal of Cleaner Production, 138*, 237–247.

Barros, R. S., & Ferreira, A. M. D. S. C. (2019). Bridging management control systems and innovation: The evolution of the research and possible research directions. *Qualitative Research in Accounting & Management, 16*(3), 342–372.

Bresciani, S., Rehman, S. U., Alam, G. M., Ashfaq, K., & Usman, M. (2022). Environmental MCS package, perceived environmental uncertainty and green performance: In green dynamic capabilities and investment in environmental management perspectives. *Review of International Business and Strategy, 33*(1), 105–126.

Chen, W. K., Nalluri, V., Hung, H. C., Chang, M. C., & Lin, C. T. (2021). Apply dematel to analyzing key barriers to implementing the circular economy: An application for the textile sector. *Applied Sciences, 11*(8), 3335.

Chen, Y., Luo, H., Chen, J., & Guo, Y. (2022). Building data-driven dynamic capabilities to arrest knowledge hiding: A knowledge management perspective. *Journal of Business Research, 139*, 1138–1154.

Denuwara, N., Maijala, J., & Hakovirta, M. (2019). Sustainability benefits of RFID technology in the apparel industry. *Sustainability, 11*(22), 6477.

Do, H., Budhwar, P., Shipton, H., Nguyen, H. D., & Nguyen, B. (2022). Building organizational resilience, innovation through resource-based management initiatives, organizational learning and environmental dynamism. *Journal of Business Research, 141*, 808–821.

Farrukh, A., Mathrani, S., & Sajjad, A. (2022). A natural resource and institutional theory-based view of green-lean-six sigma drivers for environmental management. *Business Strategy and the Environment, 31*(3), 1074–1090.

Franco-Santos, M., & Otley, D. (2018). Reviewing and theorizing the unintended consequences of performance management systems. *International Journal of Management Reviews, 20*(3), 696–730.

Guo, Z. X., Wong, W. K., Leung, S. Y. S., & Li, M. (2011). Applications of artificial intelligence in the apparel industry: A review. *Textile Research Journal, 81*(18), 1871–1892.

Henri, J. F., & Journeault, M. (2010). Eco-control: The influence of management control systems on environmental and economic performance. *Accounting, Organizations and Society, 35*(1), 63–80.

Kafuku, J. M. (2019). Factors for effective implementation of lean manufacturing practice in selected industries in Tanzania. *Procedia Manufacturing, 33*, 351–358.

Kalyar, M. N., Shoukat, A., & Shafique, I. (2020). Enhancing firms' environmental performance and financial performance through green supply chain management practices and institutional pressures. *Sustainability Accounting, Management and Policy Journal, 11*(2), 451–476.

Kamble, Z., & Behera, B. K. (2021). Upcycling textile wastes: Challenges and innovations. *Textile Progress, 53*(2), 65–122.

Khan, S. N., Hussain, R. I., Ur-Rehman, S., Maqbool, M. Q., Ali, E. I. E., & Numan, M. (2019). The mediating role of innovation between corporate governance and organizational performance: Moderating role of innovative culture in Pakistan textile sector. *Cogent Business & Management, 6*, 1–23.

Kumar, D., Pandit, P. D., Patel, Z., Bhairappanavar, S. B., & Das, J. (2019). Perspectives, scope, advancements, and challenges of microbial technologies treating textile industry effluents. In *Microbial Wastewater Treatment* (pp. 237-260). Elsevier.

Laguir, I., Gupta, S., Bose, I., Stekelorum, R., & Laguir, L. (2022). Analytics capabilities and organizational competitiveness: Unveiling the impact of management control systems and environmental uncertainty. *Decision Support Systems, 156*, 113744.

Lee, C. K. H., Choy, K. L., Law, K. M. Y., & Ho, G. T. S. (2014). Application of intelligent data management in resource allocation for effective operation of manufacturing systems. *Journal of Manufacturing Systems, 33*(3), 412–422.

Li, X., Chen, L., & Ding, X. (2019). Allocation methodology of process-level carbon footprint calculation in textile and apparel products. *Sustainability, 11*(16), 4471.

Mangenda Tshiaba, S., Wang, N., Ashraf, S. F., Nazir, M., & Syed, N. (2021). Measuring the sustainable entrepreneurial performance of textile-based small–medium enterprises: A mediation–moderation model. *Sustainability, 13*(19), 11050.

Ngai, E. W. T., Peng, S., Alexander, P., & Moon, K. K. (2014). Decision support and intelligent systems in the textile and apparel supply chain: An academic review of research articles. *Expert Systems with Applications, 41*(1), 81–91.

Pakdel, E., Naebe, M., Sun, L., & Wang, X. (2019). Advanced functional fibrous materials for enhanced thermoregulating performance. *ACS Applied Materials & Interfaces, 11*(14), 13039–13057.

Peters, J., & Simaens, A. (2020). Integrating sustainability into corporate strategy: A case study of the textile and clothing industry. *Sustainability, 12*(15), 6125.

Quesado, P., Marques, S., Silva, R., & Ribeiro, A. (2022). The balanced scorecard as a strategic management tool in the textile sector. *Administrative Sciences, 12*(1), 38.

Rahman, M. M., Bari, A. M., Ali, S. M., & Taghipour, A. (2022). Sustainable supplier selection in the textile dyeing industry: An integrated multi-criteria decision analytics approach. *Resources, Conservation & Recycling Advances, 15*, 200117.

Rehman, S. U., Mohamed, R., & Ayoup, H. (2019). The mediating role of organizational capabilities between organizational performance and its determinants. *Journal of Global Entrepreneurship Research, 9*(1), 1–23.

Rehman, S. U., Bhatti, A., Kraus, S., & Ferreira, J. J. M. (2021). The role of environmental management control systems for ecological sustainability and sustainable performance. *Management Decision, 59*(9), 2217–2237.

Tran, V. T., Ly, P. C., Ngo, N. N. T., Tran, P. H., & Nguyen, V. C. (2022). Factor affecting the implementation of responsibility accounting on firm performance–Empirical analysis of listed textile firms. *Cogent Business & Management, 9*(1), 2032912.

Ullah, H. M. K., Lejeune, J., Cayla, A., Monceaux, M., Campagne, C., & Devaux, É. (2022). A review of noteworthy/major innovations in wearable clothing for thermal and moisture management from material to fabric structure. *Textile Research Journal, 92*(17–18), 3351–3386.

Wickramasinghe, D., & Hopper, T. (2005). A cultural political economy of management accounting controls: A case study of a textile Mill in a traditional Sinhalese village. *Critical Perspectives on Accounting, 16*(4), 473–503.

Zeng, S., Hu, J., Zhang, M., Xiang, Y., Wu, J., Su, M., Zhang, Y., Shen, M., Hong, P., Huang, Z., Chen, M., Zhou, N., Hou, C., Zhou, H., Zhang, D., & Tao, G. (2022). Cooling textiles for personal thermal management. *Chinese Science Bulletin, 67*(11), 1167–1179.

Jéssica Nunes Martins has a Bachelor's degree in Accounting by ISCAL, a Master's degree in Statistics and Information Management, with a specialization in Risk Analysis and Management by Nova University of Lisbon-NOVA IMS, and currently is doing a Ph.D. in Information Management. She integrates research projects at Insight—Piaget Research Center for Ecological Human Development and provides consultant services in information management.

Enhancing Textile Enterprises' Integrated Development: Innovative Management Optimization and Monitoring System

Kateryna Boichenko

Abstract This study introduces an innovative monitoring system for the integrated development of textile enterprises, addressing a significant gap in current literature. The research focuses on the integration of management practices with technological advancements, essential for the textile industry's dynamic and evolving market. Utilizing a vector model of the innovation system, the study systematically evaluates the integration efforts of textile enterprises, particularly emphasizing Ukrainian enterprises and their integration with European networks. The methodology combines bibliometric analysis and practical application to identify trends, challenges, and opportunities in the industry. Key findings reveal the importance of synergistic integration of management and technology for enhanced competitiveness and sustainability in the global market. The study's outcomes provide a practical framework for managing integrated development in the textile industry, offering significant contributions to both theory and practice. This novel approach is instrumental in guiding textile enterprises through the complexities of integrated development, ensuring sustainable and profitable growth.

1 Introduction

The textile industry, known for its dynamic market demands and rapidly evolving technological landscape, plays a pivotal role in the global economy. In recent years, this sector has witnessed significant transformations, driven by the integration of innovative technologies and management practices (Islam et al., 2021). This integration has become a cornerstone for enterprises aiming to enhance their competitiveness and sustainability. The ability to adapt and efficiently manage these changes is critical for textile enterprises, particularly in the context of increasing global integration and collaboration (Rahmanzadeh et al., 2023; Todeschini et al., 2020). The management

K. Boichenko (✉)
Department of Business Economics and Entrepreneurship, Kyiv National Economic University named after Vadym Hetman, Beresteysky Ave, 54/1, Kyiv, Ukraine
e-mail: boichenko.kateryna@kneu.edu.ua

Lisbon Accounting and Business School, Polytechnic Institute of Lisbon (ed.), *Industry and Innovation: Textile Industry*, SDGs and Textiles,
https://doi.org/10.1007/978-3-031-57804-5_14

of these integrations and the monitoring of their effectiveness remain complex tasks, compounded by the diverse and fast-paced nature of the industry. This complexity necessitates a comprehensive approach that not only embraces the latest technological advancements but also aligns them with strategic management objectives. The effective monitoring and management of these integrations are crucial for enterprises to leverage the full potential of their collaborative efforts and to ensure a sustainable and profitable future.

Despite the acknowledged importance of innovative monitoring systems in managing integrated development in the textile sector, there exists a noticeable gap in scientific literature, particularly in the context of developing and implementing such systems. Existing research primarily focuses on general management practices and technological innovations in isolation, often overlooking the synergistic potential of integrating these two critical components. This oversight is especially evident in the context of textile enterprises, where the specific challenges and opportunities of the industry require tailored solutions. There is a pressing need for a comprehensive framework that can effectively address the nuances of integrating management practices with technological innovations, especially in the context of monitoring and evaluating the effectiveness of these integrations.

Recognizing this gap, the purpose of this study is to develop and propose an innovative monitoring system specifically designed for the integrated development of textile enterprises. This system aims to bridge the existing gap in the literature by providing a cohesive and practical framework for effectively managing and monitoring the integration of management practices and technological advancements in the textile industry. The study's focus is not only on conceptualizing this system but also on demonstrating its practical applicability and effectiveness in real-world scenarios. By doing so, it seeks to offer a valuable tool for textile enterprises to navigate the complexities of integrated development and to harness the full potential of their collaborative efforts.

The study addresses several key issues facing the textile industry in the realm of integrated development management. Through an extensive analysis of literature sources, incorporating elements of bibliometrics, the study identifies prevailing trends, challenges, and opportunities in the management of textile enterprises. It proposes a vector model of the innovation system for monitoring the effectiveness of integrated enterprise development management. This model is designed to provide a comprehensive and systematic approach to evaluate and enhance the performance of textile enterprises in their integration efforts. Furthermore, the study is applied specifically to the context of textile enterprises in Ukraine, examining their integration with European network companies. This regional focus provides valuable insights into the unique challenges and opportunities faced by these enterprises, contributing to the broader understanding of integrated development in the global textile industry.

This study not only fills a significant gap in the scientific literature but also offers practical solutions to the pressing challenges faced by textile enterprises in managing their integrated development. By proposing an innovative monitoring system and applying it to the specific context of Ukrainian textile enterprises and their integration with European networks, this study contributes significantly to the field. It provides a roadmap for textile enterprises to navigate the complexities of integrated development, ultimately enhancing their competitiveness and sustainability in the global market.

2 Literature Review

Optimising the management of integrated development of textile enterprises through innovative monitoring systems is a multifaceted field encompassing advances in technology, management strategies and industry specific problems. This literature review aims to explore various aspects of this topic based on bibliometric analyses and drawing information from a variety of academic articles and studies.

Figure 1 shows the growth in the number of publications for the keyword "integrated development" in the Scopus database under the "Business, Management and Accounting" sub-database.

Despite the absence of a clear dynamic in the growth of publication volumes in this sub-database, it is also possible to state an increase in the number of publications over the last 20 years from 64 to 165—a 2.5-fold increase. This indicates a growing interest of researchers in integrating development in the context of business management. Subsequently, we focused our bibliometric analysis specifically on this period and the Scopus scientometric sub-database.

The map in Fig. 2 shows the distribution of publications on integrated development in the Scopus database sub-database from 2003 to 2022 by the country of origin.

The obtained results demonstrate that the USA, the United Kingdom, and China are the top three leaders in terms of the number of publications −439, 279, and 158 publications respectively (Fig. 3).

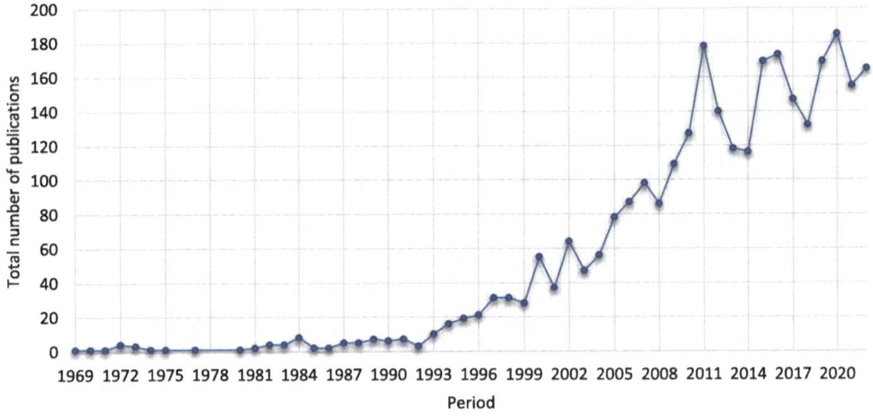

Fig. 1 Growth in the number of publications for the keyword "integrated development" in the Scopus sub-database "Business, Management and Accounting"

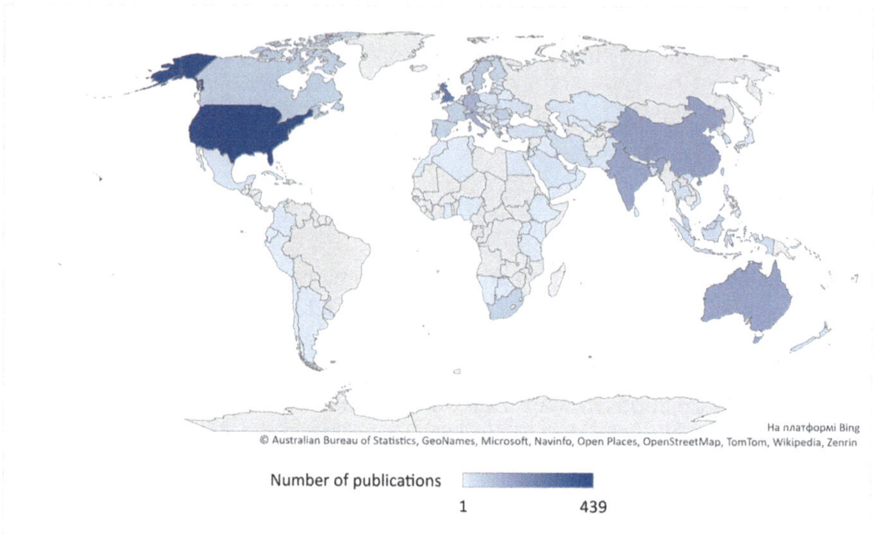

Fig. 2 Map of publication sources on integrated enterprise development

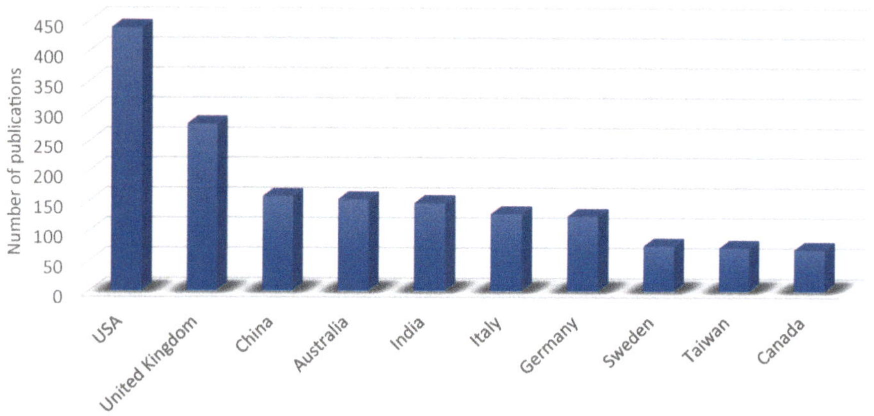

Fig. 3 Countries with the highest volume of publications for the keyword "integrated development"

Publications on integrated development in the context of business management are widespread around the world. However, some countries, such as the USA, the UK, and China, lead in the number of publications. This indicates that these countries have a strong research and innovation potential in studies on integrating enterprise development and its management. This suggests that in some regions of the world, such as Africa and some of South America, where the number of publications is relatively low, there is still a need for additional research. Therefore, researchers in these regions may need to intensify their research efforts to make a greater contribution to the knowledge base on managing integrated business development.

Also, a study was conducted on other keywords most frequently associated with publications on integrated development (Fig. 4).

The most commonly used keywords in publications on integrated development in the context of business management, particularly in the Scopus sub-database "Business, Management and Accounting," are identified (Fig. 5).

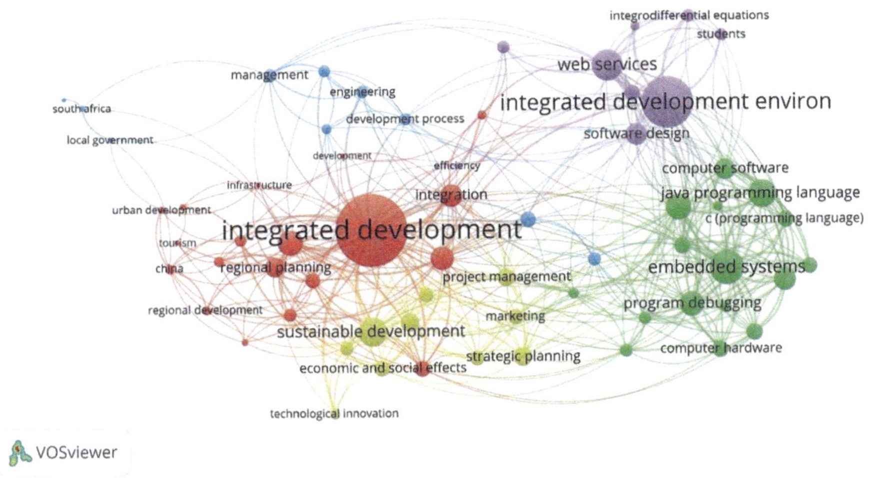

Fig. 4 Cluster analysis of publications with the keyword "integrated development" in the Scopus sub-database for 2003–2022

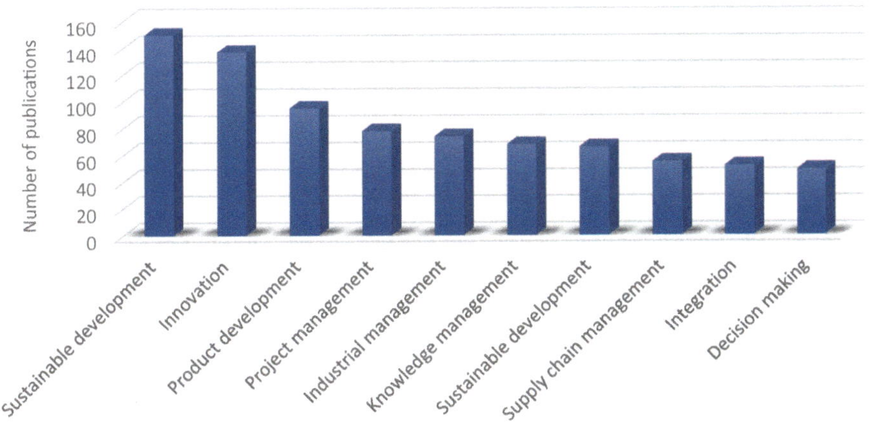

Fig. 5 Most commonly used keywords in publications on integrated development in the Scopus sub-database for 2003–2022

Thus, we can state that integrated development is most often associated with sustainable development, innovation, and product development, emphasizing and focusing attention specifically on the complexity of this concept. However, "integration" is not the most frequently used keyword regarding integrated development over the past 20 years, although it still has a significant share in research in this direction.

Figure 6 shows the distribution of "integrated development" publications in the Scopus sub-database from 2003 to 2022, grouped by knowledge fields.

Most publications were found in the subject areas of "Engineering," "Computer Science," and "Environmental Science." The subject area of "Business, Management and Accounting" is among the top five leaders for the studied period.

- the characteristics of research on integrated enterprise development include:
- multidisciplinary approach–researchers consider integrated development from the perspectives of various disciplines, such as economics, management, ecology, sociology, and others;
- stakeholder engagement–research often takes into account the interests of various stakeholders, such as customers, employees, shareholders, public organizations, and others, to achieve more comprehensive results;

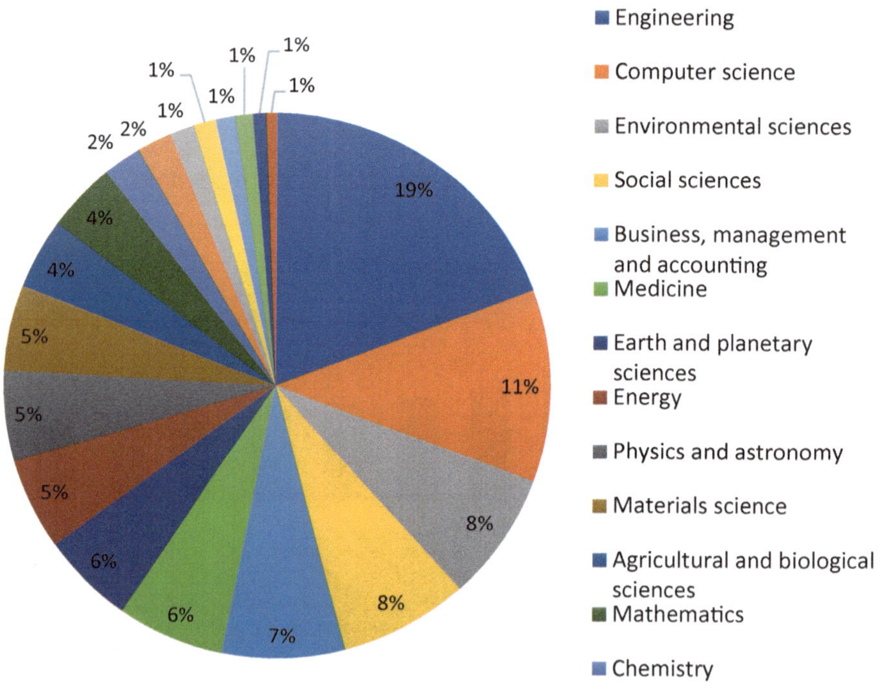

Fig. 6 Distribution of knowledge fields of publications with the keyword "integrated development" in the Scopus database for 2003–2022

- the importance of risk management–researchers emphasize the necessity of managing risks as a key aspect of integrated enterprise development. They recognize the need to consider potential risks, such as economic changes, legislative changes, environmental crises, and others, and develop strategies for their management;
- sustainability and environmental responsibility–researchers focus on the need for enterprise development in accordance with the principles of sustainable development and ensuring environmental responsibility. They explore ways to reduce environmental impact and develop green initiatives.

The integration of advanced technologies in the textile industry has been a significant focus. Innovations such as IoT (Internet of Things), AI (Artificial Intelligence), and real-time monitoring systems have revolutionized the way textile enterprises operate. A study highlights the impact of IoT in streamlining operations and enhancing production efficiency (Allioui & Mourdi, 2023; Fernández-Caramés & Fraga-Lamas, 2018). Similarly, the role of AI in predictive maintenance and quality control is emphasized in a recent publication (Hammad et al., 2023). The implementation of real-time monitoring systems has been shown to significantly improve supply chain transparency and operational responsiveness (Keung Kwok & Wu, 2009; Kumar et al., 2017).

Effective management strategies are crucial for the success of integrated development in textile enterprises. The adoption of lean management principles has been identified as a key factor in improving operational efficiency (Prasad et al., 2020; Robertsone et al., 2022). Strategic resource allocation and human resource management have been cited as critical for sustaining innovation and competitiveness (Amjad et al., 2021). Cross-functional team collaboration, as discussed by Wood and Bischoff (2020), further enhances the efficiency and innovative capacity of textile enterprises.

The textile industry faces increasing pressure to address environmental and sustainability concerns. Research on sustainable production processes and eco-friendly materials has gained prominence (de Oliveira et al., 2021; Raj et al., 2022). The implementation of circular economy principles in textile manufacturing is explored by Chae and Hinestroza (2020), Ki et al. (2020), highlighting the potential for waste reduction and resource efficiency. Additionally, the adoption of green supply chain practices is crucial for reducing the environmental impact of textile enterprises (Reche et al., 2020).

Despite advancements, the textile industry faces several challenges in implementing integrated development strategies. These include resistance to change, high implementation costs, and skill gaps (Abreu et al., 2021; Behera et al., 2021). The future of textile enterprise management lies in overcoming these barriers and leveraging technology for sustainable growth. As proposed by Hader et al. (2022), continuous innovation and adaptation are key to staying competitive in a rapidly evolving market.

The global nature of the textile industry necessitates a broader understanding of integrated development strategies. Comparative studies of textile industries in different countries provide insights into diverse management practices and technological adaptations (Simegnaw et al., 2021; Tian et al., 2022). Case studies from leading textile enterprises offer practical examples of successful implementation of innovative monitoring systems and management strategies (Herrera Almanza & Corona, 2020).

The optimization of management in the textile industry through innovative monitoring systems is a dynamic and complex field. The integration of advanced technologies, effective management strategies, sustainability considerations, and global perspectives are essential for the holistic development of textile enterprises. Future research should focus on addressing the existing challenges and exploring innovative solutions for sustainable and efficient textile production and management.

The common features of research on integrated enterprise development include:

- focus on a comprehensive approach–researchers emphasize the necessity of combining various aspects such as financial, social, ecological, and ethical, to achieve sustainable development;
- recognition of the importance of stakeholders–studies focus on considering the interests of various stakeholders and establishing a balance among them;
- practical orientation–many studies are practically oriented and strive to develop recommendations and tools for enterprises regarding the implementation of integrated development.

Researchers emphasize different aspects of integrated enterprise development, but a common feature of their studies is the practical orientation and aspiration for the real implementation of an integrated approach. They view the concept of "integrated enterprise development" as a comprehensive approach that includes economic, social, ecological, and ethical aspects, as well as interaction with stakeholders. Researchers stress the importance of risk management, sustainability, and environmental responsibility in the context of integrated enterprise development.

In summary, research on integrated enterprise development expands the understanding of the importance and impact of various aspects of development on the enterprise. They provide valuable insights into how enterprises can be resilient and efficient, maintaining a balance between economic, social, and ecological aspects of their activity. The research forms the basis for developing strategies, policies, and tools aimed at achieving integrated enterprise development.

Despite the significant scientific contribution to enterprise development, research remains fragmented in areas related to the necessity of implementing an ecosystem approach and adapting management tools based on the integration of development, considering networking and unpredictable changes in the modern business environment. The dialectic between existing theories and the needs of the contemporary textile business is gaining more significance, as enterprises seek new tools and strategies that will enable them to interact effectively in global networks, be more adaptable to changes, and generate competitive advantages in this turbulent environment. Therefore, discussions regarding the formation of an innovative system for monitoring the effectiveness of integrated textile industry enterprise management require continuation, and further research and innovations become particularly relevant and significant for the economic growth of the business.

3 Materials and Methods

This study presents the methodological approach applied in the study for analyzing the integrated development of enterprises. The research is based on three key components: BV (Business Virtualisation), SIG (Socio-Inclusive Growth), and SD (Sustainable Development). The methodology includes several stages: correlation analysis, factor analysis, vector analysis, classification, and forecasting.

In this study, a justification is proposed for the core of the paradigmatic construct as an indicator of the effectiveness of the integrated development management system of enterprises based on its concentration and harmonization, identifying its key directions and components, particularly based on the integral indicator of integrated development—IDI (Integrated Development Indicator). The need to include these components is dictated based on reverse induction. In this context, managerial influence in the context of managing business virtualization (d_{vb}), socio-inclusive growth (d_{sig}) and sustainable development (d_{sd}) of business processes is considered within such an equation. (Bashuk et al., 2023):

$$\max_{d_{sd}} aD_{sig}x_{sd} - \beta D_{sig}(t_{sd} - t_0) - d_{vb} - d_{sig} - d_{sd}$$

In this case, the following condition is mandatory, namely:

$$P_{sig}(x_{sd} = 1) = \chi_{sd}\left(1 - exp^{-\lambda_{sd}d_{sd}}\right)x_{sig},$$

$$t_{sd} = t_{sig} + a_{sd},$$

$$e_{sd} \geq 0.$$

In this case, the equation $t_{sd} = t_{sig} + a_{sd}$ has a constant form and does not depend on the management of integrated development in the context e_{sd}. In this case:

$$D_{sig}x_{sd} = 1 \cdot P_{sig}(x_{sd} = 1) = \chi_{sd}\left(1 - exp^{-\lambda_{sd}d_{sd}}\right)x_{sig},$$

Which involves transformation into an equation:

$$\max_{e_{sd} \geq 0} a\chi_{sd}\left(1 - exp^{-\lambda_{sd}d_{sd}}\right)x_{sig} - e_3,$$

This can be reflected in the solution of the first-order equation:

$$d_{sd} = \frac{1}{\lambda_{sd}} ln a \chi_{sd} \lambda_{sd}, \text{ if} a \chi_{sd} \lambda_{sd} > 1 \text{and} x_{sig} = 1$$

$$d_{sd} = 0, \text{ if} a \chi_{sd} \lambda_{sd} x_{sig} \leq .1$$

It can be assumed that the optimal management of integrated business development F_3 in the context of d_{sd} can be formulated as follows:

$$d_{sd} = F_3(e_{sig}, x_{sig}, t_{sig}) = \frac{1}{\lambda_{sd}} ln a \chi_{sd} \lambda_{sd}, \text{ if} a \chi_{sd} \lambda_{sd} > 1 \text{and} x_{sig} = 1$$

$$d_{sd} = F_3(e_{sig}, x_{sig}, t_{sig}) = 0, \text{ if} a \chi_{sd} \lambda_{sd} x_{sig} \leq 1.$$

If we assume that the effectiveness of managing integrated development does not actually depend on the effectiveness of its management at the previous link e_v, as well as on the speed t_{sig} integration with other development directions, then it can be assumed that the effectiveness of managing integrated development in the context of (d_{sd}) for ensuring further overall business development is more important $x_{sig} = 1$, than in the case of (c). Also, the optimal effectiveness of modeling in the second iteration of the proposed system can be determined in a similar way:

$$\max_{e_{sig}} a D_{vb} x_{sd} - \beta D_{vb}(t_{sig} + a_{sd} - t_0) - d_{vb} - d_{sig} - D_1 d_{sd}$$

Given the fulfilment of the preceding conditions, we can form:

$$D_{vb} x_{sd} = D_{vb}(D_{sig}(x_{sd})) = D_{vb}(\chi_{sd}(1 - exp^{-\lambda_{sd} d_{sd}}) x_{sig})$$

Based on the effectiveness of integrated development management in the context of d_{sd}, we can identify:

$$D_{vb} x_{sd} = \chi_{sd}(1 - \frac{1}{a \chi_{sd} \lambda_{sd}}) P_{vb}(x_{sig} = 1), \text{ if} a \chi_{sd} \lambda_{sd} > 1, \text{ and} D_{vb} x_{sd} = 0, \text{ if} a \chi_{sd} \lambda_{sd} \leq 1$$

$$D_{vb} d_{sd} = \frac{1}{\lambda_{sd}} ln(a \chi_{sd} \lambda_{sd}) P_{vb}(x_{sig} = 1) = \frac{1}{\lambda_{sd}} ln(a \chi_{sd} \lambda_{sd}) \chi_{sig}(1 - exp^{-\lambda_{sig} d_{sig}}) x_{vb},$$

$$\text{If} a \chi_{sd} \lambda_{sd} > 1, \text{ and} D_{vb} d_{sd} = 0, \text{ if} a \chi_{sd} \lambda_{sd} \leq 1.$$

At $a \chi_3 \lambda_3 > 1$ the search for optimal management of integrated development of enterprises in the context of e_2 can be reduced to the following form:

$$\max_{d_{sig}} (a \chi_{sd} - \frac{1}{\lambda_{sd}} - \frac{1}{\lambda_{sd}} ln(a \chi_{sd} \lambda_{sd}) \chi_{sig}(1 - exp^{-\lambda_{sig} d_{sig}})$$
$$x_{vb} - \beta D_{vb}(t_{sig} - t_{vb}) - d_{sig}$$

If the above conditions are met, the solution to the equation with respect to the variable t_{sig}, can be displayed as follows:

$$\max_{d_{sig}\geq 0}(a\chi_{sd} - \frac{1}{\lambda_{sd}} - \frac{1}{\lambda_{sd}} ln(a\chi_{sd}\lambda_{sd})\chi_{sig}(1 - exp^{-\lambda_{sig}bp_{sig}})x_{vb}$$
$$- \frac{\beta}{\mu_{sd}}(\gamma_{sig}d_{sig} - \delta_{sig}d_{vb} + a_{sig}) - d_{sig},$$

$$\max_{d_{sig}\geq 0} -(a\chi_{sd} - \frac{1}{\lambda_{sd}} - \frac{1}{\lambda_{sd}} ln(a\chi_{sd}\lambda_{sd})\chi_{sig}exp^{-\lambda_{sig}d_{sig}}x_{vb} - \left(\frac{\beta}{\mu_{sd}}\gamma_{sig} + 1\right)d_{sig}.$$

Similarly, it is possible to determine the optimal efficiency of managing the integrated development of enterprises in the context of d_{sig} when $a\chi_{sd}\lambda_{sd} \leq 1$ on the basis of the solution:

$$\max_{d_{sig}\geq 0} -\frac{\beta}{\mu_{sd}}(\gamma_{sig}d_{sig} - \delta_{sig}d_{vb} + a_{sig}) - d_{sig},$$

At the same time, $x_{sig} = 0$, so when $a\chi_{sd}\lambda_{sd} > 1$ we get the optimal management of the company's integrated development:

$$d_{sig} = F_{sig}(e_{vb}, x_{vb}, t_{vb}) = \frac{1}{\lambda_{sd}} ln \frac{\mu_{sd}\lambda_{sig}\chi_{sig}(a\chi_{sd} - (1 + ln(a\chi_{sd}\lambda_{sd}))/\lambda_{sd})}{\beta\gamma_{sig} + \mu_{sd}},$$

Provided that $\frac{\mu_{sd}\lambda_{sig}\chi_{sig}(a\chi_{sd}-(1+ln(a\chi_{sd}\lambda_{sd}))/\lambda_{sd})}{\beta\gamma_{sig}+\mu_{sd}} > 1$ and $x_{vb} = 1$;

$$d_{sig} = F_{sig}(e_{vb}, x_{vb}, t_{vb}) = 0, \text{ if } \frac{\mu_{sd}\lambda_{sig}\chi_{sig}(a\chi_{sd}-(1+ln(a\chi_{sd}\lambda_{sd}))/\lambda_{sd})}{\beta\gamma_{sig}+\mu_{sd}} x_{vb} \leq 1;$$

$$d_{sig} = F_{sig}(e_{vb}, x_{vb}, t_{vb}) = 0, \text{if} a\chi_{sd}\lambda_{sd} \leq 1$$

To complete the implementation of these assumptions, it is necessary to determine the optimal management of integrated business development at the first stage, taking into account the above-mentioned considerations. In this context, the target function for searching e_{vb} can be represented as follows:

$$\max_{d_{vb}} aD_0x_{sd} - \beta D_0(t_{sd} - t_0) - d_{vb} - D_0d_{sig} - D_0d_{sd}$$

Based on this, it can be determined:

$$D_0x_{sd} = D_0(D_{vb}(D_{sig}(x_{sd}))) = D_0(D_{vb}(\chi_{sd}(1 - exp^{-\lambda_s d_{sd}})x_{sig}));$$

$$D_0 x_{sd} = \chi_{sd}\left(1 - \frac{1}{a\chi_{sd}\lambda_{sd}}\right) D_0(P_{vb}(x_{sig} = 1))$$
$$= \chi_{sd}\left(1 - \frac{1}{a\chi_{sd}\lambda_{sd}}\right) D_0((1 - exp^{-\lambda_{sig}d_{sig}})x_{vb}),$$

if $a\chi_{sd}\lambda_{sd} > 1$ and $D_0 d_{sd} = 0$, if $a\chi_{sd}\lambda_{sd} \leq 1$.

In this way it is possible to lead the offensive:

$$D_0 x_{sd} = \chi_{sd}\left(1 - \frac{1}{a\chi_{sd}\lambda_{sd}}\right)\left(1 - \frac{\beta\gamma_{sig} + \mu_{sd}}{\mu_{sd}\lambda_{sig}\chi_{sig}(a\chi_{sd} - (1 + ln(a\chi_{sd}\lambda_{sd}))/\lambda_{sd}}\right)$$
$$P_0(x_{vb} = 1) = \chi_{vb}\chi_{sd}\left(1 - \frac{1}{a\chi_{sd}\lambda_{sd}}\right)$$
$$\left(1 - \frac{\beta\gamma_{sig} + \mu_{sd}}{\mu_{sd}\lambda_{sig}\chi_{sig}(a\chi_{sd} - (1 + ln(a\chi_{sd}\lambda_{sd}))/\lambda_{sd}}\right) exp^{-\lambda_{vb}d_{vb}} x_0,$$

If $\frac{\mu_{sd}\lambda_{sig}\chi_{sig}(a\chi_{sd} - (1+ln(a\chi_{sd}\lambda_{sd}))/\lambda_{sd})}{\beta\gamma_{sig}+\mu_{sd}} > 1$, and

$D_0 x_{sd} = 0$, if $\frac{\mu_s\lambda_{sig}\chi_{sig}(a\chi_{sd} - (1+ln(a\chi_{sd}\lambda_{sd}))/\lambda_{sd})}{\beta\gamma_{sig}+\mu_{sd}} x_{vb} \leq 1$ or $a\chi_{sd}\lambda_{sd} \leq 1$.

Thus, we will continue to have:

$$D_0(t_{sd} - t_0) = D_0(t_{sig} + a_{sd} - t_0) = D_0(t_{sig} - t_{vb}) + D_0(t_{vb} - t_0) + a_{sd}.$$

The resulting equation can be transformed:

$D_0(t_{vb} - t_0) = D_0((\gamma_{vb}d_{vb} + a_{sig})\varepsilon_{vb}) = \frac{\gamma_q d_{vb} + a_{sig}}{\mu_{vb}}$ and $D_0(t_{sig} - t_{vb}) = D_0(\gamma_{sig}e_{sig} - \delta_{sig}d_{vb} + a_{sig})\varepsilon_{sig} = D_0(D_{vb}((\gamma_{sig}e_{sig} - \delta_{sig}e_{vb} + a_{sig})\varepsilon_{sig})) = \frac{1}{\mu_{sig}} D_0(\gamma_{sig}e_{sig} - \delta_{sig}d_{vb} + a_{sig})$.

As a result, it can be argued that:

$$\frac{\mu_{sd}\lambda_{sig}\chi_{sig}(a\chi_{sd} - (1 + ln(a\chi_{sd}\lambda_{sd}))/\lambda_{sd})}{\beta\gamma_{sig} + \mu_{sd}} x_{vb} > 1,$$

$$D_0(t_{sig} - t_{vb}) = \frac{a_{sig} - \delta_{sig}d_{vb}}{\mu_{sig}}$$
$$+ \frac{\gamma_{sig}}{\mu_{sig}\lambda_{sig}} \ln \frac{\mu_{sd}\lambda_{sig}\chi_{sig}(a\chi_{sd} - (1 + ln(a\chi_{sd}\lambda_{sd}))/\lambda_{sd})}{\beta\gamma_{sig} + \mu_{sd}}$$
$$P_0(x_{vb} = 1) = \frac{a_{sig} - \delta_{sig}e_{vb}}{\mu_{sig}} + \frac{\gamma_{sig}\chi_{vb}}{\mu_{sig}\lambda_{sig}}$$
$$\ln \frac{\mu_{sd}\lambda_{sig}\chi_{sig}(a\chi_{sd} - (1 + ln(a\chi_{sd}\lambda_{sd}))/\lambda_{sd})}{\beta\gamma_{sig} + \mu_{sd}} exp^{-\lambda_{vb}d_{vb}} x_0.$$

It should also be borne in mind that if:
$\frac{\mu_{sd}\lambda_{sig}\chi_{sig}(a\chi_{sd} - (1+ln(a\chi_{sd}\lambda_{sd}))/\lambda_{sd})}{\beta\gamma_{sig}+\mu_{sd}} x_{vb} \leq 1$ or $a\chi_{sd}\lambda_{sd} \leq 1$, then:

$$D_0\left(t_{sig}-t_{vb}\right)=\frac{a_{sig}-\delta_{sig}d_{vb}}{\mu_{sig}}.$$

If $\frac{\mu_{sd}\lambda_{sig}\chi_{sig}\left(a\chi_{sd}-\left(1+ln\left(a\chi_{sd}\lambda_{sd}\right)\right)/\lambda_{sd}\right)}{\beta\gamma_{sig}+\mu_{sd}}x_{vb}>1$, we will have:

$$D_0d_{sig}=\frac{1}{\lambda_{sig}}\ln\frac{\mu_{sd}\lambda_{sig}\chi_{sig}(a\chi_{sd}-(1+ln(a\chi_{sd}\lambda_{sd}))/\lambda_{sd})}{\beta\gamma_{sig}+\mu_{sd}}$$

$$P_0(x_{vb}=1)=\frac{\chi_{vb}}{\lambda_{sig}}\ln\frac{\mu_{sd}\lambda_{sig}\chi_{sig}(a\chi_{sd}-(1+ln(a\chi_{sd}\lambda_{sd}))/\lambda_{sd})}{\beta\gamma_{sig}+\mu_{sd}}exp^{-\lambda_{vb}d_{vb}}x_0.$$

If $\frac{\mu_{sd}\lambda_{sig}\chi_{sig}\left(a\chi_{sd}-\left(1+ln\left(a\chi_{sd}\lambda_{sd}\right)\right)/\lambda_{sd}\right)}{\beta\gamma_{sig}+\mu_{sd}}x_{vb}\le 1$, or $a\chi_{sd}\lambda_{sd}\le 1$, then we will have: $D_0d_{sig}=0$.

As a result, we get this:

$$\begin{aligned}D_0d_{sd}&=\frac{1}{\lambda_{sd}}\ln(a\chi_{sd}\lambda_{sd})P_0\left(x_{sig}=1\right)=\frac{\chi_{sig}}{\lambda_{sd}}\ln(a\chi_{sd}\lambda_{sd})D_0\left(\left(1-exp^{-\lambda_{sig}d_{sig}}\right)x_{vb}\right)\\&=\frac{\chi_{sig}}{\lambda_{sd}}\ln(a\chi_{sd}\lambda_{sd})1-1-\frac{\beta\gamma_{sig}+\mu_{sd}}{\frac{\mu_{sd}\lambda_{sig}\chi_{sig}(a\chi_{sd}-(1+ln(a\chi_{sd}\lambda_{sd}))}{\lambda_s}})P_0\left(x_{sig}=1\right)\\&=\frac{\chi_q\chi_{sig}}{\lambda_{sig}}\ln(a\chi_{sd}\lambda_{sd})\left(1-\frac{\beta\gamma_{sig}+\mu_{sd}}{\frac{\mu_{sd}\lambda_{sig}\chi_{sig}(a\chi_{sd}-(1+ln(a\chi_{sd}\lambda_{sd}))}{\lambda_{sd}}}\right)exp^{-\lambda_{vb}d_{vb}}x_0),\end{aligned}$$

Provided: $\frac{\mu_{sd}\lambda_{sig}\chi_{sig}\left(a\chi_{sd}-\left(1+ln\left(a\chi_{sd}\lambda_{sd}\right)\right)/\lambda_{sd}\right)}{\beta\gamma_{sig}+\mu_{sd}}x_{vb}>1$.

If $\frac{\mu_{sd}\lambda_{sig}\chi_{sig}\left(a\chi_{sd}-\left(1+ln\left(a\chi_{sd}\lambda_{sd}\right)\right)/\lambda_{sd}\right)}{\beta\gamma_{sig}+\mu_{sd}}x_{vb}\le 1$ or $a\chi_{sd}\lambda_{sd}\le 1$, we will have: $D_0d_{sd}=0$.

Based on the reduction of the obtained equations into an objective function for determining the optimal management of the integrated development of an enterprise, taking into account d_{vb}, we have:

$\max\limits_{d_{sig}\ge 0}-Aexp^{-\lambda_{vb}d_{vb}}-Bd_{vb}$, де $B=1-\beta(\frac{\delta_{sig}}{\mu_{sig}}-\frac{\gamma_{vb}}{\mu_{vb}})$, provided: $\frac{\mu_{sd}\lambda_{sig}\chi_{sig}\left(a\chi_{sd}-\left(1+ln\left(a\chi_{sd}\lambda_{sd}\right)\right)/\lambda_{sd}\right)}{\beta\gamma_{sig}+\mu_{sd}}>1$.

Thus, we have:

$$A=-a\chi_{vb}\chi_{sd}\left(1-\frac{1}{a\chi_{sd}\lambda_{sd}}\right)\left(1-\frac{\beta\gamma_{sig}+\mu_{sd}}{\frac{\mu_{sd}\lambda_{sig}\chi_{sig}(a\chi_{sd}-(1+ln(a\chi_{sd}\lambda_{sd}))}{\lambda_3}}\right)$$

$$x_0-\beta(\frac{\gamma_{sig}\chi_{vb}}{\lambda_{sig}\mu_{sig}}\ln\frac{\mu_{sd}\lambda_{sig}\chi_{sig}(a\chi_{sd}-(1+ln(a\chi_{sd}\lambda_{sd}))/\lambda_{sd})}{\beta\gamma_{sig}+\mu_{sd}}$$

$$x_0+\frac{\chi_{vb}}{\lambda_{sig}}\ln\frac{\mu_{sd}\lambda_{sig}\chi_{sig}(a\chi_{sd}-(1+ln(a\chi_{sd}\lambda_{sd}))/\lambda_{sd})}{\beta\gamma_{sig}+\mu_{sd}}$$

$$x_0 + \frac{\chi_{vb}\chi_{sig}}{\lambda_{sig}} \ln(a\chi_{sd}\lambda_{sd})\left(1 - \frac{\beta\gamma_{sig} + \mu_{sd}}{\frac{\mu_{sd}\lambda_{sig}\chi_{sig}(a\chi_{sd} - (1 + ln(a\chi_{sd}\lambda_{sd}))}{\lambda_{sd}}})x_0\right),$$

If $\frac{\mu_{sd}\lambda_{sig}\chi_{sig}(a\chi_{sd}-(1+ln(a\chi_{sd}\lambda_{sd}))/\lambda_{sd})}{\beta\gamma_{sig}+\mu_{sd}} x_{vb} \le 1$ or $a\chi_{sd}\lambda_{sd} \le 1$, we will have: $A = 0$.

Thus, optimal management of integrated development in the case of $A\lambda_{vb}/B > 1$ and $B > 0$ can be represented using:$d_{vb} = \frac{1}{\lambda_{sd}}\ln\frac{A\lambda_{vb}}{B}$. If $A\lambda_{vb} \le B$ and ≥ 0, then the development in this direction will be $d_{vb} = 0$. Therefore, it can be stated that there will be no actual solution if $B < 0$ or if $B = 0$ and$A > 0$. However, in real business operating conditions, such a situation cannot actually arise, as all three development directions (components) of the proposed integrated development management system of enterprises cannot be effective separately.

The determination of the integral indicator of the effectiveness of integrated development is proposed to be defined in the form of a vector that extends on the plane of three ordinates, which demonstrate the development directions of the company:

$$IDI = \sqrt{D_{bv}^2 + D_{sig}^2 + D_{sd}^2}, \tag{1}$$

where D_{bv}—indicator of effectiveness in the development of business virtualization; D_{sig}—indicator of effectiveness in socio-inclusive growth of the company;D_{sd} – indicator of effectiveness in sustainable development of the enterprise.

The paradigmatic construct of an effective integrated development system for enterprises presupposes a comprehensive approach to the company's development, focusing on various aspects of its activity. The main feature of such a system is that it includes four main components that interact with each other and complement each other:

- the economic component includes planning and managing financial resources, market and competitor analysis, strategic planning, and financial support of projects;
- the organizational component is focused on the internal processes of the enterprise, including personnel management, organization of production processes, and structuring of management processes;
- the information component encompasses the development and implementation of information systems, data security, customer relationship management, and marketing strategy;
- the innovation component is oriented towards the implementation of new technologies, creation of new products and services, and the development of scientific research and developments (Boichenko et al., 2022).

Each of these components has its specific goals and tasks, but their interaction allows the creation of a comprehensive system that promotes the effective development of the enterprise. Additionally, such a system has a certain flexibility, allowing it to adapt to changes in the external environment and respond to new opportunities and challenges.

Besides the components, the integrated development system of the enterprise foresees development according to three main directions:

Virtual direction—involves the use of modern technologies in business management, implementation of electronic document management and CRM systems, use of social networks for marketing, and implementation of automated technologies.

Socio-inclusive direction—entails ensuring the accessibility of products, creating social partnership programs, organizing training for people with disabilities, and developing products for people with disabilities.

Sustainable direction–involves reducing waste, using renewable energy, implementing energy-saving systems, and introducing green technologies.

To determine the interrelations between the BV, SIG, and SD components, a correlation analysis was conducted. This allows assessing how the change in one indicator affects others. For a more detailed analysis, regression analysis was used to determine the impact of each component on the enterprise's development effectiveness.

The use of vector analysis provides a measure of the harmonization of integrated development. Evaluation occurs through determining the angle between the vectors of the components (BV, SIG, SD), reflecting their level of harmonization. To determine the effectiveness of enterprise development in this study, a vector model of diagnostics of integrated development quality is proposed. Its evaluation is proposed to be conducted through determining the level of harmonization of business integrated development. The deviation of the vector of the studied enterprises from the standard reflects the level of disharmony in business development, specifically its asymmetry in the context of integrating the determined components (Fig. 7).

The angle (γ), formed between the vectors, shows the level of disharmony in the integrated development of the enterprise:

$$\gamma = arccos\frac{BV + \mathrm{SIG} + \mathrm{SD}}{\sqrt{n}\cdot\sqrt{\mathrm{BV}^2 + \mathrm{SIG}^2 + \mathrm{SD}^2}} \tag{2}$$

where γ—the angle formed between the vectors; n is the number of indicators included in the system. The formed angle takes values in the range $0 < \gamma < \gamma_{max}$, where γ_{max}, depends on the number of indicators included in the system. In the study, three directions and their corresponding development indicators were used, so:

$$\gamma_{max} = arccos\frac{1}{\sqrt{3}} \tag{3}$$

An increase in the angle between the vectors indicates a growing level of disharmony in integrated development, so in contrast to this indicator, one can propose an indicator that helps diagnose the level of harmonization of business integrated development (HID, %):

$$HID = \left(1 - \frac{\gamma}{\gamma_{max}} - \frac{1-\frac{45}{\pi}arccos\frac{1}{\sqrt{3}}}{\gamma_{max}} \bigg/ 1 - \frac{1-\frac{45}{\pi}arccos\frac{1}{\sqrt{3}}}{\gamma_{max}}\right) \times 100\% \tag{4}$$

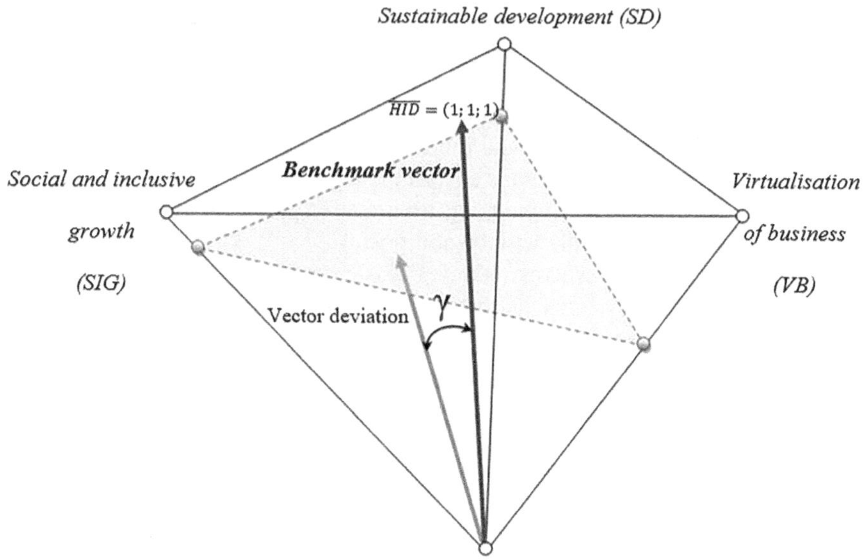

Fig. 7 Vector of harmonization of the monitoring system of integrated development management processes of enterprises

The proposed indicator of business development harmonization follows the logic that the closer the arccosine approaches 1, the higher the level of harmonization of the company's development. For diagnostic purposes, the obtained indicator can also be converted to percentages.

Cluster analysis was used to segment the companies under study based on indicators of integrated development. This allows grouping enterprises with similar characteristics and development effectiveness.

Based on the obtained data, forecasting of enterprise development is carried out. It is suggested to use indicators of the level of harmonization and integrated development to assess potential changes in the future.

It is important to consider that the obtained results depend on the quality and volume of available data. The methodology may require adaptation to specific conditions and characteristics of different industries. The study is based on a specific sample of enterprises, which may not reflect all aspects of a broader business environment. The conclusions may not be fully applicable to other companies of different sizes, industries, or geographical locations.

The presented methodology provides a comprehensive approach to the analysis of integrated enterprise development, combining several analytical tools for a deep understanding of the interaction and impact of key components BV, SIG, and SD.

4 Results

As benchmarks in this study, three key components of integrated development were used: BV (Business Virtualisation)–business virtualization (includes the indicator of increasing net profit of the enterprise through the implementation of virtual tools in activities, increasing the level of automation of business processes using virtual technologies to reduce management costs, ensuring the accessibility and security of electronic data and information thanks to virtual tools, development and implementation of new virtual products and services that meet the needs of customers and the market; SIG (Socio-Inclusive Growth)—socio-inclusive growth involves ensuring the accessibility of products, creating a social partnership program, participating in events for people with special needs, developing products for people with special needs; SD (Sustainable Development) sustainable development involves reducing the amount of waste, using renewable energy, implementing an energy-saving system, and implementing green technologies.

The proposed system includes three components: BV (Business Virtualisation), SIG (Socio-Inclusive Growth), and SD (Sustainable Development), which contribute to the development of the enterprise at different levels. BV provides business virtualization, SIG entails socio-inclusive growth, and SD promotes sustainable development. The proposed system includes the harmonization indicator of integrated development based on the BV, SIG, and SD vector, as well as modeling the impact of components on its deviation. Additionally, a cluster analysis of companies is proposed, which allows identifying groups of enterprises with similar characteristics and determining the factors that affect their development.

The result of using the proposed system for monitoring the processes of managing integrated enterprise development is an increase in the net profit of the enterprise, an increase in the level of automation of business processes, a reduction in the amount of waste, and the use of renewable energy, which ensures sustainable development. Additionally, socio-inclusive growth involves ensuring the accessibility of products for different population groups, which contributes to creating a positive image of companies and increases their market competitiveness.

Furthermore, the system includes stages of monitoring and analysis of results, which allows timely identifying problem areas and developing effective strategies for their resolution. A separate component is the use of machine learning tools for automating the processes of monitoring and data analysis. Thus, the proposed system for monitoring the processes of managing integrated enterprise development based on the vector of harmonization of integrated company development is innovative and allows effectively solving management tasks of the enterprise, taking into account key aspects such as economic stability, social responsibility, and ecological balance.

For modeling and determining the most influential factors on deviation, a correlation analysis was conducted (Fig. 8).

The obtained regression analysis indicators are presented in Table 1.

To determine the appropriateness of the proposed approach to monitoring the effectiveness of integrated development management processes of enterprises, it is

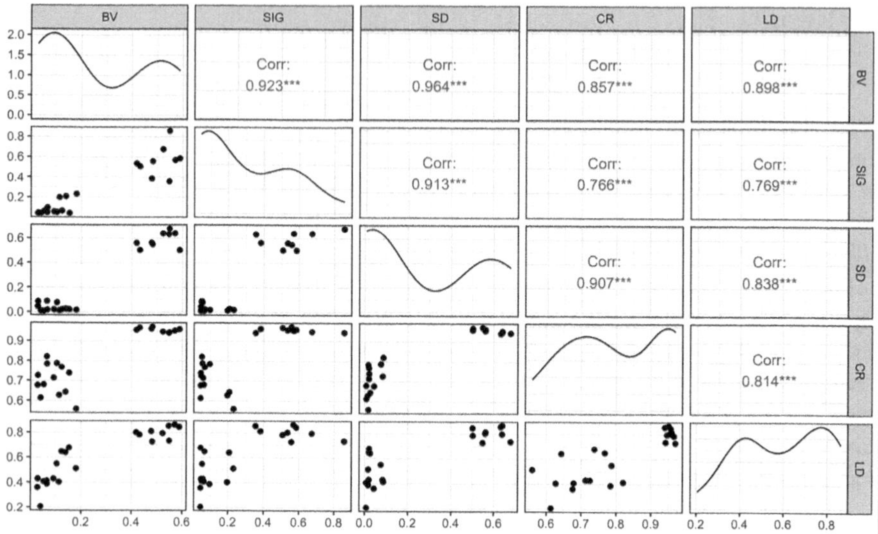

Fig. 8 Correlation analysis matrix of the effectiveness indicators of integrating development components of the companies under study

Table 1 Indicators of factor analysis of the integration of development of the companies under study

Development directions	Coefficients	Standard error	t-stat	P-value	Lower 95%	Higher 95%
Y-intercept	20.2965	3.3757	6.0125	0.0000	13.2549	27.3381
Business Virtualization (X_1)	40.8512	7.7204	5.2913	0.0000	24.7467	56.9557
Socio-Inclusive Growth (X_2)	−37.2812	11.2791	−3.3053	0.0035	−60.8090	−13.7535
Sustainable Development (X_3)	−34.2422	13.3168	−2.5714	0.0182	−62.0206	−6.4639

proposed to conduct a comparative analysis of its harmonization level for Ukrainian and European enterprises. The level of harmonization of integrated development of the studied Ukrainian and European companies as of 2022 is presented in Table 2.

The highest level of harmonization of integrated development of enterprises is characteristic of European companies—87–94%. This indicates effective management of the integration processes of these companies and a relatively minor deviation of their development vectors from the standard vector. For Ukrainian companies, we can observe a relatively low level of harmonization of integrated business development—23–46%. Therefore, it can be stated that there is low effectiveness in managing

Table 2 Indicators of the level of harmonization of integrated development of the studied companies before the implementation of the replication model

Company	BV	SIG	SD	$arcos\gamma$, degrees	$1-\frac{\gamma}{\gamma_{max}}$	Level of Harmonization of Integrated Development (HID), %
Elita	0,063	0,469	0,018	47,9	0,125	23
Kalyna	0,031	0,231	0,010	47,7	0,129	23
Edelwika	0,056	0,234	0,007	44,6	0,186	28
Kremtex	0,033	0,257	0,038	43,7	0,202	30
Kremenchuk TC	0,141	0,252	0,026	33,5	0,389	46
Mriya	0,051	0,264	0,028	42,9	0,216	31
Novovolynsk SF	0,072	0,23	0,021	39,6	0,277	36
Rivne FCF	0,072	0,253	0,021	40,8	0,255	34
Pryluky SF	0,03	0,247	0,017	47,1	0,139	24
Ballistyka	0,051	0,257	0,019	44,1	0,195	29
Kozak	0,034	0,260	0,040	43,4	0,208	30
Stritex	0,069	0,275	0,014	43,3	0,209	30
Elegant	0,011	0,228	0,008	51,4	0,062	17
H&M	0,548	0,719	0,677	6,4	0,883	90
Inditex	0,48	0,554	0,562	4,0	0,927	94
G-Star Raw	0,432	0,39	0,504	6,1	0,889	90
Vistula	0,548	0,535	0,636	4,5	0,918	93
S.Oliver	0,483	0,388	0,551	8,0	0,853	87
Impetus	0,42	0,442	0,561	7,4	0,864	88
Promod	0,59	0,517	0,504	4,0	0,926	93
Silvano Fashion	0,526	0,535	0,640	5,2	0,905	92
LPP	0,571	0,555	0,641	3,6	0,934	94

the integrated development of these companies and the presence of disharmony in business development directions. This situation requires qualitative changes, which are proposed to be implemented based on the replication model. The results obtained from monitoring the effectiveness of managing integrated development based on its level of harmonization are presented in Table 3.

The implementation of the replication model of integrated development will have a positive impact on its harmonization for all Ukrainian and partially European companies. For Ukrainian companies, we can observe an average doubling in the harmonization of integrated development. The highest result is characteristic of Kalina and Elegant–more than a threefold increase, while the minimal growth (2–4% points) is for the Novovolynsk Sewing Factory and the Kremenchuk Knitting Factory. Some European companies have a decrease in the level of harmonization of integrated

Table 3 Forecasted indicators of the level of harmonization of integrated development of the studied companies as a result of the implementation of the replication model

Company	BV	SIG	SD	$arcos\gamma$, degrees	$1 - \frac{\gamma}{\gamma_{max}}$	Level of Harmonization of Integrated Development (HID), %
Elita	0,181	0,232	0,017	32,6	0,404	47
Kalyna	0,031	0,042	0,044	8,3	0,848	87
Edelwika	0,055	0,062	0,007	30,6	0,441	51
Kremtex	0,032	0,045	0,084	22,4	0,591	64
Kremenchuk TC	0,140	0,208	0,026	31,1	0,433	50
Mriya	0,106	0,052	0,076	15,8	0,711	74
Novovolynsk SF	0,154	0,043	0,021	38,7	0,293	38
Rivne FCF	0,125	0,066	0,020	31,4	0,426	49
Pryluky SF	0,068	0,096	0,017	28,5	0,480	54
Ballistyka	0,094	0,057	0,020	27,9	0,490	55
Kozak	0,067	0,049	0,087	12,9	0,764	79
Stritex	0,115	0,197	0,013	34,8	0,364	44
Elegant	0,043	0,042	0,008	27,7	0,494	55
H&M	0,549	0,856	0,678	10,3	0,812	83
Inditex	0,478	0,383	0,563	8,8	0,839	86
G-Star Raw	0,432	0,507	0,503	4,1	0,925	93
Vistula	0,547	0,355	0,633	12,8	0,766	79
S.Oliver	0,482	0,557	0,549	3,6	0,934	94
Impetus	0,419	0,535	0,561	7,0	0,873	89
Promod	0,589	0,585	0,502	4,1	0,925	93
Silvano Fashion	0,523	0,675	0,639	6,0	0,890	90
LPP	0,571	0,569	0,640	3,2	0,942	95

development (Fig. 9)–H&M, Inditex, Vistula, and Silvano Fashion. Although for most European companies, the harmonization of integrated development increases as a result of implementing the replication model, this increase is not significant. The main factor is that the studied European network companies already have a high level of development, and additional integration does not have as significant an effect as for Ukrainian enterprises.

Since a similarity in the response of companies to the replication of integrated development has been identified, it can be assumed that enterprises may have common characteristics. This allowed for the clustering of the studied companies considering the effectiveness indicators for the recognized directions of integrated development (Fig. 10).

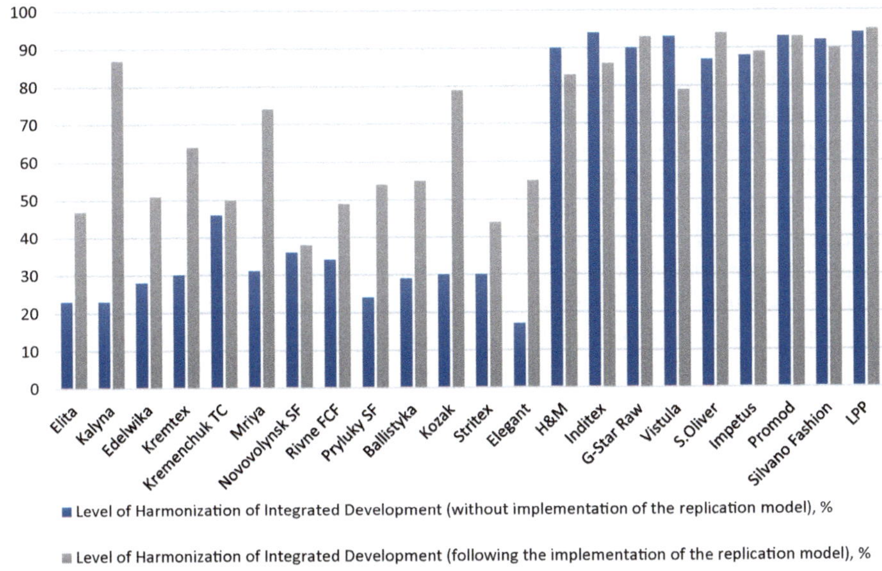

Fig. 9 Forecasted results of monitoring the harmonization of integrated development of the studied enterprises considering the implementation of the replication model

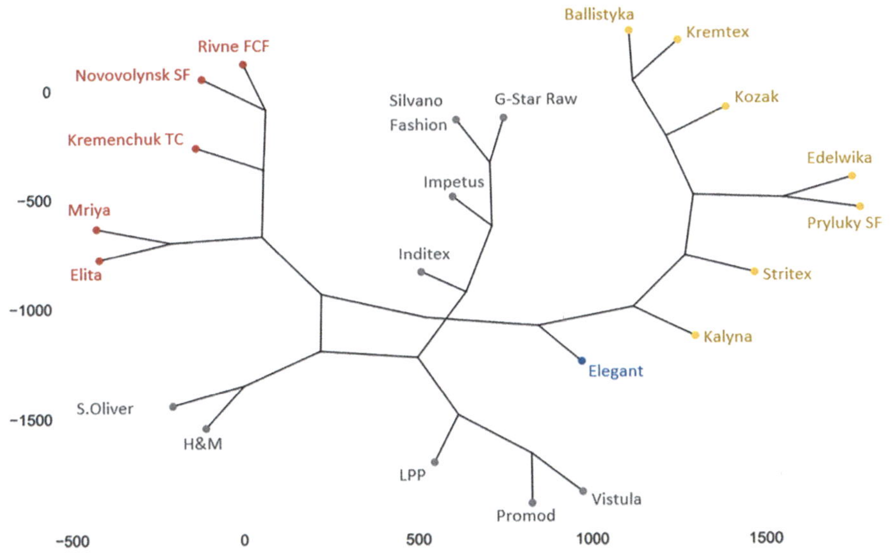

Fig. 10 Dendrogram of clustering of the studied companies by indicators of integrated development directions

As a result of clustering by indicators of integrated development directions, four clusters were formed. It is worth noting that all European companies fall into one cluster, as they have similar indicators in development directions, which integrate with a minor deviation from the standard vector of its harmonization. Domestic companies are divided into three clusters, among which one is autonomous—Elegant. This company is singled out into an autonomous cluster, as it has the lowest level of development in terms of business virtualization, as well as a fairly low level of sustainable enterprise development. The diagnostics of the effectiveness of managing integrated development confirms this, as among all the studied companies, Elegant has the lowest indicator of harmonization of integrated development—only 17%. The clustering of the studied companies by the method k-NN, based on the determined indicators of harmonization of integrated development and the indicator of overall business development, is presented in Fig. 11.

European companies form one cluster, as they are distinguished by a high level of development and business harmonization, compared to Ukrainian ones. Ukrainian enterprises form two other clusters: the yellow cluster includes companies with similar input parameters, in particular, a level of overall development above 0.5, and the grey cluster includes all others. The next step is to study the regular changes in the harmonization of integrated development for companies in the identified clusters (Fig. 12).

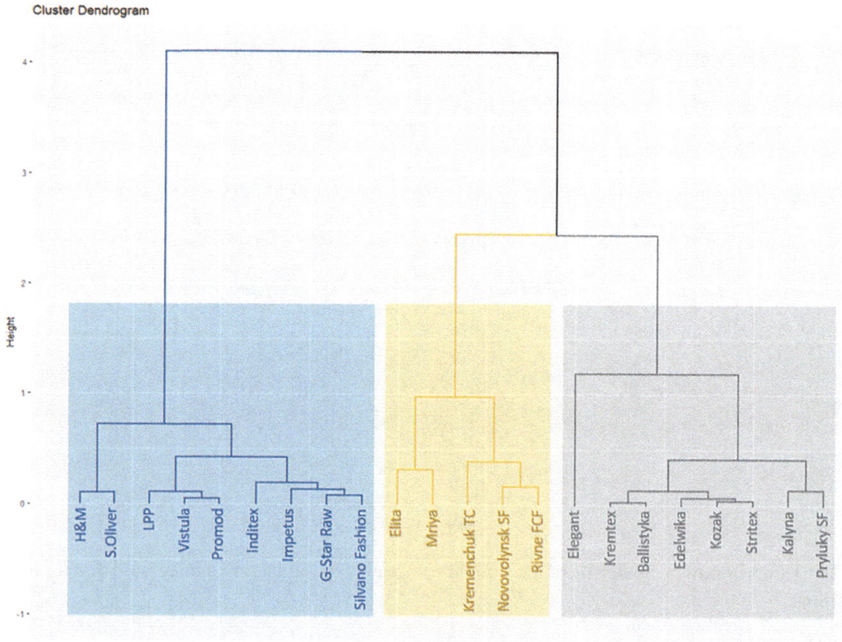

Fig. 11 Hierarchical dendrogram of clustering of the studied companies by the level of harmonization and overall business development

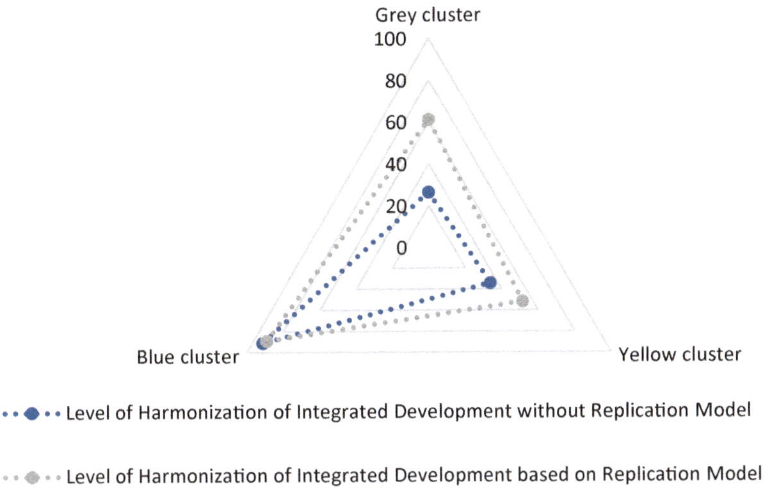

Fig. 12 Forecasted level of harmonization of integrated development for individual clusters based on the implementation of the replication model

For the blue cluster of European companies, the highest average level of harmonization of integrated development is characteristic. The replication of integrated components does not have a significant impact on the change in the harmonization of development directions. For the yellow cluster, there is an average increase in the harmonization of integrated development by 18% points. Ukrainian companies in this cluster had a level of harmonization of integrated development below average, and the implementation of the replication model has a sufficient positive impact on its growth. Ukrainian enterprises of the grey cluster had the lowest level of development and harmonization of its integrated directions, but the use of the replication model allows eliminating disharmony and increasing HL almost threefold. This confirms the effectiveness of the proposed replication model primarily for companies with a low level of development, as its use contributes to a significant increase in key indicators used for monitoring the effectiveness of managing integrated business development.

5 Conclusion

The conducted modeling proves that the proposed system for monitoring the effectiveness of integrated development processes based on its harmonization indicator not only allows assessing its level but also forming suggestions for correcting asymmetries in integration across development directions. The advantages of this system include the ability to promptly track the results of implementing strategic development directions of the company in terms of its social, ecological, and economic components. The system also enables the company to respond quickly to changes in the external environment, ensure the improvement of product and service quality, increase customer satisfaction, reduce costs, and enhance profitability.

The system for monitoring the processes of managing integrated enterprise development based on the Harmonization Vector of integrated company development has several other advantages. First and foremost, it allows enterprises to quickly respond to changes in the economic environment and ensure sustainable development by harmonizing the three key development directions. Moreover, the system enables the evaluation of the effectiveness of different development directions and the analysis of the effectiveness of implementing new technologies and launching new products.

This research contributes to the theoretical understanding of integrated enterprise development by introducing a novel harmonization vector. The practical implications of this study are significant, offering enterprises a comprehensive tool for evaluating and adjusting their development strategies across key dimensions of business virtualization, socio-inclusive growth, and sustainable development.

The findings reveal that the implementation of the replication model significantly enhances the harmonization of integrated development, particularly for enterprises with previously low levels of harmonization. This underscores the importance of a balanced approach to integrating various development aspects, ensuring economic viability while also considering social and environmental responsibilities.

The study acknowledges certain limitations, primarily its reliance on a specific set of enterprises, which may not fully represent the broader spectrum of business environments. The model's effectiveness may vary across different industries and geographic locations, necessitating adaptations to local contexts.

Future research should focus on expanding the model's applicability across diverse industries and regions. There is also a need for longitudinal studies to assess the long-term impacts of the replication model on enterprise development. Further exploration into the integration of advanced technologies, such as artificial intelligence and machine learning, into the monitoring system could provide deeper insights and more robust decision-making tools for managing integrated enterprise development. This research provides a valuable framework for businesses to enhance their integrated development strategies, adapting to the dynamic economic landscape while fostering sustainability, inclusivity, and technological advancement.

References

Abreu, M. C. S. D., Ferreira, F. N. H., Proenca, J. F., & Ceglia, D. (2021). Collaboration in achieving sustainable solutions in the textile industry. *Journal of Business & Industrial Marketing, 36*(9), 1614–1626.

Allioui, H., & Mourdi, Y. (2023). Exploring the full potentials of IoT for better financial growth and stability: A comprehensive survey. *Sensors, 23*(19), 8015.

Amjad, F., Abbas, W., Zia-Ur-Rehman, M., Baig, S. A., Hashim, M., Khan, A., & Rehman, H. U. (2021). Effect of green human resource management practices on organizational sustainability:

The mediating role of environmental and employee performance. *Environmental Science and Pollution Research, 28*, 28191–28206.

Bashuk, I., Dantas, R. M., Mata, M. N., Boichenko, K., Rita, J. X., & Correia, A. B. (2023). The impact of business process modelling on the companies' efficiency under Covid-19 conditions. *Economic Research-Ekonomska Istraživanja, 36*(3), 2183418.

Behera, M., Nayak, J., Banerjee, S., Chakrabortty, S., & Tripathy, S. K. (2021). A review on the treatment of textile industry waste effluents towards the development of efficient mitigation strategy: An integrated system design approach. *Journal of Environmental Chemical Engineering, 9*(4), 105277.

Boichenko, K., Gherghina, ŞC., Abreu, A., Mata, M. N., & Martins, J. M. (2022). Towards financing system of integrated enterprise development in the time of COVID-19 outbreak. *International Journal of Financial Studies, 10*(3), 50.

Chae, Y., & Hinestroza, J. (2020). Building circular economy for smart textiles, smart clothing, and future wearables. *Materials Circular Economy, 2*(1), 2.

de Oliveira, C. R. S., da Silva Júnior, A. H., Mulinari, J., & Immich, A. P. S. (2021). Textile re-engineering: Eco-responsible solutions for a more sustainable industry. *Sustainable Production and Consumption, 28*, 1232–1248.

Fernández-Caramés, T. M., & Fraga-Lamas, P. (2018). Towards the Internet of smart clothing: A review on IoT wearables and garments for creating intelligent connected e-textiles. *Electronics, 7*(12), 405.

Hader, M., Tchoffa, D., El Mhamedi, A., Ghodous, P., Dolgui, A., & Abouabdellah, A. (2022). Applying integrated blockchain and big data technologies to improve supply chain traceability and information sharing in the textile sector. *Journal of Industrial Information Integration, 28*, 100345.

Hammad, M., Jillani, R. M., Ullah, S., Namoun, A., Tufail, A., Kim, K. H., & Shah, H. (2023). Security framework for network-based manufacturing systems with personalized customization: An industry 4.0 approach. *Sensors, 23*(17), 7555.

Herrera Almanza, A. M., & Corona, B. (2020). Using social life cycle assessment to analyze the contribution of products to the sustainable development goals: A case study in the textile sector. *The International Journal of Life Cycle Assessment, 25*, 1833–1845.

Islam, M. M., Perry, P., & Gill, S. (2021). Mapping environmentally sustainable practices in textiles, apparel and fashion industries: A systematic literature review. *Journal of Fashion Marketing and Management: An International Journal, 25*(2), 331–353.

Keung Kwok, S., & Wu, K. K. (2009). RFID-based intra-supply chain in textile industry. *Industrial Management & Data Systems, 109*(9), 1166–1178.

Ki, C. W., Chong, S. M., & Ha-Brookshire, J. E. (2020). How fashion can achieve sustainable development through a circular economy and stakeholder engagement: A systematic literature review. *Corporate Social Responsibility and Environmental Management, 27*(6), 2401–2424.

Kumar, V., Koehl, L., Zeng, X., & Ekwall, D. (2017). Coded yarn based tag for tracking textile supply chain. *Journal of Manufacturing Systems, 42*, 124–139.

Prasad, M. M., Dhiyaneswari, J. M., Jamaan, J. R., Mythreyan, S., & Sutharsan, S. M. (2020). A framework for lean manufacturing implementation in Indian textile industry. *Materials Today: Proceedings, 33*, 2986–2995.

Rahmanzadeh, S., Pishvaee, M. S., & Govindan, K. (2023). Emergence of open supply chain management: The role of open innovation in the future smart industry using digital twin network. *Annals of Operations Research, 329*(1), 979–1007.

Raj, A., Gyaneshwar, A., Chadha, U., Chadha, A., Asija, A., Abrol, A., Priya, S., Alam, A., Joshi, A., Bastola, K., Selvaraj, S. K., & Hadidi, H. (2022). Green manufacturing via machine learning enabled approaches. *International Journal on Interactive Design and Manufacturing (IJIDeM)*, 1–30.

Reche, A. Y. U., Junior, O. C., Estorilio, C. C. A., & Rudek, M. (2020). Integrated product development process and green supply chain management: Contributions, limitations and applications. *Journal of Cleaner Production, 249*, 119429.

Robertsone, G., Mezinska, I., & Lapina, I. (2022). Barriers for Lean implementation in the textile industry. *International Journal of Lean Six Sigma, 13*(3), 648–670.
Simegnaw, A. A., Malengier, B., Rotich, G., Tadesse, M. G., & Van Langenhove, L. (2021). Review on the integration of microelectronics for e-textile. *Materials, 14*(17), 5113.
Tian, J., Coreynen, W., Matthyssens, P., & Shen, L. (2022). Platform-based servitization and business model adaptation by established manufacturers. *Technovation, 118*, 102222.
Todeschini, B. V., Cortimiglia, M. N., & de Medeiros, J. F. (2020). Collaboration practices in the fashion industry: Environmentally sustainable innovations in the value chain. *Environmental Science & Policy, 106*, 1–11.
Wood, G., & Bischoff, C. (2020). Challenges and progress in integrating knowledge: Cases from clothing and textiles in South Africa. *Journal of Knowledge Management, 24*(1), 32–55.

Kateryna Boichenko is an Associated Professor of Kyiv National Economic University named after Vadym Hetman. She graduated with B.S. and M.S. in Enterprise Economics, and a Ph.D. in Economics and Business Management (by type of economic activity). Her current research interests are empirical research in Business Economics, Application of Optimization Methods to Integrated Development, Smart Business models, and Risk-Management.

Export Performance of the Textile Industry: Influence of Financial Resources and Productivity

Isabel Soares de Moura

Abstract This study aims to examine the effect of certain factors on the export performance of the Portuguese textile industry. Considering the strong representation of European markets in Portuguese exports, we also aim to determine whether geographic diversification has an impact on export performance. Utilizing a multi-sectoral sample of 106 Portuguese small and medium-sized enterprises (SMEs) exporting textiles, this research adopts a quantitative methodological approach, applying a panel regression model to test our hypothesis after ensuring the reliability and validity of the data collected for this study. The study provides several findings for export-oriented firms, revealing that financial resources, productivity, and size are significant determinants of export performance. The findings resulting from empirical testing of the theoretical model provide new and important information for the Portuguese textile sector.

1 Introduction

SMEs constitute the vast majority of European firms, representing 99% of EU-27 firms (European Commission, 2019). In many economic sectors, more than 50% of the EU's export volume is sold by SMEs and around 87% of EU SMEs export (Cernat et al., 2020). Many authors report that exports are the preferred way of entering the foreign market for SMEs, using the flexibility of their structure to better adapt their strategies to the specific needs of the markets (Sousa & Bradley, 2009; Wang et al., 2017). In addition to these arguments, it should be added that, with a small involvement of resources and a limited risk of exposure, exports offer a high degree of flexibility in international sales (Chen et al., 2016; Lages & Montgomery, 2004; Sousa, 2016). SMEs are experiencing a new competitive dynamism, in which their competitiveness is very dependent on their ability to grow internationally.

I. S. de Moura (✉)
Universidade Lusófona, Campo Grande 376, 1749-024 Lisboa, Portugal
e-mail: isabel.moura@ulusofona.pt

Lisbon Accounting and Business School, Polytechnic Institute of Lisbon (ed.),
Industry and Innovation: Textile Industry, SDGs and Textiles,
https://doi.org/10.1007/978-3-031-57804-5_15

In this context, Portugal is an interesting country to study given the small existing domestic market, which has encouraged firms to follow an international orientation.

Portugal's opening to the outside world has been a vital condition for the country's affirmation and for the creation of a more favourable position in the international context. Therefore, the present study focuses on Portugal and it purposes to offer a useful instrument both for current or potential exporters, and for financing decision-makers, as the export characteristic is an attribute that qualifies for financing attribution (IAPMEI, 2022). It will also make it possible to reduce the information asymmetry that still characterizes the SME scope, designing conditions that encourage firms to participate in international markets. Furthermore, it is important to have greater knowledge about the determinants that can contribute to the export success of SMEs, as considered by Chen et al. (2016).

Therefore, we consider the following research problem. What are the determinants that drive Portuguese textile exports?

The structure of the study is as follows. First, grounded on a review of the existing literature, we develop a set of theoretically hypotheses for empirical testing. Next it defines the research design of the empirical Portuguese textile study. Then, we presented the findings. We conclude with a discussion of the findings, some of the study's limitations, and implications for practice and theory.

2 Literature Review

The foundations of Resources Based View (RBV) seem to have emerged with Penrose (1959), in her book The Theory of Growth of the firm. According to Penrose, the optimal growth of a firm involves a balance between the exploitation of existing resources and the development of new ones. According to RBV (Barney, 1991; Wernerfelt, 1984), firms build competitive advantages based on the set of resources they have. This theory combines four attributes of resources that can bring sustainable and specific advantages to the firm: having value, being rare, having little mobility and not being easily replaceable (Barney, 1991). Dhanaraj and Beamish (2003) identify three other sets of resources that cover the domain of a firm's resources, namely, management and organizational resources, business resources and technological resources. These quantitative and qualitative resources promote the firm's performance and, consequently, its export performance. The literature on exporting has addressed various issues related to resources, suggesting that firms with greater financial resources are more likely to pursue international expansion opportunities (Dhanaraj & Beamish, 2003; Penrose, 1959). Consequently, successful export performance can be attributed to a firm's acquisition and effective utilization of a set of resources.

The Resource-Based View (RBV) emphasizes the resources and capabilities that firms develop, combine, and deploy in the process of creating offerings for their current and potential markets. It defines resources as existing knowledge or individual skills that can be utilized to enhance organizational processes. Thus, they can lead to a positive response from customers, obtaining competitive advantages and meeting performance objectives (Kaleka, 2012). Resources are stocks of tangible and intangibles and capabilities are organizational processes that transform resources into strategic results, show the firm's ability to respond quickly to environmental changes and achieve innovative forms of competitive advantages and superior performance (Teece, 2014; Teece et al., 1997). Several other resource-based theoretical models have emerged in the literature as viable models to explain the main drivers of the firm's internationalization process, in general, and the firm's export strategy and performance, in particular. Fahy (2002), in his resource-based model of competitive advantages, relates not only firm-level resources, skills and performance, but also includes the specific resources of each country, of origin and destination. More recently, Chen et al. (2016) argue that RBV should not be seen individually, especially in SMEs, but should be seen together with contingency theory, complementing internal resources and capabilities with the surrounding environment, thus achieving a better explanation of successful export performance, a consequence of the alignment of organizational and internal (resources and capabilities) and external factors. In other words, the important resources are not only internal, but also external to the firm. Wheeler et al. (2008) applied a resource-based framework, through the integration and interpretation of the determinants of export performance, identified in 30 export performance studies carried out in the United Kingdom, during the period 1990–2003. In this study, the firm's export performance was measured using the EXPERF scale by Zou et al. (1998).

Exports are the preferred mode of entry into foreign markets for SMEs, as with minimal resource involvement and limited exposure risk, it offers a strong degree of flexibility in international sales. Several authors (Golovko & Valentini, 2011; Lages & Montgomery, 2004) have stated that SMEs use the flexibility of their structure to better adapt their strategies to the specific needs of external markets, with the commitment of the owner or top management being essential to create a good relationship with your customers.

Since SMEs have an important position in the Portuguese economy, and it is necessary to understand how to improve their export performance, it is important to understand which are the most important resources for this purpose. We considered that this study could offer a useful instrument both for the managers of exporting firms, for those responsible for defining public support and for Portuguese banking, in general, which has been committed, in recent years, to position as one of the players in supporting Portuguese exports.

2.1 Financial Resources

The literature is clear in its references that access to financing is essential to enable export operations (Griffith, 2011) and is a critical factor in driving successful sales to foreign markets (Morgan et al., 2004).

Several studies show that SMEs have more financing restrictions than large firms, as many do not even have access to formal financing (Beck et al., 2006, 2008). Beck et al. (2008) showed that the size of firms is important to understand the financing pattern. Empirically, they demonstrated that small businesses use less external financing, especially bank financing. And this situation was more evident in countries with weaker financial and legal systems. Harvie et al. (2013), in their study with SMEs from eight Asian countries, revealed that external financing is crucial for national SMEs, which are fast growing but still not mature, also suggesting that access to financing increases the innovation capabilities of SMEs and the possibility of export.

The ability of firms to secure additional loans from financial institutions is often correlated with their size, given that larger firms can offer more collateral for financing compared to smaller counterparts. Consequently, the authors concluded that firms with greater access to financing sources are more likely to engage in exporting and tend to export more.

Numerous studies have delved into the issue of constraints in access to financing and their repercussions on exports. Wagner (2019) concluded that there exists a positive and statistically significant estimated effect between heightened credit restrictions on industrial firms and exports across five countries, including Portugal, Denmark, Finland, France, and Ireland. The author based their findings on data from the Survey on Access to Finance of Enterprises (SAFE), a questionnaire conducted by the EC and the European Central Bank (ECB), wherein constraints are gauged through reported problems in credit access by enterprises across 25 European countries. Additionally, Padmaja and Sasidharan (2021), in their research involving Indian firms, revealed that financing restrictions exert a significant impact on export sales. They observed that many firms encounter difficulties entering foreign markets due to insufficient resources to cover fixed costs inherent to exports.

Various causes underlie the existence of these constraints, particularly in SMEs. These include projects with low quality, business plans that lack sustainability for financiers, and the pronounced asymmetry of information (Abor et al., 2014).

Nevertheless, SMEs often lack sufficient internal resources and frequently turn to external financing options through debt or equity (St-Pierre et al., 2018). Despite this, it is not always straightforward for these firms to access external financing (Bodlaj et al., 2020), presenting challenges to both exports and innovation. Some firms may find it necessary to defer growth and export objectives due to financing restrictions (Riding et al., 2012; Griffith & Czinkota, 2012). However, there is evidence suggesting that exporting firms tend to have easier access to financing compared to their non-exporting counterparts. Manole and Spatareanu (2010) explored the financial constraints faced by SMEs engaged in export activities in the context of the Czech Republic. Their findings indicated that exporting firms exhibit greater liquidity and face fewer credit restrictions than non-exporting firms. This is attributed to the former's higher productivity and more robust cash flows, enabling them to secure preferential financing. Notably, industries with pronounced external financing needs may also find it easier to access bank financing due to their status as exporters.

A study with German firms, Goldbach and Nitsch (2014) examined whether export-oriented SMEs rely more on bank financing than non-exporting SMEs and whether bank financing can increase the export probability of German SMEs.

Wagner (2019) provided a summary of various studies investigating the absence or limited access to financial resources and its impact on export performance. The majority of these studies conclude that there is indeed a significant impact of financial resources on export performance. However, it's worth noting that some authors present different results, even in studies involving Portuguese firms. For instance, Pacheco (2017) found that less financially leveraged SMEs demonstrate a greater orientation towards exports.

SMEs play a fundamental role in economic growth, job creation, regional and local development and social cohesion. Access to finance is a critical prerequisite for the development and growth of SMEs.

Therefore, as can be seen from the literature, the availability of export financial resources is important to improve export performance results. Thus, hypothesis 1 theorizes:

H1: The higher the firm's level of financial resources, the higher its export performance.

2.2 Assets

As mentioned in the literature, based on the RBV, assets are firm resources, which can take the form of intangible resources and tangible resources, and which can determine the success and performance of firms (Abu Bakar & Ahmad, 2010). Several authors have contributed to the literature, showing evidence that intangible resources are important determinants of firms' success (Barney, 1991) and, specifically, firms' export success (D'Angelo et al., 2013; Chen et al., 2019). Abu Bakar and Ahmad (2010) also concluded that tangible resources are relevant to the firm's superior performance. The latter authors state that all assets that are acquired take their place as assets on the firm's balance sheet. Tangible resources include capital, access to capital, and location, such as location of buildings, warehouse, and other facilities. Intangible resources consist of knowledge, brands, patents, entrepreneurial orientation, proactivity, innovation and the ability to face risks.

Based on the theory that all assets contribute to the creation of value and firm performance, hypothesis 2 tests the following:

H2: The firm's asset level positively influences export performance.

2.3 *Productivity*

In the literature on export performance, productivity has been considered a variable with a strong relationship with export performance (Boehe & Jiménez, 2015; Faria et al., 2020). Faria et al (2020), in a study with Portuguese firms in the wine sector, concluded that productivity is a positive driver of export performance at firm level. Boehe and Jiménez (2015), who worked with Brazilian exporters, argued that productivity underpins the export strategy and the level of commitment to export markets.

Also, Onkelinx et al. (2016) concluded that productivity and investment in human capital are essential for the success of SMEs and for firms to achieve a higher level of internationalization. The aforementioned authors framed this theme in the debate that has existed, in international economics literature, about the causal link between productivity and internationalization, trying to understand whether the most productive firms decide to move towards exporting or, on the other hand, whether it is through learning obtained from the export experience that there is productivity growth. Reis and Forte (2016) corroborated this relationship of variables, having highlighted that labor productivity is important for firms and political decision-makers to face the need to address their policies towards increasing productivity, with the essential objective of improving competitiveness in foreign markets. Onkelinx et al. (2016) concluded that productivity positively affects the entire organization, including export performance, bringing insights from the Theory of Firm Resources, as they consider that investment in human capital (through remuneration, training, skills and knowledge of people), increases productivity, which, in turn, increases export intensity.

H3: The firm's productivity level positively influences export performance.

2.3.1 Geographic Diversification

In international literature, and focusing on MNCs, the relationship between geographic diversification, defined as the number of markets in which the firm operates, and firm performance has been the subject of numerous studies, as referred to by Hennart (2007). Existing studies with SMEs are, in fact, essentially about exports, but in this area there has not been broad attention to geographic diversification, as a determinant of export performance. Sousa et al. (2008) state that only four studies had focused on this topic. Even today, this relationship remains an underexplored area in international marketing literature (Bodlaj et al., 2020).

Most studies on export diversification strategy use the number of markets to which firms export as a measure. And this is even the most used metric for geographic diversification (Katsikeas et al., 2000; Cieślik et al., 2012; Cos et al., 2019; Benito-Osorio et al., 2020). This research concluded that firms with greater profitability and larger size tend to have a greater export intensity and a greater degree of geographic diversification.

The literature does not always point in the same direction regarding the relationship between geographic diversification strategy and export performance. Brouthers et al. (2009) found that the export sales of small firms, predominantly directed to a single market, an extreme form of concentration, lead to better export performance. In contrast, Cos et al. (2019), in their study with Spanish exporters, concluded that there is a positive relationship between a strong level of geographic diversification and better export performance, even stating that this relationship is linear and not curvilinear. The same authors had already suggested, in a previous study, that this relationship was influenced by the size of the firm, concluding in a U-shaped relationship for medium-sized firms and a linear and negative relationship for small firms, noting that, as the number of countries to which a firm exports increases, there is an increase in costs that exceeds the firm's incremental benefits (Benito-Osorio et al., 2016).

This study investigates previous results in the literature, through the following hypothesis:

H4: The greater the number of destination markets, the greater the export performance.

2.4 Export Performance

In the last three decades several researchers showed in their literature review works the existing evolution of the determinants of export performance (Beleska-Spasova, 2014; Chen et al., 2016; Katsikeas et al., 1996, 2000; Sousa, 2004; Sousa et al., 2008; Wheeler et al., 2008; Zou & Stan, 1998). In all these investigations, the central idea in all these investigations is that the success of export performance is at the center of strategic decisions and the decision-making process, both for firm decision-makers and for public and corporate policy makers.

In the investigation by Chen et al. (2016), which focused on 124 export performance studies, several clues were presented on the determinants that managers and public decision-makers should most take into consideration, with a view to export success. After analysing the various factors that moderate the relationships between firm resources and export performance, recommendations were made to managers and policy makers. For example, technology-intensive industries must focus on developing innovation capacity to improve their competitive position in international markets, as derived from the work of Pla-Barber and Alegre (2007). On the other hand, they must also think comprehensively, not only considering their internal capacity, but also the institutional environment, cultural diversity, psychological distance and the dynamism of the export market, as follows from the research by Sousa and Lengler (2009). For policy makers, the authors (Chen et al., 2016) suggested that government support for exporters should be adapted to the characteristics of the firms, with special attention to SMEs, so that they can overcome resource limitations, which tends to be a reality in these types of firms, in order to improve their capabilities, for successful international expansion of exports.

As seen, achieving successful export performance is a central factor in the strategy of firm decision-makers. For firms, the success of export performance indicates the extent to which the firm's objectives, economic and non-economic, are achieved in an international environment, reflecting the adequacy of the chosen export strategy. In literature review studies it is possible to confirm that there are numerous export performance indicators. Katsikeas et al. (2000) analysed 42 different indicators and Sousa et al. (2008) discovered 40 different determinants of export performance grouped into internal factors and external factors, moderating and control variables.

The dimensions of export performance measurement have also evolved over the years. In the literature, we find objective and subjective measures. The main objective measures include those mentioned by Sousa (2004), when compiling export performance studies between 1998 and 2004. Of the 43 studies analysed, the author classified the objective measures into three categories. In the first category, it considered measures related to sales, in the second it added measures related to results and profitability and, in the third, those related to market indicators. It should be noted that these indicators can also be subject to subjective measurement (depending on how they are collected).

In this study, all measures used are objective measures, due to the type of data collected, which is in line with what is recommended by Sousa (2004), when arguing that, for SMEs, objective measures of export performance are more adequate than subjective ones.

The selection of the unit of analysis is important for a correct assessment of export performance. A study at the firm level aims to identify the determinants of success that describe its global export activity, while a study at the export venture level focuses on the determinants of the performance of the product/market combination (Oliveira et al., 2012; Sousa, 2004). As an export performance unit of analysis, the firm level was used in this investigation, which, according to Sousa (2004), is suitable for this study with the available data. Reinforcing the suitability of our measurement unit, Chen et al. (2016) found that three quarters of studies use the firm level as the unit of analysis, advancing even though, for many firms, the venture level is a completely unknown concept and that these firms measure export performance by the volume of exports in the last few years. Sousa (2004), in the clues he left for future research, considered that dynamic studies are the most correct for measuring export performance, and researchers should use the past and present performance. Also, Katsikeas et al. (2000) defended the importance of longitudinal studies to ensure increased effectiveness of performance measures. In fact, the work of Aaby and Slater (1989) had already shown concern with this topic. Our model was built based on a dynamic view, with performance being measured over three years.

The size and the age of the firm were used in this research as control variables, as they are important predictors of export performance and are widely used in the literature as this type of variable. In some studies by Boehe and Jiménez (2015), Calof (1994) and Chen et al. (2019), firm size has a positive and significant relationship with export performance. Bekteshi (2020) also concluded that dimension is an important variable of export activity, due to the economies of the scale it allows. Benito-Osorio et al. (2020), in their research with Spanish SME firms, obtained results with several models, in which the variable size was always significant in the relationship with export performance. Firm age is used in this study as a control variable, as it is an important predictor of export performance and widely used in the literature as a control variable (Chen et al., 2019; Sousa & Bradley, 2009). In the literature, the firm's age is combined with business experience and has a positive impact on export activities (Chen et al., 2019; Majocchi et al., 2005). In their study of Italian industrial firms, Majocchi et al. (2005) argued that the longer a firm stays in a country, the greater its knowledge about market relations and this knowledge can be used to enter new markets and increase the costumer base. The same authors concluded there is a positive relationship between firm age and export performance. Other authors present inconsistent results on the relationship between firm age and internationalisation (e.g., Wagner, 2015; Wang et al., 2017). Some researchers find no correlation, while others find a positive sign or even a negative relationship, highlighting the growing role of born globalsfirms that carry out external activities soon after their creation (Baronchelli & Cassia, 2014; Cavusgil & Knight, 2015). Zou and Stan (1998) had also found several studies with non-significant or negative relationships.

3 Research Method

3.1 Sample and Data Collection

To answer the study's research question, a quantitative approach was used, as in other studies (Dai et al., 2019). This empirical study is drawn from SMEs Portuguese exporters' textile sector. Considering the slight magnitude of the Portuguese market, it is expected that firms are pressured to globalize their activities to take advantage of economies of scale. The financial data was accessed through the Simplified Business Information (IES) database, an annual repository of tax and accounting information from firms registered in Portugal. Data were collected from firms that exported between 2017 and 2019. This study included firms with at least 10% of export intensity (Bank of Portugal criteria).

3.2 Portuguese Context

The Portuguese textile and clothing industry are one of the oldest and most traditional in the country and represents a crucial business sector. It originated during the Industrial Revolution and industrialization intensified in the nineteenth century. Its quality, design, innovation and technology have led to international recognition. Today, ITV maintains a significant role in the Portuguese economy, emphasizing innovation, sustainability and collaboration with research institutions (Pereira, 2021). The Textile and Clothing Industry is one of the most important industries for the Portuguese economy. It represents 10% of total Portuguese exports, 20% of Manufacturing Industry employment and 9% of manufacturing production. Portugal has around 6,300 firms operating in all sub-sectors of the textile and clothing industry, the majority are small and medium-sized entreprises, all well known for their flexibility and rapid response, know-how and innovation (ATP, 2023). They are mainly located in the north of Portugal (Porto, Braga, Guimarães and Famalicão).

Divided into textiles, clothing, home textiles and other made-up textile articles, the sector is supported by associations such as ATP and ANIVEC, as well as competence centers including CENIT, CITEVE and *Associação Selectiva Moda*. The presence of firms at the national and international fairs and events, such as *ModaLisboa* and Portugal Fashion, is key to increasing exports by promoting their international reputation (AICEP, 2018).

Exports in the textile industry have been growing for several years, leading to a geographical increase in Portuguese enterprises. With a remarkable export capacity, reaching 80% of production and exporting to 189 countries, the sector contributes significantly to the economy, accounting for 3% of national GDP Bank of Portugal (2023).

With the war in Europe and instability in the energy market, widespread inflation and rising interest rates, demand for non-essential goods was affected, resulting in a slowdown in demand and a 7% drop in production in the sector in 2022 (ATP, 2023). With the effect of the pandemic, in 2020, exports fell by 11%, however in 2022, export figures reached 6.1 billion euros, surpassing the 2019 figure of 5.2 billion euros by 17% (ATP, 2023).

3.3 Statistical Analysis

After the selection and organisation of the database, we use econometric software Gretl 2023 (GNU Regression, Econometric, and Time-series Library) for descriptive analyses, regression analyses, and parameter estimation.

The nature of panel data suggests the use of linear models that simultaneously incorporate information from the firms in the sample and the different years within the time horizon. The data is panel or longitudinal, as it has two dimensions: a cross-sectional dimension, where there are 106 different firms for each year, and a time dimension, where the same firm is evaluated over three years, between 2017 and 2019.

The use of panel data offers several advantages, according to Hsiao (2007), allowing a more realistic study with better estimators and effect measures that cannot be observed with other methodologies, such as cross-sectional or time studies.

- Enables an increase in degrees of freedom.
- Considers a larger number of observations (318 in this study).

- Allows control of potential individual heterogeneity among observed firms.
- Extracts more information from the data, yielding greater variability and efficiency.
- By studying the repeated cross-section over observations, panel data is more suitable for studying dynamic evolution.
- Thus, we address the limitations pointed out by Sousa et al. (2008) regarding the absence of longitudinal studies and the construction of dynamic models.

3.4 *Measures*

Our study variables were based on objective and quantitative measures used in existing literature. To quantify financial resources, we utilized accounting data for both current and non-current financing. Current financing represents remunerated debt contracted by the firm and paid in the short term, while non-current financing refers to credit obtained in the medium and long term. This approach aligns with the metric used by Kaleka (2012). Assets, denoted as firm resources, were measured using accounting data of total assets, encompassing tangible and intangible assets, consistent with existing literature (Abu Bakar & Ahmad, 2010) and as another measure of the firm's resources (Bodlaj et al., 2020). The firm's productivity was gauged by the ratio of sales to the number of employees, a metric employed by various authors (Ganotakis & Love, 2012; Iyer, 2010; Reis & Forte, 2016). Geographic diversification was measured by the number of markets to which firms exported, in line with studies by Katsikeas et al. (2000), Cos et al. (2019), and Benito-Osorio et al. (2020).

As for control variables, age and size were considered—variables widely used in export performance literature (Chen et al., 2016). Firm size was quantified in terms of the number of employees in the respective year (Leonidou et al., 2002; Katsikea et al., 2005), while firm seniority was measured by the number of years in operation, spanning from the year of its establishment to 2019 (Benito-Osorio et al., 2020; Leonidou et al., 2002). The metric employed for export performance was an objective measure based on financial data. Therefore, export sales, one of the most commonly used metrics, was utilized in this study, as mentioned by Sousa (2004).

4 Results

During the analysis period (from 2017 to 2019), the average volume of exports per firm amounted to around 5,720.0 thousand euros, with a median of 3,880.0 thousand euros. The standard deviation is somewhat high, demonstrating the volatility of the variable between firms over the three years (Table 1).

Table 2 shows the results of the descriptive statistics for the explanatory and control variables. The results obtained for financial resources average 1970,000€

Table 1 Descriptive statistics for the dependent variable–Export performance

Variable	Unit	Obs	Average	Median	SD	Min	Max
Export performance	10^3 Euro	318	5 720	3 880	5 040	132	39 700

Note N = 570; T = 3

and confirm wide variations between firms. The assets of the sample firms average around 6 million euros, but vary between a minimum of 303 thousand euros and a maximum of 35.8 million euros. With regard to the number of market destination, the highest incidence (median) was found to firms that export to four markets. At the other end, there are firms that export to 37 countries around the world, thus diversifying the risk of their sales. This result is in line with the country's data. Portugal concentrates its exports on four main European markets (Spain, France, Germany and the United Kingdom) and the United States of America, which represents the main destination outside Europe. The average age of the firms is 30 years, which confirms the maturity of this sector. Over the three years of the analysis, the average size was 80 employees and the median was 70 employees, medium-sized enterprises. The average value for productivity was 105,000€, with very different ranges, varying from a minimum of 17,000€ to a maximum of 494,000€.

Considering the organization of panel data, we first assessed the options for constructing the regression model to use: the fixed effects model, the random effects model, or the pooled ordinary least squares (OLS) model. Chow and Hausman tests (Hausman 1978) were conducted to choose among these various models. The OLS model was considered the most suitable due to the coefficients obtained in determining the variables that determine the export performance, like in similar studies, such as Pandey and Sahu (2019). The impact of different determinants of export performance, obtained through the OLS method, is presented in Table 3. The F-statistic results, used to test the overall significance of the regression model, were significant at 1% significance level. Given the significance of the 'F' value, the prediction equation is considered well-suited for the export performance of Portuguese SMEs. The 'p' value of the model is 0.00, suggesting that the model is highly significant. Additionally, with a Durbin-Watson coefficient close to 2, it was confirmed to be free from autocorrelation, as per Wooldridge (2019). The R^2 value was 0.696,

Table 2 Descriptive statistics for the independent and control variables

Variable	Unit	Obs	Average	Median	SD	Min	Max
Financial resources	10^3 Euro	318	1970	1300	2260	2	11,500
Assets	10^3 Euro		6020	4180	5090	303	35,800
Productivity	10^3 Euro		105	75	83	17	494
Geograph. diversif	N.countries		4	4	5	0	37
Age	N.years		30	30	16	3	79
Size	N.employees		80	70	45	9	236

indicating that 69.6% of the variation in the export sales of the sample firms was explained by the six selected variables used in the study. The adjusted coefficient of determination for the regression allows us to conclude that 68,9% of the variance in the dependent variable is determined by the effects of independent variables, highlighting the statistical evidence of the model's good fit to the data.

The model results are as follows:

$$\begin{aligned} \text{EXPORT PERFORMANCE}_{i,t} &= 0,345^{***}\text{FINANCIALRESOURCES}_{i,t} \\ &+ 45,048^{***}\text{PRODUCTIVITY}_{i,t} \\ &+ 58,935^{***}\text{SIZE}_{i,t} + u_{i,t} \end{aligned} \tag{1}$$

According to the hypotheses proposed previously, the factors determining firms' exports and the control variables were evaluated using objective and quantitative analyses accompanied by existing literature.

This study shows that of six relationships three are statistically significant. First, we find evidence supporting the financial resources on export performance. The financial resources accessible to firms are key determinants of exports for Portuguese textile SMEs. The outcomes derived from the statistical model reveal that financial resources exhibit a positive and significant coefficient in their association with export performance ($\beta = 0.365$; $\rho = 0.009$), providing support for H1. To elucidate, for every 10 thousand euros increase in external financing for firms, export sales witnessed a corresponding increase of 3,650€. Moving on to the results for the asset variable, no statistically significant relationship was identified with export performance ($\beta = 0.079$; $\rho = 0.785$), contradicting the hypothesis that the firm's asset level positively influences export performance. Export sales appear to remain unaffected by the augmentation of assets dedicated to the business activity. Furthermore, the results establish that worker productivity exerts a positive and significant impact on the

Table 3 Export performance estimators

Variables		Coeff	Stand. Errors	t-ratio	*p value*	
Const		−2760,10	626,307	−4,407	<0,0001	a
Financial resources		0,365	0,141	2,595	0,0099	a
Assets		0,071	0,089	0,792	0,429	
Productivity		45,048	2,674	16,850	<0,0001	a
Geographic diversification		10,277	37,611	0,273	0,7848	
Size		58,935	6,684	8,817	<0,0001	a
Age		−13,1446	10,442	−1,259	0,209	
R^2		0,696				
R^2 Adjust		0,689		P(F)	0,000	
F		101,191				

Notes [a]represents statistically significant relationships at a significance level of 1%

export performance of Portuguese SMEs ($\beta = 45.048$; $\rho < 0.0001$). These findings align with the conclusions drawn by Onkelinx et al. (2016), asserting that labor productivity and investments in human capital play a crucial role in the success of SME internationalisation. However, in contrast, geographic diversification of exports did not demonstrate statistical significance ($\rho = 0.785$), failing to support H4, which posited a positive relationship between the number of destination markets and export sales. This result contradicts the conclusions reached by many authors and challenges the idea that sales vary with the geographic scale of operations. In other words, increasing the number of export destination countries does not necessarily lead to a proportional increase in the volume of exports. Wheeler et al. (2008) suggested that exporters' involvement might differ based on the number of destination markets, and the study's findings indicate that the presumed economies of scale and competitive advantages associated with a greater number of export markets were not confirmed in our research.

Fifth, two control variables were also included: firm size, measured as the number of employees and firm age. Firm age has no significative effect on export performance.

But there is statistical evidence that firm size has a significant effect on the dependent variable. This effect may be associated with possible economies of scale and in line with the conclusions of other studies, such as Lejpras (2019), Chen et al. (2016) and Pinho and Prange (2016). It was clear that there is a positive contribution to the relationship between firm size and export performance, which means that SMEs achieve better export performance compared to smaller firms.

Seniority is not statistically significant in the export performance of Portuguese SMEs ($p = 0.209$), not allowing us to conclude about the impact of increasing seniority on improving export performance.

This result we arrived at may have been influenced by the use of a single industry, in line with Bekteshi (2020) who concluded that the effects of firm age may depend on the type of business, in a study on exporting SMEs in Kosovo.

5 Discussion

Successful exports are a result of efficient use of firm resources and capabilities that generate international competitiveness. This research was carried out in a mature sector, the textile sector in Portugal, so the results have to be analysed in this context. We concluded that financial resources, productivity and size are key drivers of SMEs' export strategy and growth, and that these firms are indeed capable of attaining competitive advantage.

A positive and significant coefficient was found for financial resources, allowing us to conclude that there is a positive relationship between financial resources and the export performance of textile SMEs. This result is in line with Huang and Liu (2017), who argued that firms with more access to financing have greater export capacity, and with Goldbach and Nitsch (2014), who suggested that bank credit lines support firms to further expand their exports, due to greater flexibility they provide.

As suggested by the latter authors, exporting SMEs obtain, on average, 15% more bank credit than non-exporting firms and, therefore, additional bank financing can increase firms' export sales. This investigation is consistent with the conclusions of Goldbach and Nitsch (2014), who argued that the reduction of financial barriers in international business leads to an increase in the level of confidence of firms, which leads to them being able to become more involved in business activities like export. The results of our research brought insights into the importance of access to finance in the success of Portuguese SMEs' export operations, consistent with Griffith's (2011) approach, and greater involvement with exports correlated with access to finance.

The results did not validate expectations about the relationship between firm assets and export performance. Although the coefficient was positive, the relationship was not statistically significant. This result did not allow us to confirm the position of Abu Bakar and Ahmad (2010), not allowing us to demonstrate that an increase in the level of assets is decisive for an increase in export sales. The geographic diversification of exports, measured by the number of export markets, presented a positive coefficient, but did not prove to be significant in the relationship with export performance. These results are in line with other studies that also demonstrated that there is no significant relationship between these variables (Ruzo et al., 2011). On the other hand, several authors have confirmed the positive and significant existence of this relationship. Cos et al. (2019) argued for a better level of export performance for firms that reach a wider number of foreign markets. Also, Wheeler et al. (2008), Bartlett and Ghoshal (1987) suggested that exporters' involvement may differ according to the number of destination markets and that the greater the number of markets, the more economies of scale they can achieve, with more competitive advantages. The measure used can partly justify its non-significance, as defended by Sun and Govind (2018), who concluded that there was a positive relationship between the aforementioned variables, but using another metric for geographic diversification, the number of continents.

This study proves that worker productivity has a positive and significant impact on the export performance of Portuguese SMEs. These results confirm the conclusions obtained in the literature, such as those of Onkelinx et al. (2016), who argued that labour productivity and investments in human capital are crucial to the success of SME internationalization. Also, López-Rodríguez et al. (2018) concluded that there is a positive relationship between qualified workers with advanced training and greater export performance. All these results demonstrate that human resources are crucial for product quality and improvements in market power, therefore, greater productivity leads to better export performance.

Regarding the control variables, the relationship between firm size and export performance is positive. The size of SMEs proved to have a positive and significant effect on the volume of exports, associated with possible economies of scale and in line with the conclusions of several studies, such as that of Bekteshi (2020), Benito-Osorio et al. (2020), Lejpras (2019); Chen et al. (2016); Ogasavara et al. (2016); Eickelpasch and Vogel (2011); Ruzo et al. (2011); Sousa et al. (2008). Majocchi et al. (2005) also argue that size has a positive effect on performance and that exporting requires a strong commitment of resources and organizational processes that can vary significantly with size and according to resource availability. On the contrary, Schlegelmilch and Crook (1988) concluded a non-linear relationship in the shape of an inverted U, in which export intensity reached its maximum value when the firm size reached 1439 workers. However, the authors reported that only a small number of firms exceeded this size. And, on the other hand, it concerns application to large firms, which is outside the scope of our investigation.

Surprisingly, the coefficient of seniority is positive but not significant on export performance. This result is contradictory to the conclusions of a vast literature that argues that, as firms get older, they tend to develop management skills and accumulate knowledge and experience about the competitive environment, which can positively impact their engagement capabilities in foreign markets, as argued by several authors: Behmiri et al. (2019) and Villar et al. (2014). The effect of the seniority variable reinforces the controversial results in the literature (Lejpras, 2019). Our result is in line with empirical work that did not find a significant relationship between seniority and export performance, Faria et al. (2020) on SME exporters in the Portuguese wine sector.

Thus, we can answer the research question by stating that financial resources are what most determine the export performance of Portuguese SMEs, that is, they positively influence export sales, as well as productivity, which represents a capacity that influences in a direct export.

6 Conclusion

This study seeks to contribute to the development of the literature on factors that influence SMEs' export performance through a robust empirical study. The central context of this research is on SMEs, which constitute the vast majority of firms in Portugal, as in most European economies.

Understanding the effects of decisions made by management in selecting strategic orientations is crucial and highly relevant to both theory and practice. Moreover, our intent is also to contribute to the ongoing scholarly conversation on the value of intangibles as strategic resources to SMEs.

This study emphasizes the resources theory. It is acknowledged that strategy involves intentional and emergent initiatives adopted by management, encompassing resources and capabilities utilized to enhance business performance (Nag et al., 2007). To stay competitive, firms need to assess which strategic factors provide them with an advantage over their competitors. The findings contribute to clarifying the impact of financial resources, productivity, and size on the export performance of SMEs.

Acknowledgement It is my pleasure to express my acknowledgment to my Students - Luís Antunes Barbosa, Manuel David Loureiro and Ana Filipa Pedro - who supported me in the Portuguese context and in the final revision of this book chapter.

References

Aaby, N. E., & Slater, S. F. (1989). Management influences on export performance: A review of the empirical literature 1978–88. *International Marketing Review, 6*(4). https://doi.org/10.1108/EUM0000000001516

Abor, J. Y., Agbloyor, E. K., & Kuipo, R. (2014). Bank finance and export activities of small and medium enterprises. *Review of Development Finance, 4*(2), 97–103.

Abu Bakar, L. J., & Ahmad, H. (2010). Assessing the relationship between firm resources and product innovation performance: A resource-based view. *Business Process Management Journal, 16*(3), 420–435.

AICEP. (2018). PortugalGlobal, Indústria têxtil e de vestuário uma referência a nível mundial, 113.

ATP. (2023). Estatísticas—Caracterização. Accessed 15 December 2023. https://atp.pt/ptpt/estatisticas/caraterizacao/

Barney, J. (1991). Firm resources and sustained competitive advantage. *Journal of Management, 17*(1), 99–120.

Bank of Portugal. (2023). Dashboards Quadros do Setor, Retrieved December 15, 2023, from https://www.bportugal.pt/QS/qsweb/Dashboards.

Bartlett, C. A., & Ghoshal, S. (1987). Managing across borders: New strategic requirements. *Sloan Management Review, 28*(4), 7–17. https://doi.org/10.4324/9781315199689-19

Baronchelli, G., & Cassia, F. (2014). Exploring the antecedents of born-global companies' international development. *International Entrepreneurship and Management Journal, 10*(1), 67–79. https://doi.org/10.1007/s11365-011-0197-9

Behmiri, B., N., Rebelo, J. F., Gouveia, S., & António, P. (2019). Firm characteristics and export performance in Portuguese wine firms. *International Journal of Wine Business Research, 31*(3), 419–440. https://doi.org/10.1108/IJWBR-07-2018-0032

Beck, T., Demirgüç-Kunt, A., & Maksimovic, V. (2006). The determinants of financing obstacles. *Journal of International Money and Finance, 25*(6), 932–952.

Beck, T., Demirgüç-Kunt, A., & Maksimovic, V. (2008). Financing patterns around the world: Are small firms different? *Journal of Financial Economics, 89*(3), 467–487.

Bekteshi, S. A. (2020). Firm size related to export performance. *International Journal of Economics and Business Administration, 8*(1), 51–61.

Beleska-Spasova, E. (2014). Determinants and measures of export performance–comprehensive literature review. *Jcebi, 1*(1), 63–74.

Benito-Osorio, D., Colino, A., Guerras-Martín, L. Á., & Zúñiga-Vicente, J. Á. (2016). The international diversification-performance link in Spain: Does firm size really matter? *International Business Review, 25*(2), 548–558.

Benito-Osorio, D., Colino, A., Guerras-Martín, L. Á., & Zúñiga-Vicente, J. Á. (2020). The combined effects of product and geographical diversification on performance: Evidence in manufacturing SMEs. *BRQ Business Research Quarterly, 23*(2), 91–106.

Bodlaj, M., Kadic-Maglajlic, S., & Vida, I. (2020). Disentangling the impact of different innovation types, financial constraints and geographic diversification on SMEs' export growth. *Journal of Business Research, 108*(October), 466–475.

Boehe, D. M., & Jiménez, A. (2015). How does the geographic export diversification–performance relationship vary at different levels of export intensity? *International Business Review, 25*(6), 1262–1272.

Brouthers, L. E., Nakos, G., Hadjimarcou, J., & Brouthers, K. D. (2009). Key factors for successful export performance for small firms. *Journal of International Marketing, 17*(3), 21–38.

Calof, J. L. (1994). The relationship between firm size and export behaviour revisited. *Journal of International Business Studies, 25*(2), 367–387.

Cavusgil, S. T., & Knight, G. (2015). The born global firm: An entrepreneurial and capabilities perspective on early and rapid internationalization. *Journal of International Business Studies, 46*(1), 3–16.https://doi.org/10.1057/jibs.2014.62

Cernat, L., Jakubiak, M., & Preillon, N. (2020). The role of SMEs in extra-EU exports: Key performance indicators. European Commission.

Chen, J., Sousa, C. M. P., & He, X. (2016). The determinants of export performance: A review of the literature 2006–2014. *International Marketing Review, 33*(5), 626–670.

Chen, J., Sousa, C. M. P., & He, X. (2019). Export market re-entry: Time-out period and price/ quality dynamisms. *Journal of World Business, 54*(2), 154–168. https://doi.org/10.1016/j.jwb.2019.01.001

Cieślik, J., Kaciak, E., & Welsh, D. H. B. (2012). The impact of geographic diversification on export performance of small and medium-sized enterprises (SMEs). *Journal of International Entrepreneurship, 10*(1), 70–93. https://doi.org/10.1007/s10843-012-0084-7

Cos, P., Colom, A., & Cabasés, A. (2019). Geographic export diversification: Determinants and their link with export performance. *Thunderbird International Business Review, 61*(2), 397–411.

Dai, M., Nucci, F., Pozzolo, A. F., & Xu, J. (2019). Access to finance and the exchange rate sensitivity of exports. Centro Studi Luca D'Agliano Development Studies Working Paper, 442.

D'Angelo, A., Buck, T., Majocchi, A., & Zucchella, A. (2013). Geographical pathways for SME internationalization: Insights from an Italian sample. *International Marketing Review, 30*(2), 80–105.

Dhanaraj, C., & Beamish, P. W. (2003). A Resource-Based Approach to the Study of Export Performance. *Journal of Small Business Management, 41*(3), 242–261.

Eickelpasch, A., & Vogel, A. (2011). Determinants of the export behaviour of German business services companies. *Service Industries Journal, 31*(4), 513–526. https://doi.org/10.1080/02642069.2010.504304

European Comission (2019). An open, sustainable and assertive trade policy, accessed 13 November 2023

Faria, S., Rebelo, J., & Gouveia, S. (2020). Firms' export performance: A fractional econometric approach. *Journal of Business Economics and Management, 21*(2), 521–542.

Ganotakis, P., & Love, J. H. (2012). Export propensity, export intensity and firm performance: The role of the entrepreneurial founding team. *Journal of International Business Studies, 43*(8), 693–718. https://doi.org/10.1057/jibs.2012.16

Goldbach, S., & Nitsch, V. (2014). Extra credit: Bank finance and firm export status in Germany. *World Economy, 37*(7), 883–891.

Golovko, E., & Valentini, G. (2011). Exploring the complementarity between innovation and export for SMEs growth. *Journal of International Business Studies, 42*(3), 362–380.

Griffith, D. A. (2011). Insights into gaining access to export financing: Understanding export lenders' ideal exporter profile. *Journal of World Business, 46*(1), 84–92.

Griffith, D. A., & Czinkota, M. R. (2012). Release the constraints: Solving the problems of export financing in troublesome times. *Business Horizons, 55*(3), 251–260.

Harvie, C., Narjoko, D., & Oum, S. (2013). Small and Medium Enterprises' Access to Finance: Evidence from Selected Asian Economies. ERIA Discussion Paper Series, 1–54.

Hausman, J. A. (1978). Specification tests in econometrics. *Econometrica, 46*, 1251–1271.

Hennart, J. F. (2007). The theoretical rationale for a multinationality-performance relationship. *Management International Review, 47*(3), 423–452.

Hsiao, C. (2007). Panel data analysis-advantages and challenges. *Test, 16*(1), 1–22. https://doi.org/10.1007/s11749-007-0046-x

Huang, X., & Liu, X. (2017). How do firms finance their exports?–Evidence from China. *Journal of International Trade and Economic Development, 26*(2), 154–173.

IAPMEI. (2022). Portal do Financiamento. Retrieved January 12, 2024.

Iyer, K. (2010). The determinants of firm-level export intensity in New Zealand agriculture and forestry. *Economic Analysis and Policy, 40*(1), 75–86. https://doi.org/10.1016/S0313-5926(10)50005-5

Kaleka, A. (2012). Studying resource and capability effects on export venture performance. *Journal of World Business, 47*(1), 93–105.

Katsikea, E. S., Theodosiou, M., Morgan, R. E., & Papavassiliou, N. (2005). Export market expansion strategies of direct-selling small and medium-sized firms: Implications for export sales management activities. *Journal of International Marketing, 13*(2), 57–92.

Katsikeas, C. S., Leonidou, L. C., Neil, A. M., Constantine, S., & Leonidas, C. (2000). Firm-level export performance assessment review, evaluation, and development. *Academy of Marketing Science. Journal, 28*(4), 493–511.

Katsikeas, C. S., Piercy, N. F., & Ioannidis, C. (1996). Determinants of export performance in a European context. *European Journal of Marketing, 30*(6), 6–35.

Lages, L. F., & Montgomery, D. B. (2004). Export performance as an antecedent of export commitment and marketing strategy adaptation. *European Journal of Marketing, 38*(9/10), 1186–1214.

Lejpras, A. (2019). Determinants of export performance: Differences between service and manufacturing SMEs. *Service Business, 13*(1), 171–198.

Leonidou, L. C., Katsikeas, C. S., & Samiee, S. (2002). Marketing strategy determinants of export performance: a meta-analysis. *Journal of Business Research, 55*(1), 51–67. https://doi.org/10.1016/S0148-2963(00)00133-8

López-Rodríguez, J., Dopico, D. C., & Del Castillo Puente, A. M. (2018). Export performance in Spanish wineries: The role of human capital and quality management system. *European Journal of International Management, 12*(3), 311–333. https://doi.org/10.1504/ejim.2018.091372

Majocchi, A., Bacchiocchi, E., & Mayrhofer, U. (2005). Firm size, business experience and export intensity in SMEs: A longitudinal approach to complex relationships. *International Business Review, 14*(6), 719–738.

Manole, V., & Spatareanu, M. (2010). Exporting, capital investment and financial constraints. *Review of World Economics, 146*(1), 23–37. https://doi.org/10.1007/s10290-009-0040-3

Morgan, N. A., Kaleka, A., & Katsikeas, C. S. (2004). Antecedents of export venture performance: A theoretical model. 68(January), 90–108.

Nag, R., Hambrick, D., & Chen, M. (2007). What is strategic management, really? Inductive derivation of a consensus definition of the field. *Strategic Management Journal, 955*, 935–955.

Ogasavara et al. (2016). Experience, resources and export market performance: The pivotal role of international business. *International Marketing Review, 33*(6), 867–893. https://doi.org/10.1108/IMR-10-2013-0247

Oliveira, J. S., Cadogan, J. W., & Souchon, A. (2012). Level of analysis in export performance research. *International Marketing Review, 29*(1), 114–127.

Onkelinx, J., Manolova, T. S., & Edelman, L. F. (2016). The human factor: Investments in employee human capital, productivity, and SME internationalization. *Journal of International Management, 22*(4), 351–364.

Pacheco, L. (2017). Financial structure effects on export intensity and diversification: The case of Portuguese industrial firms. *International Journal of Globalisation and Small Business, 9*(4), 252–276.

Padmaja, M., & Sasidharan, S. (2021). Financing constraints and exports: Evidence from India. *Journal of Economics and Finance, 45*(1), 118–145.

Pandey, K. D., & Sahu, T. N. (2019). Debt financing, agency cost and firm performance: Evidence from India. *Vision, 23*(3), 267–274.

Penrose, E. T. (1959). *The theory of the growth of the firm (Oxford)*. Basil Blackwell.

Pereira, B. (2021). *Analysis of the productivity determinants in the Portuguese textile and clothing industry*. Master Thesis, University of Minho.

Pinho, J. C., & Prange, C. (2016). The effect of social networks and dynamic internationalization capabilities on international performance. *Journal of World Business, 51*(3), 391–403.

Pla-Barber, J., & Alegre, J. (2007). Analysing the link between export intensity, innovation and firm size in a science-based industry. *International Business Review, 16*(3), 275–293. https://doi.org/10.1016/j.ibusrev.2007.02.005

Reis, J., & Forte, R. (2016). The impact of industry characteristics on firms' export intensity. *International Area Studies Review, 19*(3), 266–281.

Riding, A., Orser, B. J., Spence, M., & Belanger, B. (2012). Financing new venture exporters. *Small Business Economics, 38*(2), 147–163.

Ruzo, E., Losada, F., Navarro, A., & Díez, J. a. (2011). Resources and international marketing strategy in export firms: Implications for export performance. *Management Research Review, 34*(5), 496–518. https://doi.org/10.1108/01409171111128698

Schlegelmilch, B. B., & Crook, J. N. (1988). Firm-level determinants of export intensity. *Managerial and Decision Economics, 9*(4), 291–300.

Sousa. (2004). Export performance measurement: An Evaluation of the Empirical Research in the Literature. *Academy of Marketing Science Review*, 2004(09).

Sousa, C. M. P., & Bradley, F. (2009). Effects of export assistance and distributor support on the performance of SMEs: The case of Portuguese export ventures. *International Small Business Journal, 27*(6), 681–701.

Sousa, & Lengler, J. (2009). Psychic distance, marketing strategy and performance in export ventures of Brazilian firms. *Journal of Marketing Management, 25*(5–6), 591–610. https://doi.org/10.1362/026725709X461876

Sousa, C. M. P., Martínez-López, F. J., & Coelho, F. (2008). The determinants of export performance: A review of the research in the literature between 1998 and 2005. *International Journal of Management Reviews, 10*(4), 343–374.

Sousa. (2016). A review of the empirical research on export channel selection between 1979 and 2015. *International Business Review, 26*(2), 303–323. https://doi.org/10.2991/cesses-18.2018.156

St-Pierre, J., Sakka, O., & Bahri, M. (2018). External financing, export intensity and inter-organizational collaborations: Evidence from Canadian SMEs. *Journal of Small Business Management, 56*, 68–87.

Sun, W., & Govind, R. (2018). Geographic diversification, product diversification, and firm cash flow volatility: The moderating effect of firm dynamic capability. *Journal of Strategic Marketing, 26*(5), 440–461.

Teece, D. J. (2014). A dynamic capabilities-based entrepreneurial theory of the multinational enterprise. *Journal of International Business Studies, 45*(1), 8–37.

Teece, D. J., Pisano, G., & Shuen, A. (1997). Dynamic capabilities and strategic management. *Strategic Management Journal, 18*(7), 509–533.

Villar, C., Alegre, J., & Pla-Barber, J. (2014). Exploring the role of knowledge management practices on exports: A dynamic capabilities view. *International Business Review, 23*(1), 38–44. https://doi.org/10.1016/j.ibusrev.2013.08.008

Wagner, J. (2015). Credit constraints and the extensive margins of exports: First evidence for German manufacturing. *Economics, 9*, 1–18.

Wagner, J. (2019). Access to finance and exports–Comparable evidence for small and medium enterprises from industry and services in 25 European countries. *Open Economies Review, 30*(4), 739–757.

Wang, X., Chen, A., Wang, H., & Li, S. (2017). Effect of export promotion programs on export performance: Evidence from manufacturing SMEs. *Journal of Business Economics and Management, 18*(1), 131–145.

Wernerfelt, B. (1984). A resource-based view of the firm. *Strategic Management Journal, 5*, 171–180.

Wheeler, C., Ibeh, K., & Dimitratos, P. (2008). UK export performance research: Review and implications. *International Small Business Journal, 26*(2), 207–239.

Wooldridge, J. M. (2019). Introduction econometrics-A modern approach (7th ed.). Cengage Learning, Inc.

Zou, S., et al. (1998). The EXPERF scale: a cross-national generalized export performance measure. *Journal of International Marketing, 6*(3), 10.

Zou, S., & Stan, S. (1998). The determinants of export performance: A review of the empirical literature between 1987 and 1997. *International Marketing Review, 15*(5), 333–356.

Isabel Soares de Moura has a Bachelor's degree in Management, a Master's degree in Management Sciences, and is a Ph.D. student in the Global Management, Strategy and Entrepreneurship (ISCTE-IUL). She is a senior economist at Caixa Geral de Depósitos (CGD–Banking). Her research interests include the measurement and impact of financial resources on international activities, and innovation and their links to export performance. She has also advised government members (Economy and Finance).

Sustainability Innovation in the Textile Industry

Budi Harsanto, Joval Ifghaniyafi Farras, and Dimas Indradi

Abstract The textile industry is known for being labor-intensive and has a significant impact on the environment and society. The aim of this paper is to identify key themes of sustainability innovation in the textile industry and to map research in this area. The methodology used is bibliometric approach, with a total of more than 2,500 Scopus-indexed documents analyzed using the bibliometric method. More than sixty percent of the articles are peer-reviewed journal articles. Publications exhibit an increasing trend from year to year, with the main outlets including, for example, the Journal of Cleaner Production, and Journal of Silk. Conceptually, four quadrants of the thematic map are identified such as niche, emerging, declining, and motor themes. These include basic themes consisting of sustainability, innovation, circular economy, and nanotechnology; motor themes include the economy; niche themes consisting of green innovation, smart textiles, and electrospinning; as well as emerging themes, namely competitiveness and technological innovation.

1 Introduction

Sustainability innovation is a combination of two main streams: eco-innovation and social innovation (Hansen & Große-Dunker, 2013). Sustainability innovation in the Textile Industry has received significant attention recently. For instance, Salo et al. (2020) explored the motivation for carrying out eco-innovation in the textile industry in Nordic countries, while Wanniarachchi et al. (2020) carried out research on the

B. Harsanto (✉)
Faculty of Economics and Business, Universitas Padjadjaran, Bandung, Indonesia
e-mail: budi.harsanto@unpad.ac.id

J. I. Farras
Digital Marketing Program, Universitas Padjadjaran, Bandung, Indonesia
e-mail: joval.ifghaniyafi@unpad.ac.id

D. Indradi
Master in Management Science Program, Universitas Padjadjaran, Bandung, Indonesia
e-mail: dimas22011@mail.unpad.ac.id

Lisbon Accounting and Business School, Polytechnic Institute of Lisbon (ed.), *Industry and Innovation: Textile Industry*, SDGs and Textiles,
https://doi.org/10.1007/978-3-031-57804-5_16

sustainability and innovation of textile handloom in Sri Lanka. Another study is by Di Chiacchio et al. (2023), which examines sustainability innovation, particularly from an eco-innovation perspective in the textile industry, connecting it with knowledge management. Additionally, L. Zhang et al. (2022) conducted a study highlighting technological innovation and its role in advancing the textile life cycle.

The increasing attention to the textile industry is not surprising, given its known substantial environmental and social impact. For instance, environmentally, this industry is recognized for its high consumption of resources, including water, electricity, and oil, often leading to overconsumption (Macchion et al., 2015). Furthermore, the manufacturing process generates pollution, involving chemicals, gas emissions, and even extending to the end-of-use phase in the form of textile waste. Additionally, the industry typically operates at super-high volumes, resulting in significant challenges after use (Morlet et al., 2017). The numerous challenges faced by the textile industry prompt both practitioners and academics to seek innovative solutions, aiming to maintain economic sustainability while addressing environmental and social impacts simultaneously.

Given the extensive and well-established nature of research in this field, it is essential to undertake efforts to review it systematically, aiming to comprehend the patterns and interconnections of various concepts presented in various published studies. Recently, Harsanto et al. (2023) conducted a systematic literature review focusing on sustainability innovation practices in the textile industry, providing a valuable contribution to the existing knowledge in this domain. This paper, taking a different approach, seeks to expound upon previous publications in the field of sustainability innovation using bibliometric methods and concentrating on metadata from various existing studies. Consequently, the research aims to identify key themes of sustainability innovation in the textile industry and map the research landscape in this area.

The paper is arranged as follows. This session explains the research background and the next section contains the methodology, including search methods and bibliometric analysis techniques. After that, the next section presents the results of the analysis, followed by a discussion of the results in the next section. The last section consists of conclusions, including limitations and future directions.

2 Methodology

Bibliometric is a valuable tool in research methods, offering quantitative analysis of academic literature and scholarly communication. It involves applying statistical methods to analyze patterns in publication, citations, and usage (Hicks et al., 2015). The use of bibliometrics in the field of business and management has gained significant attention, particularly in the context of analyzing emerging research trends and evaluate the implications of various factors on business (Cillo et al., 2019; Harsanto & Firmansyah, 2023; Verma & Sharma, 2019).

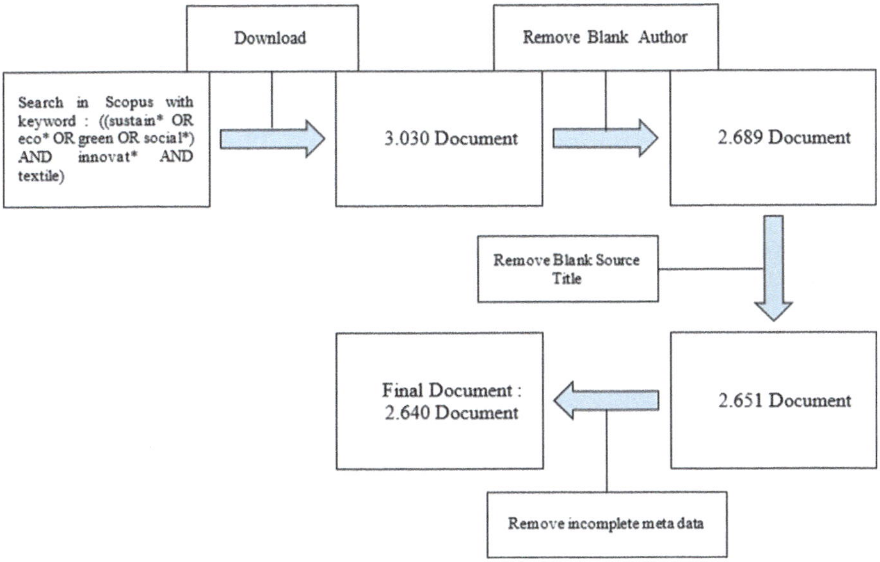

Fig. 1 Document production

The data used comes from the Scopus database. The utilization of the Scopus database for bibliometric analysis is firmly established and widely acknowledged in academic literature (Ioseliani et al., 2023; Olatinwo & Joubert, 2022; Sweileh, 2021; Zyoud et al., 2022). We conducted a literature search on January 5, 2024, with the keywords ((sustain* OR eco* OR green OR social*) AND innovat* AND textile)). The data obtained was 3,030 documents which after the filtering process became 2,640 documents. We decided to analyze all documents to get a more comprehensive picture of this theme. We conducted text analysis using Biblioshiny and R-Studio software developed by Aria and Cuccurullo (2017). Text analysis is a method to determine the number of words that often appear related to a particular field of science. The data collection procedure in this research can be seen in Fig. 1.

3 Results and Discussion

3.1 Document Type

The total number of documents we examined is 2,640 in the period 1965–2024 which is divided into five document types, article (1664), book (62), book chapter (243), conference paper (411), and review (260). Based on the grouping results, the most widely used document type is article with a percentage of 63.03% and the least document type is book with a percentage of 2.35% (see Table 1).

Table 1 Document type

Description	Results	%
Main information about data		
Timespan	1965–2024	
Document types		
Article	1664	63.03
Book	62	2.35
Book chapter	243	9.20
Conference paper	411	15.57
Review	260	9.85
Total	**2640**	

The analysis was conducted using Biblioshiny Software with R-Studio. The analyzed text undergoes additional analysis in searches related to frequently occurring words in the context of textile industry sustainability innovation. This is aimed at incorporating references from global studies to enhance the understanding of the development of Sustainability Innovation in the textile Industry.

A visual mapping chart of 2640 documents in the textile industry's sustainability innovation context will be shown to further elaborate on the results of the metanalysis. The outcomes of the key word mapping analyze serve as the foundation for correlating significant or distinctive terms found within these documents. Mapping is a procedural approach that facilitates the identification of knowledge elements, along with their structure, dynamics, interconnections, and interactions.

3.2 Source Analysis

The quantity of research articles released by individual journals, categorized according to their relevance to the topic of sustainability innovation within the textile industry, is illustrated in Fig. 2. We decided to use a bar chart to display the list of journals that published the most articles, along with a range of the number of articles published. The more intense the color, the greater the quantity and pertinence of research topics, along with the volume of articles published by the journal. The Journal of Cleaner Production stands at the forefront with 50 published articles, attributed to its strong alignment with the discussed theme. Following closely in the second position is Sustainability.

In this study, we discuss the development of journal resources on the theme of textile industry sustainability innovation. The curve in Fig. 3 outlines the progression of each journal's annual presence from 1990 to 2024, providing insights into whether the publication frequency related to textile industry sustainability innovation has risen or fallen over the research period. The curve indicates that research focused on

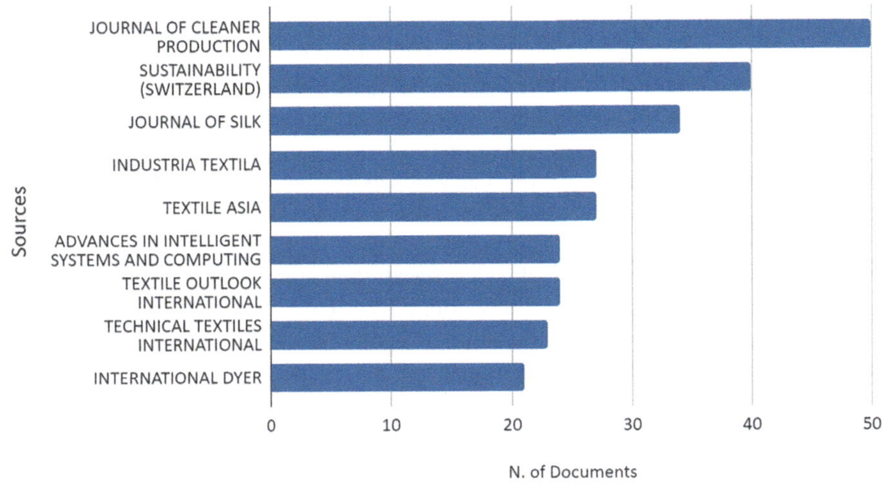

Fig. 2 Most relevant sources

textile industry sustainability innovation tends to undergo varying levels of growth in publications.

It is evident from the line diagram that the two journals with the most publications both have the greatest number. The Journal of Cleaner Production was first published in 2004 and has since grown steadily. Beginning in 2016, the number of publications with a sustainability innovation subject has increased dramatically, reaching a peak in 2023. Second, until 2022, Sustainability (Switzerland) is going to grow rapidly and fluctuate. It was initially introduced in 2013.

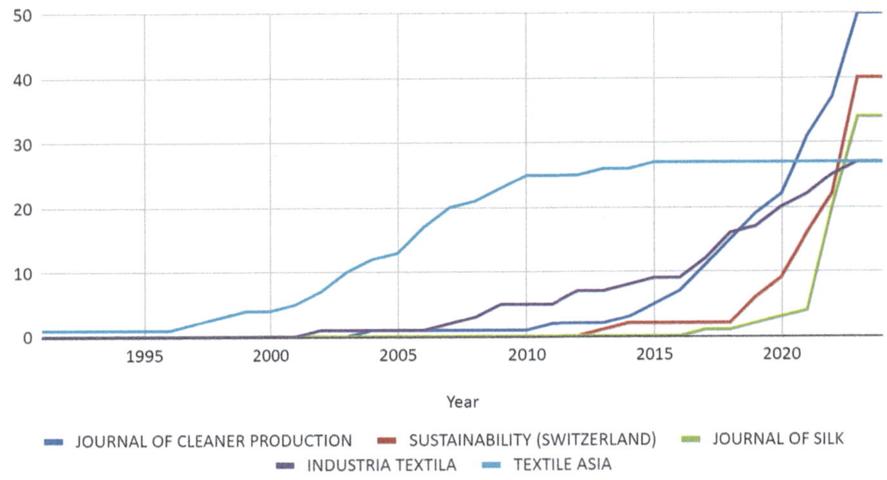

Fig. 3 Source production over time

In this study, we tried to measure the productivity of each author who published their articles in the context of textile industry sustainability innovation (Fig. 4). It can be seen in the figure that there are top authors in the period 1999 to 2023. The productivity of each author during a certain time period is depicted in the figure by the red line. In addition, the circle on the red line displays the total number of publications released in that specific year. Two authors with extensive backgrounds in the field of sustainable innovation in the textile industry are Watkins P (1999–2019) and Li Y (2001–2023).

Figure 5 depicts the countries of the corresponding authors who published articles on textile industry sustainability innovation. The quantity of articles published by the corresponding authors from that country is represented by the size of the circle, and the average relevance score of those articles is indicated by the color of the circle. The map indicates that China, India, and Italy account for the majority of the corresponding writers of the papers about the sustainability and innovation of the textile sector.

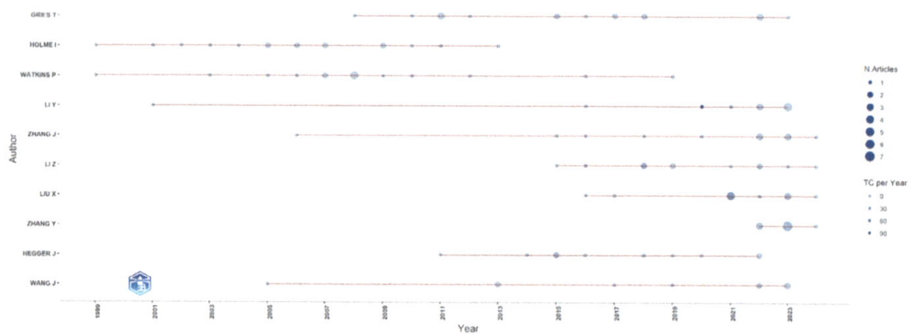

Fig. 4 Authors production over time

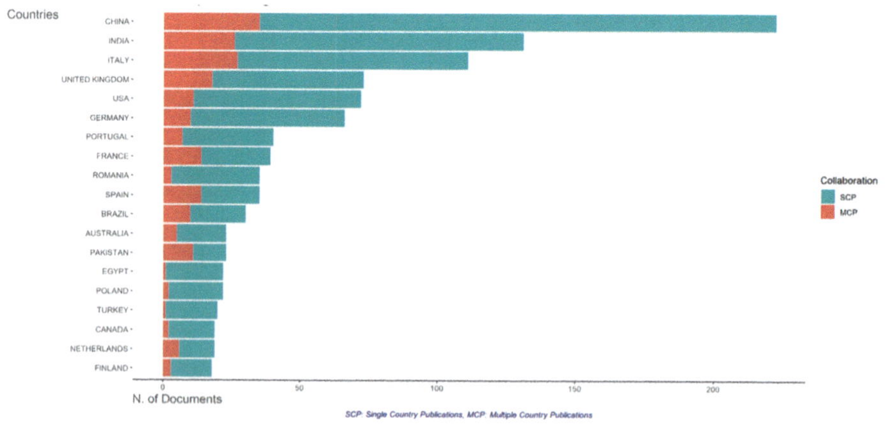

Fig. 5 Corresponding author's countries

China is one of the countries that attracts attention in the textile industry because of its large impact and scale on the global market (Huang, 2022). The development of the textile industry in China is driven by exports of textiles and finished garments which have an impact on economic growth for the country and the textile industry globally (Liu & Zhao, 2022). Not only China, the textile industry in India also contributes significantly to gross domestic product, export earnings, industrial production through apparel exports (Muniyasamy et al., 2020). In addition, India focuses on strengthening its traditional textiles, which is the focus of the current research (Deepshikha, 2019). From the MSME sector, Italy is one of the countries that focuses on the sector by emphasizing environmental sustainability (Bakos et al., 2020). More focused, the Italian fashion industry plays an important role as an international fashion landscape centered on the city of Milan (Merlo & Polese, 2006). Several Italian fashion centers in the city of Milan have organized International Exhibitions with the theme of Textiles and Clothes, and these exhibitions have become one of Italy's history in the textile industry and made it a specialist country in the textile industry (Wubs & Maillet, 2017).

The size of the circles for those nations is relatively considerable, indicating that a substantial number of papers have been published by corresponding authors from these countries. This implies that the publications published by corresponding authors from these nations are extremely significant to the topic of textile industry sustainability innovation.

Figure 6 provides a comprehensive compilation detailing the top 10 most globally cited documents that explore textile industry sustainability innovation. These documents serve as key pillars of influence, making significant contributions to the academic landscape. The table meticulously presents detailed information, including authors' names, publication years, and the citation counts for each document.

Research conducted by Chen et al. (2020) at the Department of Bioengineering, University of California, Los Angeles, has garnered 583 citations. The research

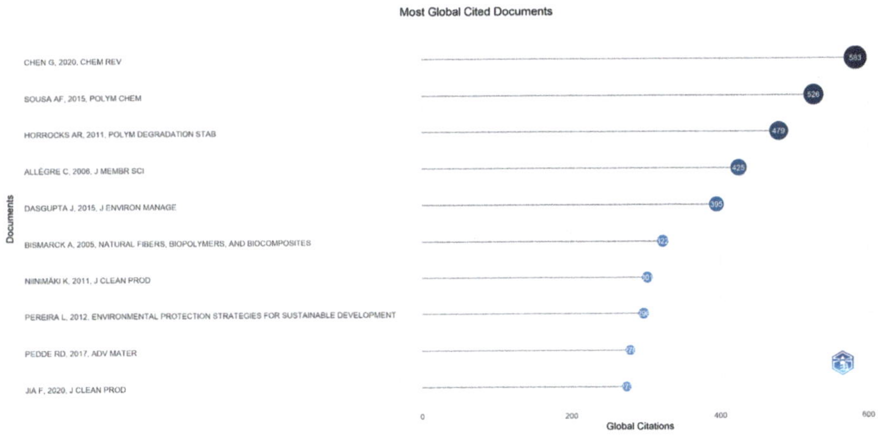

Fig. 6 Most global cited documents

Fig. 7 Word cloud

focuses on employing smart textiles to address the energy crisis and environmental pollution. Furthermore, research by Sousa et al. (2015) from the Aveiro Institute of Materials (cited 526 times) investigates novel and sustainable procedures employing renewable-based polymers, particularly polyesters generated from 2, 5-furandicarboxylic acid. Furthermore, research conducted by Horrocks (2011) from the Institute for Research and Innovation, University of Bolton, with 479 citations, delves into the potential of nanocomposite fibers and the use of nanoparticulate fillers to enhance fire performance in textile coatings.

Figure 7 visually represents a word cloud highlighting key terms related to textile industry sustainability innovation. Each word in the cloud is sized according to its frequency in the analyzed documents, with larger words indicating higher occurrence.

Based on Fig. 7, it can be seen that the most prevalent terms associated with textile industry sustainability innovation are: Sustainability with a frequency of 148 times (14%), Innovation with a frequency of 121 times (12%), circular economy with a frequency of 97 times (9%), and fashion with a frequency of 52 times (5%). Furthermore, this study found that research with the theme of textile industry sustainability innovation is mostly conducted in the fashion sector. In line with what is described in Fig. 6 (Most globally cited documents), the top three authors discuss the importance of innovation in the textile industry in achieving sustainability.

Thematic map is each cluster represented on a specific plot. There are two parts to the thematic map, the Y-axis relates to density and the X-axis relates to centrality. Centrality refers to the significance of a theme within the broader research domain, while Density can be interpreted as an indicator of the theme's level of advancement (Aria & Cuccurullo, 2017; Aria et al., 2022; Cobo et al., 2011; Rusydiana et al., 2023).

There are four categories in the thematic map analysis based on density and centrality (Fig. 8). The explanation of each category in the quadrant is as follows (Aria et al., 2022): (1) Niche themes, though strongly developed, remain on the periphery

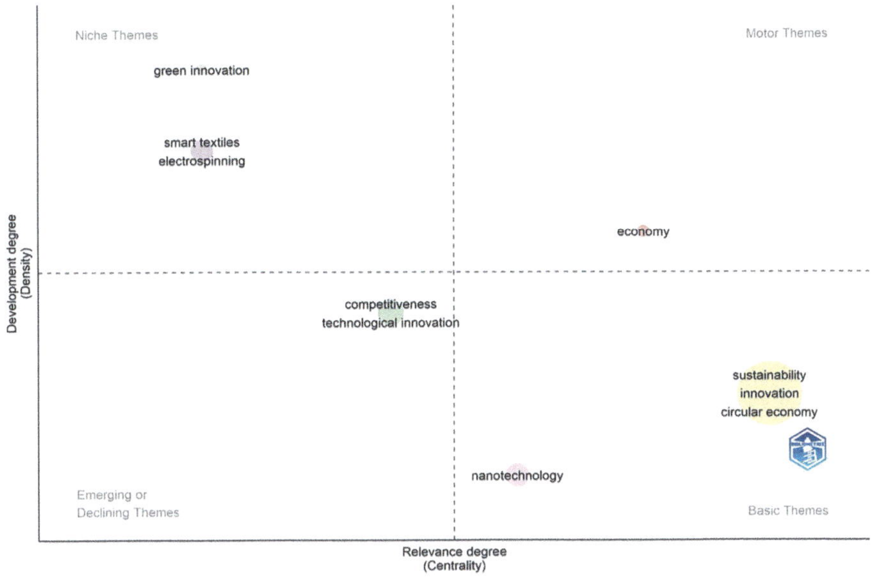

Fig. 8 Thematic map

of the domain under examination, as evidenced by lower centrality and higher density values. (2) Emerging or declining themes within the domain are marked by lower centrality and density values, indicating incomplete development or marginal interest. (3) Motor themes within the domain are characterized by elevated centrality and density values, contributing substantially to the well-constructed conceptual framework. (4) Basic themes crucial to the domain, with high centrality and low-density values, have significant relevance and span various areas within it.

In this study, according to Fig. 8, the first quadrant is Niche Themes, themes in this quadrant are green innovation, smart textiles, and electrospinning. The textile and garment industry's transition towards sustainable development and green innovation is multifaceted, as highlighted by various studies. Tran et al. (2023) emphasize the significance of green human capital, green relational capital and green structural capital in enhancing innovation. Khan et al. (2023a, b) and Zhou et al. (2023) discuss the role of green innovation and green human resource management in linking environmental performance and human resource management. On the financial side, Wang (2023) highlighted green financial policies and the need for special attention for the highly polluting textile industry. Badi et al. (2022) investigated sustainability performance indicators considering the implications of industrialization and population growth. Feng (2023) underscored the role of smart manufacturing in promoting corporate sustainability through green innovation. Xiaoyi et al. (2023) emphasized the role of organizational green culture in enhancing green innovation and business sustainability. Furthermore, SMEs can improve their sustainability performance by adopting new business strategies and responsive practices (Chau et al., 2023). In addition, Ali et al. (2021) highlighted the importance of green intellectual capital in driving green innovation adoption in manufacturing SMEs. Scafuto et al. (2021) emphasized the relevance of project management in driving green innovation in the context of sustainable business. Yang and Wang (2020) investigated supplier selection for green innovation adoption in sustainable supply chain management in the Chinese textile manufacturing industry. Dulal et al. (2022) provided recommendations for the textile industry to undergo green innovation through restructuring, use of clean energy, advocacy of green consumption principles, and improvement of recycling systems. Huo et al. (2022) provides insights on apparel, textile, and Fashion labels emphasizing Corporate Social Responsibility (CSR) with the aim of enhancing their performance in green innovation. Wakeford et al. (2017) provide valuable insights into innovation dynamics within the sector, especially in the context of driving green industrialization and sustainable practices.

In the development of smart textile, Meganathan et al. (2023) investigated the utilization of photocatalytic carbon nitride or polypyrrole coated cotton fabrics for smart textile applications, showcasing combined self-cleaning and antibacterial properties, thereby contributing to the evolution of functional textiles through the integration of these features. Additionally, López-Larraz et al. (2023) introduced a groundbreaking development in smart textiles by incorporating an EEG sensor layer within a garment, enabling accurate measurement of brain activity and offering potential implications for healthcare monitoring and human–computer interaction. Furthermore, Khan et al. (2023a, b) put forth a sustainable approach to smart textiles by proposing the recycling of textile waste into innovative carbon black. On the other hand, Dulal et al. (2022) emphasized the shift to sustainable e-textiles by using bio-based and biodegradable bioplastics such as polyhydroxyalkanoates (PHA), polylactic acid (PLA), polyhydroxybutyrate (PHB), and polybutylene succinate (PBS) to create eco-friendly technologies. These materials are promising for creating environmentally friendly technologies. Likewise, Tat et al. (2022) highlighted the importance of smart textiles for health and sustainability, focusing on increasing contact-separation space and omnidirectional interaction through the use of two-component spun yarns or connecting layers. Both articles emphasize the importance of sustainability and environmental friendliness in the development of wearable electronic textiles. In addition, Janeni and Adassooriya (2023) discuss the development of eco-friendly smart textiles from nanomaterials, providing additional insight into the utilization of eco-friendly materials in textile engineering.

Electrospinning, as explored by Stoyanova et al. (2023), enhances medicinal efficacy, ensuring increased bioavailability of extracts at the targeted site. Metwally et al. (2023) contributes to sustainable fabric production through electrospinning using recycled polyamide 6 waste. Osali et al.'s. (2023) study delves into the chemistry of electrospun nanomembranes, providing insights into nano-nonwoven fabric properties. Uddin et al. (2022) highlight the potential applications in water purification.

The next quadrant is Emerging or Declining Themes, themes in this quadrant are competitiveness and technological innovation. In this theme we can comprehensively understand competitiveness in the fashion and textile industry. It covers aspects such as corporate social responsibility (Battaglia et al., 2014), export competitiveness patterns (Fetscherin et al., 2012), policies supporting modern design and fashion (Wysokińska, 2015), comparative advantage in technical textiles, and factors influencing the sustainable productivity of SMEs (Chi et al., 2005). The synthesis of these references offers insights into the multifaceted nature of competitiveness within the fashion and textile industry, encompassing social, economic, and trade perspectives. On the other hand, from the point of view of technological innovation, it is necessary to emphasize the urgency of transformative measures in the fashion and textile industry. Blomsma et al. (2023) highlighted the need for circular disruption in response to growing concerns about unsustainable practices; this is in line with Ruan et al. (2022), and focuses on eco-efficiency analysis, underscoring the urgency for sustainable management in the textile industry. Furthermore, Shen et al. (2021) exploration of smart servitization underscores industrial transformation through the integration of smart industries and services. This is followed by Majumdar and Shaw (2023) and involves analyzing the impact of foreign direct investment on innovation in Indian textile firms. In addition, Zhou et al. (2022) and Pandeng et al. (2022) provide insights into the relationship between technological innovation, environmental performance and carbon emissions in China's textile industry, emphasizing the need for a holistic approach that integrates circular disruption, sustainable practices, innovation and technological advancement for a more environmentally sustainable and economically viable future in the textile sector. Based on the existing research the theme is currently on the rise.

The next quadrant is Motor Themes, themes in this quadrant need to be developed and are important for further research. Themes in this quadrant include economy and we, Lo (2021) identified the alignment of sustainable, economic and eco-friendly trends in the textile industry, emphasizing the importance of environmentally friendly processes and materials. In line with this, Aileni et al. (2020) provide insights into the economic aspects and the need for continuous innovation and skills development in the textile industry, particularly in the European context. More specifically, Bloomfield and Borstrock (2018) proposes the innovation of 3d printed textile introduced with Modeclix, where this system offers a cost-effective, customizable and scalable solution for the production of high-value and bespoke products, with the potential for reconfiguration and reuse, thus promoting sustainability and encouraging shifts in consumer behavior. On the other hand, utilizing supercritical water oxidation technology for the treatment of highly concentrated textile wastewater and sludge is economically viable (Zhang & Chen, 2017). Innovations made in order to increase economic profits while improving product quality can be done by using bio-based natural textiles and high performance fibers techniques, Alex et al. (2005) proved that businesses that make such innovations will lead the market fabrics and network with higher value and better profits on their own.

The last quadrant is Basic Themes which contains important themes in the current research. Themes included in this quadrant are sustainability, innovation, circular economy, and nanotechnology. The textile industry is becoming more aware of the need for sustainable practices, driven by increasing public consciousness and a sense of social responsibility towards environmental issues (Chen & Burns, 2006). Pollution challenges are prevalent in the industry, necessitating the adoption of sustainable measures to mitigate the environmental impact of textile production (Sidian et al., 2023). With the textile sector ranking as the second most polluting industry globally (Negrete & López, 2020), there is a pressing call for a shift to a circular economy-based textile model for enhanced sustainability (Ruiz et al., 2022). Effective collaboration and integrated sustainable supply chain management are identified as pivotal elements in achieving sustainable solutions within the textile industry (Abreu et al., 2020; Shahi et al., 2020). Moreover, the integration of sustainable practices is not only crucial for current operations but also holds significance for future fashion education, with expectations for sustainability to be incorporated into fashion design curriculum (Purnama et al., 2022). The environmental impact of textiles, particularly in countries like China, is a major concern, leading to the emergence of various technologies aimed at reducing water usage and pollution in the textile industry (Chen et al., 2019). The recycling and reprocessing of textile waste are gaining global importance for economic, environmental, and sustainability reasons (Özdil et al., 2020). However, challenges persist, including substantial sulfur dioxide and emissions of nitrogen oxide during production phases (Guo et al., 2023). Sustainability is a crucial issue in the fashion business, drawing significant attention from consumers and underscoring the importance of integrating sustainability into fashion brands (Pero et al., 2020; Shen et al., 2014). In the innovation perspective, the study by Filin et al. (2020) highlights the impact of real options on innovative projects in textile production, showing that innovative approaches can lead to changes in the range of textile products and overall supply. In addition, research by Larsson (2018) underlines the importance of utilizing digital innovation to create sustainable systems for the apparel industry, emphasizing the role of digital tools in minimizing waste and identifying opportunities for sustainable development in the textile and apparel industry. The introduction of environmental protection concepts in textile education, as suggested by Pu et al. (2021), is essential for implementing sustainable development strategies in the textile industry. This is in line with the findings of Macchion et al. (2016), which emphasized the influence of environmental practices, including supply chain practices, on innovation performance in the fashion industry. Furthermore, the study by Liu and Zhao (2022) emphasizes the importance of innovation in the textile industry for sustainable development, especially in the context of China's textile industry policy. This is further supported by Xu et al. (2018), who identified the need for innovative environmental services in the textile industry in China. One aspect that is currently considered in the textile industry is circular economy, where some circular economy practices have been carried out in this sector. For instance, the utilization of disposed textile materials and the incorporation of circular business models have been acknowledged as catalysts facilitating the adoption of circular economy practices by SMEs (Rizos et al., 2016). The significance of textile recycling in contributing to the circular economy has been emphasized, particularly focusing on the socio-economic benefits associated with this recycling approach (Filho et al., 2019). The transition towards a circular economy in the fashion and textile industry is intricately linked to the recycling and reuse of consumer waste, highlighting the imperative to shift from a linear to a circular economic model (Huang, 2022). Furthermore, the introduction of circular business models and sustainable design is suggested as a strategy to bolster the circular economy within the textile and clothing sector (Daňo et al., 2020). In addition to the emphasis on recycling and reuse, the extraction of cellulose from textile waste is identified as a feasible opportunity for advancing the circular economy in the textile sector (Neto et al., 2022). Moreover, the efficient management of post-consumer textile waste and the recycling of textile waste are recognized as indispensable components of the circular economy in the textile industry (Antonov et al., 2021; Santos & Campos, 2021). The necessity for effective post-consumer textile waste management, guided by principles of the circular economy, is underscored as a crucial measure to address the environmental impact associated with textile production and waste generation (Shamsi, 2020). In this study, our findings indicate that the incorporation of nanotechnology can enhance product quality for businesses. For example, the integration of nanomaterials into textile substrates is motivated by the increasing demand for durable, functional, and sustainable apparel (Yetisen et al., 2016). Nanotechnology facilitates the production of high-performance textiles endowed with improved characteristics such as enhanced breathability, flexibility, and aesthetic appeal, all while preserving the inherent properties of the fabric (Asif & Hasan, 2018). The application of nanotechnology in textiles spans various domains, encompassing the development of intelligent sportswear for improved athletic performance, wearable electronic textiles for biomedical and healthcare purposes, and the functionalization of cotton fabrics (Barman et al., 2022; Elmaaty et al., 2022; Hassabo et al., 2022; Syduzzaman et al., 2023) Furthermore, nanotechnology has played a pivotal role in the creation of multifunctional textiles, including stretchable and wearable solar textiles, thereby contributing to the advancement of the textile industry (Chowdhury, 2023; Salman et al., 2019). The distinctive and valuable attributes of nanotechnology have expedited its integration into the textile industry, fostering rapid advancements in textile finishing and the formulation of innovative textile structures (Farouk, 2019; Patra & Gouda, 2013).

4 Conclusion

Theoretical implications

This study focuses on bibliometric research in the field of textile industry sustainability innovation. The main objective is to explore and visually map the trends seen in the published scientific literature. There are four quadrants, of the four quadrants that are theoretically predicted to develop are emerging quadrants consisting of competitiveness and technological innovation. This quadrant emphasizes the importance of technological innovation, competitiveness, and transformation in the textile industry. It underscores the need for technological innovations such as nanotechnology, smart textiles to improve the competitiveness of companies while still paying attention to business sustainability that pushes towards a circular economy.

Managerial implications

Given the importance of textile industry sustainability innovation, managers must strategically prioritize and integrate sustainability practices into their innovation initiatives. This requires aligning organizational goals with the themes of sustainability, innovation, circular economy and technology identified in the literature. The emphasis on technology as a key theme in sustainability innovation suggests that managerial efforts should be directed towards investing in and adopting cutting-edge technologies in textile production processes (Chen et al., 2020). This could include the exploration of innovative materials, energy-efficient manufacturing, and sustainable technologies (nanotechnology and smart textiles) to improve overall efficiency and reduce environmental impact. Furthermore, the identification of circular economy as a recurring theme indicates the importance of adopting the principles of circularity in textile production. Managers should look for ways to minimize waste, promote recycling, and design products with a focus on longevity and recyclability to align with the circular economy model. Recognizing the linkages between sustainability, innovation and other key themes, managers should actively seek collaboration opportunities. Forming strategic partnerships with research institutions, technology providers and industry peers can foster knowledge exchange, accelerate innovation and contribute to the collective advancement of sustainability in the textile sector. Furthermore, in order to successfully implement sustainability practices, it is critical to ensure that the workforce is well-informed and trained in the principles of sustainability and innovation. Managers should consider implementing training programs to raise awareness, educate employees on sustainability practices and foster a culture of continuous improvement. The dynamic nature of sustainability innovation suggests the importance of continuous monitoring of industry trends. Managers should stay abreast of emerging themes and adjust organizational strategies to stay ahead of sustainability practices in the textile industry.

Limitations and future research

While this study provides valuable insights into sustainability innovation within the textile industry, there are certain limitations that warrant acknowledgment. The bibliometric approach focuses primarily on existing scientific literature, potentially excluding emerging trends not yet well-represented in academic publications. Additionally, the reliance on theoretical perspectives may not fully encapsulate the practical challenges faced by industry practitioners. Future research could address these limitations by incorporating more diverse data sources, and engaging industry stakeholders directly for a more comprehensive understanding of textile industry sustainability innovation.

Building on the current study's findings, future research avenues could explore the dynamic interplay between economics and sustainability practices in the textile industry. Investigating the impact of evolving costs on the adoption of sustainable technologies and circular economy principles would provide a nuanced understanding of the industry's response to external pressures. Furthermore, longitudinal studies could track the evolution of sustainability trends over time, offering insights into the industry's adaptive capacity and the effectiveness of implemented sustainability strategies. Exploring the role of consumer behavior and preferences in driving sustainability initiatives within the textile sector could also be a fruitful area of investigation. Finally, comparative analyses across different industries or sectors could provide valuable benchmarks for assessing the effectiveness of sustainability innovations in the textile industry.

References

de Abreu, M. C. S., Ferreira, F. N. H., Proença, J. F., & Ceglia, D. (2020). Collaboration in achieving sustainable solutions in the textile industry. *Journal of Business and Industrial Marketing*. https://doi.org/10.1108/jbim-01-2020-0041

Aileni, R. M., Chiriac, L., Surdu, L., & Radulescu, R. I. (2020). Field research and methodologies for textile industry innovation in the context of European economy. *Industria Textila, 71*(01), 12–16.

Alex, R., Kessler, R., Kohler, R., Mayer, G., & Tubach, M. (2005). Sustainability and profitability through intelligent value chain management in bast fibre processing. *Journal of Natural Fibers, 1*(3), 67–75.

Ali, W., Wen, J., Hussain, H., Khan, N. A., Younas, M. W., & Jamil, I. (2021). Does green intellectual capital matter for green innovation adoption? Evidence from the manufacturing SMEs of Pakistan. *Journal of Intellectual Capital, 22*(5), 868–888.

Antonov, M., Hussain, A., Kamboj, N., & Podgurski, V. (2021). Circular economy approach to recycling technologies of post-consumer textile waste in Estonia: A review. *Proceedings of the Estonian Academy of Sciences*.

Aria, M., & Cuccurullo, C. (2017). Bibliometrix: An R-tool for comprehensive science mapping analysis. *Journal of Informetrics, 11*(4), 959–975.

Aria, M., Cuccurullo, C., D'Aniello, L., Misuraca, M., & Spano, M. (2022). Thematic analysis as a new culturomic tool: The social media coverage on COVID-19 pandemic in Italy. *Sustainability, 14*(6), 3643.

Asif, A. K. M. A. H., & Hasan, Md. Z. (2018). Application of nanotechnology in modern textiles: A review. *International Journal of Current Engineering and Technology*. https://doi.org/10.14741/ijcet/v.8.2.5

Badi, I., Muhammad, L. J., Abubakar, M., & Bakır, M. (2022). Measuring sustainability performance indicators using FUCOM-MARCOS methods. *Operational Research in Engineering Sciences: Theory and Applications, 5*(2), 99–116.

Bakos, J., Siu, M., Orengo, A., & Kasiri, N. (2020). An analysis of environmental sustainability in small & medium-sized enterprises: Patterns and trends. *Business Strategy and the Environment, 29*(3), 1285–1296.

Barman, J., Tirkey, A., Batra, S., Paul, A. A., Panda, K., Deka, R., & Babu, P. J. (2022). The role of nanotechnology based wearable electronic textiles in biomedical and healthcare applications. *Materials Today Communications*.

Battaglia, M., Testa, F., Bianchi, L., Iraldo, F., & Frey, M. (2014). Corporate social responsibility and competitiveness within SMEs of the fashion industry: Evidence from Italy and France. *Sustainability, 6*(2), 872–893.

Blomsma, F., Bauwens, T., Weissbrod, I., & Kirchherr, J. (2023). The 'need for speed': Towards circular disruption—What it is, how to make it happen and how to know it's happening. *Business Strategy and the Environment, 32*(3), 1010–1031.

Bloomfield, M., & Borstrock, S. (2018). Modeclix. The additively manufactured adaptable textile. *Materials Today Communications, 16*, 212–216.

Chau, K. Y., Huang, T., Moslehpour, M., Khan, W., Nisar, Q. A., & Haris, M. (2023). Opening a new horizon in green HRM practices with big data analytics and its analogy to circular economy performance: An empirical evidence. *Environment, Development and Sustainability*. https://doi.org/10.1007/s10668-023-03725-9

Chen, G., Li, Y., Bick, M., & Chen, J. (2020). Smart textiles for electricity generation. *Chemical Reviews, 120*(8), 3668–3720.

Chen, H. L., & Burns, L. D. (2006). Environmental analysis of textile products. *Clothing and Textiles Research Journal*. https://doi.org/10.1177/0887302x06293065

Chen, L., Caro, F., Corbett, C. J., & Ding, X. (2019). Estimating the environmental and economic impacts of widespread adoption of potential technology solutions to reduce water use and pollution: Application to China's textile industry. *Environmental Impact Assessment Review*. https://doi.org/10.1016/j.eiar.2019.106293

Chi, T., Kilduff, P., & Dyer, C. (2005). An assessment of US comparative advantage in technical textiles from a trade perspective. *Journal of Industrial Textiles, 35*(1), 17–37.

Chowdhury, K. P. (2023). *Nanomaterials for multifunctional textiles*. https://doi.org/10.21741/9781644902288-8

Cillo, V., Petruzzelli, A. M., Ardito, L., & Del Giudice, M. (2019). Understanding sustainable innovation: A systematic literature review. *Corporate Social Responsibility and Environmental Management, 26*(5), 1012–1025.

Cobo, M. J., López-Herrera, A. G., Herrera-Viedma, E., & Herrera, F. (2011). An approach for detecting, quantifying, and visualizing the evolution of a research field: A practical application to the Fuzzy Sets Theory field. *Journal of Informetrics, 5*(1), 146–166.

Daňo, F., Drábik, P., & Hanuláková, E. (2020). Circular business models in textiles and apparel sector in Slovakia. *Central European Business Review*. https://doi.org/10.18267/j.cebr.226

Deepshikha, Y. P. (2019). Expressions of traditional textiles of India. *International Journal of Affective Engineering, 18*(2), 101–107.

Di Chiacchio, L., Martínez-Caro, E., Cegarra-Navarro, J. G., & Garcia-Perez, A. (2023). Eco-innovation in the textile industry: The strategic importance of data privacy and environmental knowledge management. *Kybernetes*. https://doi.org/10.1108/K-02-2023-0222

Dulal, M., Afroj, S., Ahn, J., Cho, Y., Carr, C., Kim, I.-D., & Karim, N. (2022). Toward sustainable wearable electronic textiles. *ACS Nano, 16*(12), 19755–19788.

Elmaaty, T. A., Elsisi, H., Elsayad, G. M., Elhadad, H. H., & Plutino, M. R. (2022). Recent advances in functionalization of cotton fabrics with nanotechnology. *Polymers*. https://doi.org/10.3390/polym14204273

Farouk, A. (2019). Multi-functionalization of cellulosic fabrics using nanotechnology. *International Journal of Chemtech Research*. https://doi.org/10.20902/ijctr.2019.120102

Feng, T. (2023). Do Intelligent manufacturing concerns promote corporate sustainability? Based on the perspective of green innovation. *Sustainability, 15*(14), 10958.

Fetscherin, M., Alon, I., Johnson, J. P., & Pillania, R. K. (2012). Export competitiveness patterns in Indian industries. *Competitiveness Review: An International Business Journal, 22*(3), 188–206.

Filho, W. L., Ellams, D., Han, S. L., Tyler, D., Boiten, V. J., Paço, A., Moora, H., & Balogun, A. (2019). A review of the socio-economic advantages of textile recycling. *Journal of Cleaner Production*. https://doi.org/10.1016/j.jclepro.2019.01.210

Filin, S. A., Yakushev, A. Z., Myrkhalykov, Z., Velikorossov, Vladimir. V, & Chaikovskaya, L. A. (2020). Innovative projects with real options in textile production: evaluation of their effect. In *E3s Web of Conferences*. https://doi.org/10.1051/e3sconf/202017605010

Guo, Y., Zhu, L., Ye, X., Wang, X., Xu, Y., Gao, Q., & Wang, L. (2023). Toxicity impact assessment of nitrogen oxide and sulfur dioxide emissions in China's textile industry with chemical footprint method. *Aatcc Journal of Research*. https://doi.org/10.1177/24723444231161744

Hansen, E. G., & Große-Dunker, F. (2013). *Sustainability-Oriented Innovation*. https://ssrn.com/abstract=2191679

Harsanto, B., & Firmansyah, E. A. (2023). A twenty years bibliometric analysis (2002 – 2021) of business economics research in ASEAN. *Cogent Business & Management, 10*(1).

Harsanto, B., Primiana, I., Sarasi, V., & Satyakti, Y. (2023). Sustainability innovation in the textile industry: A systematic review. *Sustainability, 15*(2), 1549.

Hassabo, A. G., Elmorsy, H., Gamal, N., Sediek, A., Saad, F., Hegazy, B., & Othman, H. (2022). Applications of nanotechnology in the creation of smart sportswear for enhanced sports performance: Efficiency and comfort. *Journal of Textiles Coloration and Polymer Science*. https://doi.org/10.21608/jtcps.2022.181608.1147

Hicks, D., Wouters, P., Waltman, L., de Rijcke, S., & Rafols, I. (2015). Bibliometrics: The Leiden Manifesto for research metrics. *Nature, 520*(7548), 429–431.

Horrocks, A. R. (2011). Flame retardant challenges for textiles and fibres: New chemistry versus innovatory solutions. *Polymer Degradation and Stability, 96*(3), 377–392.

Huang, X. (2022). How is fashion going: Structured literature review for circular economy in fashion and textile industry. *Frontiers in Business Economics and Management*. https://doi.org/10.54097/fbem.v3i3.358

Huo, C., Hameed, J., Zhang, M., Bin Mohd Ali, A. F., & Amri Nik Hashim, N. A. (2022). Modeling the impact of corporate social responsibility on sustainable purchase intentions: Insights into brand trust and brand loyalty. *Economic Research-Ekonomska Istraživanja, 35*(1), 4710–4739.

Ioseliani, A. D., Orekhovskaya, N. A., Svintsova, M. N., Panov, E. G., Skvortsova, E. M., & Bayanova, A. R. (2023). Bibliometric analysis of articles on digital educational environments. *Contemporary Educational Technology*. https://doi.org/10.30935/cedtech/13100

Janeni, J., & Adassooriya, N. M. (2023). Development of eco-friendly smart textiles from nanomaterials. In *Advances in Smart Nanomaterials and their Applications* (pp. 325–335). Elsevier. https://doi.org/10.1016/B978-0-323-99546-7.00002-1

Khan, A., Awais, M., & Mohsin, M. (2023a). Recycling textile waste into innovative carbon black and applications to smart textiles: A sustainable approach. *Biomass Conversion and Biorefinery*. https://doi.org/10.1007/s13399-023-04475-4

Khan, W., Nisar, Q. A., Roomi, M. A., Nasir, S., Awan, U., & Rafiq, M. (2023b). Green human resources management, green innovation and circular economy performance: the role of big data analytics and data-driven culture. *Journal of Environmental Planning and Management*, 1–26. https://doi.org/10.1080/09640568.2023.2189544.

Larsson, J. (2018). Digital innovation for sustainable apparel systems. *Research Journal of Textile and Apparel*. https://doi.org/10.1108/rjta-02-2018-0016

Liu, Y., & Zhao, H. (2022). Quantitative evaluation of policy based on PMC index model: A case study of China's textile industry policy. *Mathematical Problems in Engineering*. https://doi.org/10.1155/2022/1870185

Lo, C. H. (2021). Degumming silk by CO $_2$ supercritical fluid and their dyeing ability with plant indigo. *International Journal of Clothing Science and Technology, 33*(3), 465–476.

López-Larraz, E., Escolano, C., Robledo-Menéndez, A., Morlas, L., Alda, A., & Minguez, J. (2023). A garment that measures brain activity: Proof of concept of an EEG sensor layer fully implemented with smart textiles. *Frontiers in Human Neuroscience, 17*. https://doi.org/10.3389/fnhum.2023.1135153

Macchion, L., Moretto, A., Caniato, F., Caridi, M., Danese, P., & Vinelli, A. (2015). Production and supply network strategies within the fashion industry. *International Journal of Production Economics, 163*, 173–188.

Macchion, L., Moretto, A., Caniato, F., Caridi, M., Danese, P., Spina, G., & Vinelli, A. (2016). Improving innovation performance through environmental practices in the fashion industry: The moderating effect of internationalisation and the influence of collaboration. *Production Planning & Control*. https://doi.org/10.1080/09537287.2016.1233361

Majumdar, D., & Shaw, S. (2023). Innovation under the threat of FDI: A study of Indian firms. *Journal of the Knowledge Economy*. https://doi.org/10.1007/s13132-023-01477-5

Meganathan, P., Selvaraj, L. M., Subbaiah, S., Subramanian, V., Pitchaimuthu, S., & Srinivasan, N. (2023). A synergistic self-cleaning and antibacterial studies of photocatalytic carbon nitride/polypyrrole coated cotton fabrics for smart textile application. *Cellulose, 30*(17), 11211–11230.

Merlo, E., & Polese, F. (2006). Turning fashion into business: The emergence of Milan as an international fashion hub. *Business History Review, 80*(3), 415–447.

Metwally, B. S., Zayed, A. M., Rashed, S. A., El-Sheikh, M. N., & Hamouda, A. S. (2023). Sustainable nano-nonwoven fabric production from recycled polyamide 6 waste via electrospinning: Controlling characteristics and comprehensive analytical study. *Advanced Materials Technologies, 8*(18).

Morlet, A., Opsomer, R., Herrmann, S., Balmond, L., Gillet, C., & Fuchs, L. (2017). *A new textiles economy: Redesigning fashion's future.* https://www.ellenmacarthurfoundation.org/a-new-textiles-economy

Muniyasamy, A., Sivaporul, G., Gopinath, A., Lakshmanan, R., Altaee, A., Achary, A., & Velayudhaperumal Chellam, P. (2020). Process development for the degradation of textile azo dyes (mono-, di-, poly-) by advanced oxidation process-Ozonation: Experimental & partial derivative modelling approach. *Journal of Environmental Management, 265*, 110397.

Negrete, J. D. C., & López, V. N. (2020). A sustainability overview of the supply chain management in textile industry. *International Journal of Trade Economics and Finance.* https://doi.org/10.18178/ijtef.2020.11.5.673

de Oliveira Neto, G. C., Teixeira, M. M., Souza, G. L. V., Arns, V. D., & Tucci, H. N. P. (2022). Assessment of the eco-efficiency of the circular economy in the recovery of cellulose from the shredding of textile waste. *Polymers.* https://doi.org/10.3390/polym14071317

Olatinwo, S. O., & Joubert, T.-H. (2022). Deep learning for resource management in internet of things networks: A bibliometric analysis and comprehensive review. *IEEE Access.* https://doi.org/10.1109/access.2022.3195898

Osali, S., Ghiyasi, Y., Esfahani, H., Jose, R., & Ramakrishna, S. (2023). Electrospun nanomembranes at the liquid–liquid and solid–liquid interface-A review. *Materials Today, 67*, 151–177.

Özdil, N., Kayseri, G. Ö., & Mengüç, G. S. (2020). *Investigation of sound absorption characteristics of textile materials produced from recycled fibers.* https://doi.org/10.5772/intechopen.92792

Pandeng, S., Lin, H., Jianlei, Z., & Longdi, C. (2022). The impact of technological innovation from domestic innovation, import and FDI channels on carbon dioxide emissions of China's textile industry. *Industria Textila, 73*(04), 426–431.

Patra, J. K., & Gouda, S. (2013). Application of nanotechnology in textile engineering: An overview. *Journal of Engineering and Technology Research.* https://doi.org/10.5897/jetr2013.0309

Pero, M., Arrigo, E., & Fionda-Douglas, A. (2020). Sustainability in fashion brands. *Sustainability.* https://doi.org/10.3390/su12145843

Pu, C., Boshan, G., & Chen, X. (2021). Introduction of environmental protection concept in textile major education. *Journal of Educational Theory and Management.* https://doi.org/10.26549/jetm.v5i1.6471

Purnama, R., Tajuddin, R. M., & Shariff, S. M. (2022). *Examining students' attitudes towards sustainable fashion design curriculum.* https://doi.org/10.2991/assehr.k.220601.052

Rizos, V., Behrens, A., van der Gaast, W., Hofman, E., Ιωάννου, A., Kafyeke, T., Flamos, A., Rinaldi, R., Papadelis, S., Hirschnitz-Garbers, M., & Topi, C. (2016). Implementation of circular economy business models by small and medium-sized enterprises (SMEs): Barriers and enablers. *Sustainability.* https://doi.org/10.3390/su8111212

Ruan, X., Ding, N., & Yang, J. (2022). Dual-level eco-efficiency analysis in sustainable management: An empirical study on textile manufacturers and supply chains in China. *Journal of Environmental Management, 304*, 114227.

Ruiz, J.-R. R., Cantero, R., Riba-Mosoll, P., & Puig, R. (2022). Post-consumer textile waste classification through near-infrared spectroscopy, using an advanced deep learning approach. *Polymers.* https://doi.org/10.3390/polym14122475

Rusydiana, A. S., Irfany, M. I., As-Salafiyah, A., & Tieman, M. (2023). Halal supply chain: A bibliometric analysis. *Journal of Islamic Marketing, 14*(12), 3009–3032.

Salman, A., Metwally, F. A., El-Bisi, M. K., & Emara, G. (2019). Applications of Nanotechnology and Advancements in Smart Wearable Textiles: An overview. *Egyptian Journal of Chemistry.* https://doi.org/10.21608/ejchem.2019.18223.2112

Salo, H. H., Suikkanen, J., & Nissinen, A. (2020). Eco-innovation motivations and ecodesign tool implementation in companies in the Nordic textile and information technology sectors. *Business Strategy and the Environment, 29*(6), 2654–2667.

Santos, P. S. D., & Campos, L. M. de S. (2021). Practices for garment industry's post-consumer textile waste management in the circular economy context: An analysis on literature. *Brazilian Journal of Operations & Production Management*. https://doi.org/10.14488/bjopm.2021.004

Scafuto, I. C., de Araújo, V. A. A., dos Moreiras, A. A., & Kniess, C. T. (2021). Project management relationship to green innovation processes in sustainable fabric companies. *Sustainability in Debate, 12*(3), 13–26

Shahi, S. K., Shiva, A., & Dia, M. (2020). Integrated sustainable supply chain management and firm performance in the Indian textile industry. *Qualitative Research in Organizations and Management an International Journal*. https://doi.org/10.1108/qrom-03-2020-1904

Shamsi, M., & Chaudhary, A. (2020). Need of post-consumer textile waste management via circular economy. *International Journal for Modern Trends in Science and Technology*. https://doi.org/10.46501/ijmtst0609s32

Shen, B., Zheng, J., Chow, P.-S., & Chow, K.-Y. (2014). Perception of fashion sustainability in online community. *Journal of the Textile Institute*. https://doi.org/10.1080/00405000.2013.866334

Shen, L., Sun, C., & Ali, M. (2021). Path of smart servitization and transformation in the textile industry: A case study of various regions in China. *Sustainability, 13*(21), 11680.

Sidian, Y., Ghazali, R., & Tajuddin, R. M. (2023). Integrating sustainable concepts into textile design courses: An effective teaching practices. *International Journal of Global Optimization and Its Application*. https://doi.org/10.56225/ijgoia.v2i1.159

Sousa, A. F., Vilela, C., Fonseca, A. C., Matos, M., Freire, C. S. R., Gruter, G.-J.M., Coelho, J. F. J., & Silvestre, A. J. D. (2015). Biobased polyesters and other polymers from 2,5-furandicarboxylic acid: A tribute to furan excellency. *Polymer Chemistry, 6*(33), 5961–5983.

Stoyanova, N., Nachev, N., & Spasova, M. (2023). Innovative bioactive nanofibrous materials combining medicinal and aromatic plant extracts and electrospinning method. *Membranes, 13*(10), 840.

Sweileh, W. M. (2021). Bibliometric analysis of peer-reviewed literature on antimicrobial stewardship from 1990 to 2019. *Globalization and Health*. https://doi.org/10.1186/s12992-020-006 51-7

Syduzzaman, Md., Hassan, A., Anik, H. R., Akter, M., & Islam, M. R. (2023). Nanotechnology for high-performance textiles: A promising frontier for innovation. *Chemnanomat*. https://doi.org/10.1002/cnma.202300205

Tat, T., Chen, G., Zhao, X., Zhou, Y., Xu, J., & Chen, J. (2022). Smart textiles for healthcare and sustainability. *ACS Nano, 16*(9), 13301–13313.

Tran, T. D., Huan, D. M., Phan, T. T. H., & Do, H. L. (2023). The impact of green intellectual capital on green innovation in Vietnamese textile and garment enterprises: Mediate role of environmental knowledge and moderating impact of green social behavior and learning outcomes. *Environmental Science and Pollution Research, 30*(30), 74952–74965.

Uddin, Z., Ahmad, F., Ullan, T., Nawab, Y., Ahmad, S., Azam, F., Rasheed, A., & Zafar, M. S. (2022). Recent trends in water purification using electrospun nanofibrous membranes. *International Journal of Environmental Science and Technology, 19*(9), 9149–9176.

Verma, N., & Sharma, V. (2019). A literature review on energy value stream mapping (EVSM). *International Journal of Advanced Science and Technology, 27*(1), 1–8.

Wakeford, J. J., Gebreeyesus, M., Ginbo, T., Yimer, K., Manzambi, O., Okereke, C., Black, M., & Mulugetta, Y. (2017). Innovation for green industrialisation: An empirical assessment of innovation in Ethiopia's cement, leather and textile sectors. *Journal of Cleaner Production, 166*, 503–511.

Wang, M. L. (2023). Effects of the green finance policy on the green innovation efficiency of the manufacturing industry: A difference-in-difference model. *Technological Forecasting and Social Change, 189*, 122333.

Wanniarachchi, T., Dissanayake, K., & Downs, C. (2020). Improving sustainability and encouraging innovation in traditional craft sectors: The case of the Sri Lankan handloom industry. *Research Journal of Textile and Apparel, 24*(2), 111–130.
Wubs, B., & Maillet, T. (2017). Building competing fashion textile fairs in Europe, 1970–2010. *Journal of Macromarketing, 37*(1), 25–39.
Wysokińska, Z. (2015). Policy of active support for modern design and fashion in the textile and clothing sector and its influence on regional competitiveness (in the Global, European, and Regional Contexts). *Fibres and Textiles in Eastern Europe, 23*(6(114)), 14–22.
Xiaoyi, Z., Yang, H., Kumar, N., Bhutto, M. H., Kun, W., & Hu, T. (2023). Assessing Chinese textile and apparel industry business sustainability: The role of organization green culture, green dynamic capabilities, and green innovation in relation to environmental orientation and business sustainability. *Sustainability, 15*(11), 8588.
Xu, C., Cheng, H., & Liao, Z. (2018). Towards sustainable growth in the textile industry: A case study of environmental policy in China. *Polish Journal of Environmental Studies*. https://doi.org/10.15244/pjoes/79720
Yang, Y., & Wang, Y. (2020). Supplier selection for the adoption of green innovation in sustainable supply chain management practices: A case of the Chinese textile manufacturing industry. *Processes, 8*(6), 717. https://doi.org/10.3390/pr8060717
Yetisen, A. K., Qu, H., Manbachi, A., Butt, H., Dokmeci, M. R., Hinestroza, J. P., Skorobogatiy, M., Khademhosseini, A., & Yun, S. H. (2016). Nanotechnology in textiles. *ACS Nano*. https://doi.org/10.1021/acsnano.5b08176
Zhang, K. Q., & Chen, H. H. (2017). Environmental performance and financing decisions impact on sustainable financial development of Chinese environmental protection enterprises. *Sustainability (Switzerland), 9*(12).
Zhang, L., Leung, M. Y., Boriskina, S., & Tao, X. (2022). Advancing life cycle sustainability of textiles through technological innovations. *Nature Sustainability, 6*(3), 243–253.
Zhou, S., Tiruneh, W. A., & Legese, M. A. (2023). The effect of corporate social responsibility on environmental performance: The mediating role of green innovation and green human resource management. *International Journal of Emerging Markets*. https://doi.org/10.1108/IJOEM-02-2022-0211
Zhou, Y., Agyemang, A. O., Adam, I. O., & Twum, A. K. (2022). Assessing the impact of technological innovation on environmental and financial performance of Chinese textile manufacturing companies. *International Journal of Technology, Policy and Management, 22*(4), 369.
Zyoud, S. H., Shakhshir, M., Koni, A., Shahwan, M., Jairoun, A. A., & Al-Jabi, S. W. (2022). Olfactory and gustatory dysfunction in COVID-19: A global bibliometric and visualized analysis. *Annals of Otology Rhinology & Laryngology*. https://doi.org/10.1177/00034894221082735

Budi Harsanto is a Lecturer at Universitas Padjadjaran, Bandung, Indonesia. He earned a Ph.D. in Management from the University of Liverpool, UK. His research interests are in Sustainability Innovation, and Operations and Supply Chain Management.

Joval Ifghaniyafi Farras is Lecturer at Universitas Padjadjaran, especially in the Digital Marketing Vocational Education Program and Islamic Economics. Student of the Doctoral Program in Management Science (DIM) at Universitas Padjadjaran with a concentration in Halal Supply Chain Management. His research areas are mainly in Halal Supply Chain, Operations Management and Digital Marketing.

Dimas Indradi is a Master Student at Universitas Padjadjaran, focusing on Operation Management in Science Management with concentration and research Thesis in Lean Six Sigma especially in the Textile Industry.

GPSR Compliance

The European Union's (EU) General Product Safety Regulation (GPSR) is a set of rules that requires consumer products to be safe and our obligations to ensure this.

If you have any concerns about our products, you can contact us on ProductSafety@springernature.com

In case Publisher is established outside the EU, the EU authorized representative is:

Springer Nature Customer Service Center GmbH
Europaplatz 3
69115 Heidelberg, Germany

Batch number: 10761599

Printed by Printforce, the Netherlands